Asymptotic Methods for Investigating Quasiwave Equations of Hyperbolic Type

Mathematics and Its Applications

Managing Editor:

M. HAZEWINKEL

Centre for Mathematics and Computer Science, Amsterdam, The Netherlands

Volume 402

Asymptotic Methods for Investigating Quasiwave Equations of Hyperbolic Type

by

Yu. Mitropolskii

G. Khoma

International Mathematical Centre,
Ukrainian Academy of Sciences,
Kiev, Ukraine

and

M. Gromyak

Pedagogical University,
Ternopil, Ukraine

SPRINGER-SCIENCE+BUSINESS MEDIA, B.V.

A C.I.P. Catalogue record for this book is available from the Library of Congress.

ISBN 978-94-010-6426-2 ISBN 978-94-011-5752-0 (eBook)
DOI 10.1007/978-94-011-5752-0

This is a revised and updated translation of the original work
published under the same title by Naukova Dumka in 1991.
Translated from the Russian by Andrei Khruzin.

Printed on acid-free paper

Table of Contents

PREFACE

The theory of partial differential equations is a wide and rapidly developing branch of contemporary mathematics. Problems related to partial differential equations of order higher than one are so diverse that a general theory can hardly be built up. There are several essentially different kinds of differential equations called elliptic, hyperbolic, and parabolic. Regarding the construction of solutions of Cauchy, mixed and boundary value problems, each kind of equation exhibits entirely different properties.

Cauchy problems for hyperbolic equations and systems with variable coefficients have been studied in classical works of Petrovskii, Leret, Courant, Gording. Mixed problems for hyperbolic equations were considered by Vishik, Ladyzhenskaya, and that for general two-dimensional equations were investigated by Bitsadze, Vishik, Gol'dberg, Ladyzhenskaya, Myshkis, and others.

In last decade the theory of solvability on the whole of boundary value problems for nonlinear differential equations has received intensive development. Significant results for nonlinear elliptic and parabolic equations of second order were obtained in works of Gvazava, Ladyzhenskaya, Nakhushev, Oleinik, Skripnik, and others. Concerning the solvability in general of nonlinear hyperbolic equations, which are connected to the theory of local and nonlocal boundary value problems for hyperbolic equations, there are only partial results obtained by Bronshtein, Pokhozhev, Nakhushev.

As known, the solvability problem for periodic boundary value problem belongs to the basic problems in the theory of nonlinear hyperbolic equations. The development of the theory of boundary value problems for both ordinary and partial differential equations is prompted first of all by the practical needs with regard to solving various problems in physics, celestial mechanics, radio- and electrical engineering, etc. On the other hand, methods for investigating boundary value problems are insufficiently elaborate. In the theory of hyperbolic equations, the question concerning the existence of periodic solutions of boundary value problems for the quasiwave second order equation is of special interest. A particular difficulty arisen in investigating the above problems for hyperbolic linear and nonlinear equations is connected to small denominators. A number of important results have been established here by Vishik, M. M. Lavrent'yev, Leret, Lions, Skripnik, and others. The methods used to obtain these results mainly include those of nonlinear functional analysis, implicit function theory, variational methods as well as the small parameter method, Leret-Schauder fixed point principle, and the theory of monotone operators.

A significant place in the theory of differential equations is occupied by the methods for approximate solution of nonlinear differential equations. Notably, for nonlinear differential equations with small parameter these methods include the asymptotic methods of Anosov, Arnol'd, Grebenikov, Lomov, Oleinik, Tikhonov, and others; and for nonlinear wave partial differential equations with small parameter these are the asymptotic methods of nonlinear mechanics.

There are two major approaches in applying the asymptotic methods of nonlinear mechanics to investigating nonlinear wave partial differential equations, which describe distributed parameter systems.

The first approach uses reduction of a boundary value problem to a countable system of ordinary differential equations with the help of Fourier series. Then two main methods can be employed to investigate solutions of countable systems of ordinary differential equations in standard form. The first one makes use of the generalised first and second Bogolyubov principal theorems for countable systems; the second one shortens these systems with the help of the averaging method.

The second approach builds up asymptotic expansions of solutions of boundary value problems. The first approximation to an arbitrary single- or multi-frequency solution of a perturbed nonlinear boundary value problem describing oscillations in a weakly nonlinear distributed parameter system is represented as an asymptotic expansions containing one term or a sum of terms that correspond to normal modes of oscillations of the nonperturbed system. Furthermore, the amplitudes and phases as slow functions of time are determined from the system of ordinary differential equations of the first approximation. This approach proves to be quite fruitful when studying various nonlinear boundary value problems described by partial differential equations under both linear and nonlinear boundary conditions. As a result, the investigation of the single- and multi-frequency modes of oscillations in distributed systems reduces to that of discrete systems with a small number of parameters involved.

However, a series of problems related to wave propagation and interaction in boundary-free weakly nonlinear systems as well as boundary problems for large spatial intervals comparing to the wave length cannot be reduced to the interaction of a few normal oscillations of fixed spatial structure; and it becomes difficult to apply the above approch to them. It should be stressed once again that systems of partial differential equations of hyperbolic type, describing various wave processes, are of great interest for both mathematics and its applications, the distinguished place being occupaied by the one-dimensional ones.

The present book is devoted to futher development of the asymptotic theory of investigation of solutions of second order partial differential equations of hyperbolic type.

We evolve a new research direction in the theory of constructive numerical methods for studying solutions of a wide classes of periodic boundary value problems for both wave ordinary differential equations of second order and wave equations of hyperbolic type. We also suggest and justify a systematic approach to constructing and applying asymptotic methods for studying solutions of boundary value problems for nonlinear wave partial differential equations with small parameter.

CHAPTER 1

EXISTENCE THEOREMS FOR HYPERBOLIC EQUATIONS

The present chapter gives some grounds for the averaging method and the Krylov-Bogolyubov expansions to be effective in studying nonstationary vibration processes in systems which can be described by the second order partial differential equations of hyperbolic type. The chapter also establishes the weakest sufficient condition for a solution of these equations to exist and be unique and considers its representation as a Fourier series.

1.1. Preliminary remarks

The rigorous mathematical treatment of both the averaging method and the Krylov-Bogolyubov expansions for systems of ordinary differential equations can be found in [3–5, 7, 12–16, 19, 20, 42, 43, 50, 51, 53, 57–60, 63, 66, 69, 100–102, 123, 124, 160].

The existence condition for a solution of nonlinear ordinary differential equations is one of the prerequisites for such a rigorous treatment of the averaging method. (As for nonlinear partial differential equations of hyperbolic type, the corresponding existence condition has not been studied enough yet). We have in mind the question about the condition for a solution of the mixed boundary value problem [11, 23, 52, 83–85, 147]

$$u_{tt} - a^2 u_{xx} = \lambda u + \varepsilon f(x, t, u, u_t, u_x); \tag{1.1}$$

$$u(0, t) = u(\pi, t) = 0; \tag{1.2}$$

$$u(x, 0) = \varphi(x), \quad u_t(x, 0) = \psi(x) \tag{1.3}$$

to exist in a rectangular $\overline{Q} = \{0 \leq x \leq \pi, 0 \leq t \leq T\}$ and be representable as the Fourier series

$$u(x, t, \varepsilon) = \sum_{k=1}^{\infty} z_k(t, \varepsilon) \sin kx. \tag{1.4}$$

This is related to the fact that the basic idea of applying the averaging method to studying the mixed problem (1.1)–(1.3) consists in using the theory of Fourier series. Indeed, before proceeding to transformations of equation (1.1), one considers the nonperturbed equation

$$u_{tt} - a^2 u_{xx} = \lambda u, \tag{1.5}$$

which can be obtained from (1.1) by setting $\varepsilon = 0$, with the same boundary (1.2) and initial (1.3) conditions. Assuming $(ak)^2 - \lambda > 0$, $k = 1, 2, 3, \ldots$, and using the Fourier method, one finds then the solution of equation (1.5) as the series

$$u(x, t, 0) = \sum_{k=1}^{\infty} (A_k \cos \omega_k t + B_k \sin \omega_k t) \sin kx, \tag{1.6}$$

where $\omega_k = \sqrt{(ak)^2 - \lambda}$ are the frequencies of normal oscillations, A_k and B_k certain constants determined by the initial condition (1.3).

In the presence of perturbation ($\varepsilon \neq 0$), one may assume that the shapes of normal tone oscillations are determined, with sufficient accuracy, by the same functions sin kx, because the parameter ε is small. Taking into account solution (1.6) of nonperturbed equation (1.5), the solution of perturbed equation (1.1) is sought as series (1.4), where $z_k(t,\varepsilon)$ are the functions to be determined.

The above leads us to a natural question about the weakest condition imposed on the functions φ, ψ, f which ensures the mixed boundary value problem (1.1)–(1.3) admitting a continuous (classical $u \in C^2$) solution in the form (1.4). This question is very hard and has not been solved completely. To illustrate the point, we consider the linear mixed boundary value problem for the hyperbolic equation

$$u_{tt} - u_{xx} + q(x)\, u(x,t) = 0 \tag{1.7}$$

with potential

$$q(x) \in C([0,\pi]) \tag{1.8}$$

and boundary and initial conditions (1.2), (1.3), in the class of functions $u(x,t) \in C^2(\overline{Q})$ where

$$\overline{Q} = \{(x,t) \in \mathbb{R}^2 : 0 \le x \le \pi, \quad 0 \le t \le T\}. \tag{1.9}$$

The problem formulation implies that the following conditions have to be fulfilled

$$\varphi(x) \in \overset{0}{C}{}^2([0,\pi]), \quad \varphi''(0) = \varphi''(\pi) = 0; \tag{1.10}$$

$$\psi(x) \in \overset{0}{C}{}^1([0,\pi]); \tag{1.11}$$

where $\overset{0}{C}{}^k([0,\pi]) = \{h(x) \in C^k([0,\pi]) : h(0) = h(\pi) = 0\}$.

The question of solvability of the problem (1.7)–(1.9) became a subject for textbooks after [110] was first published in 1922. Nevertheless, the existence of a solution in the limiting case (1.10) and (1.11) is still an open question. The known conditions sufficient for the above problem to be solvable consist in some requirements to the functions $q(x)$ $\varphi(x)$ and $\psi(x)$ in addition to (1.8), (1.10) and (1.11):

$$q(x),\ \varphi''(x) \in \text{Lip}\,[0,\pi],\ \psi''(x) - (R)\text{– is integrable on}[0,\pi]; \tag{1.12}$$

$$\varphi(x) \in \overset{0}{C}{}^3([0,\pi]), \quad \psi(x) \in \overset{0}{C}{}^2([0,\pi]) \tag{1.13}$$

$$q(x) \in C^1([0,\pi]) \tag{1.14}$$

These conditions were found by Steklov [110, p. 223], Petrovskii [90, p. 210] and Levitan [55, pp. 108 and 115]. The conditions (1.12) and (1.13) were obtained with the help of the Fourier method, and condition (1.14) with the Riemann method, the operation of term-by-term differentiation of formal series representing the unknown solution being essentially used. Obviously, giving up that differentiation has to result in the conditions (1.12)–(1.14) becoming weaker. Following this way, a preliminary result has been obtained in [156]. The theorem below is its main statement about the linear problem (1.7)–(1.9), (1.2) and (1.3).

Theorem 1.1. [156]. *For the classical solution $u(x,t)$ of the mixed problem (1.7)–(1.9), (1.2) and (1.3) to exist it is necessary and sufficient that the initial functions $\varphi(x)$ and $\psi(x)$ satisfy the conditions (1.10) and (1.11). The solution $u(x,t)$ is then unique and represented by the Fourier series*

$$u(x,t) = \sum_{n=1}^{\infty} \left(\Phi_n \cos \omega_n t + \frac{\Psi_n}{\omega_n} \sin \omega_n t \right) y_n(x), \tag{1.15}$$

where ω_n and $y_n(x)$ are the eigenvalues and eigenfunctions of the Sturm-Liouville boundary value problem

$$-y'' + q(x)\, y = \omega^2 y, \quad y(0) = y(\pi) = 0.$$

This theorem was proved with the help of a new approach to justification of the Fourier method, which uses the asymptotic estimates of the eigen functions

$$y_n(x) = \sqrt{\frac{2}{\pi}} \sin nx + f_n(x), \quad f_n(x) = O(n^{-1}) \tag{1.16}$$

and their second derivatives.

Considering now the equation with small parameter

$$u_{tt} - u_{xx} + \varepsilon q(x) u = 0, \tag{1.17}$$

instead of (1.7), we can conclude, on the basis of representation (1.16), that not every mixed problem (1.1)–(1.3) admits a solution in the form of the Fourier series (1.4). Many mathematicians were proving the existence of a solution in the mixed problem (1.1)–(1.3) with the help of the Fourier method. In the works [21, 76, 78, 79], the existence of a solution of the periodic boundary value problem

$$u_{tt} - u_{xx} = f(x, t, u, u_t, u_x), \quad x \in \mathbb{R}, \quad t \in \mathbb{R}, \quad u(0, t) = u(\pi, t) = 0,$$

$$t \in \mathbb{R}, \quad u(x, t + T) = u(x, t), \quad x \in \mathbb{R}, \quad t \in \mathbb{R}, \tag{1.18}$$

was proved for the first time with the help of a simple modification of the d'Alambert formula, which allows the expressions containing infinite series to be avoided. This also makes it possible to introduce a continuously generelized or simply continuous solution of the problem (1.18).

1.2. Homogeneous mixed problem

We shall consider the homogeneous mixed problem

$$\begin{gathered} u_{tt} - a^2 u_{xx} = 0, \quad 0 < x < \pi, \quad t \geq 0, \\ u(0, t) = u(\pi, t) = 0, \quad t \geq 0, \\ u(x, 0) = \tilde{\varphi}(x), \quad u_t(x, 0) = \tilde{\psi}(x), \quad 0 \leq x \leq \pi. \end{gathered} \tag{1.19}$$

As known [115], the problem (2.1) can be solved by the Fourier method and its solution can be represented by the Fourier series

$$u(x, t) = \sum_{k=1}^{\infty} \left(a_k \cos kat + \frac{b_k}{ka} \sin kat \right) \sin kx, \tag{1.20}$$

where

$$a_k = \frac{2}{\pi} \int_0^{\pi} \varphi(x) \sin kx dx, \quad k = 0, 1, 2, \ldots,$$

$$b_k = \frac{2}{\pi} \int_0^{\pi} \psi(x) \sin kx dx, \quad k = 1, 2, \ldots, \tag{1.21}$$

Here $\varphi(x)$ and $\psi(x)$ are the odd 2π-periodic continuations of the functions $\tilde{\varphi}(x)$ and $\tilde{\psi}(x)$ from the interval $[0, \pi]$ to the real line $\mathbb{R}$.

It was established that $u(x,t) \in C^2$ if $\tilde{\varphi}(x) \in \overset{0}{C}{}^3([0,\pi])$, $\tilde{\psi}(x) \in \overset{0}{C}{}^2([0,\pi])$.

We shall prove that $u(x,t) \in C^2$ if $\tilde{\varphi}(x) \in \overset{0}{C}{}^2$, $\tilde{\psi}(x) \in \overset{0}{C}{}^1$. To do this, consider the following Cauchy problem

$$
\begin{gathered}
u_{tt} - a^2 u_{xx} = 0, \quad x \in \mathbb{R}, \quad t \geq 0, \\
u(0,t) = \varphi(x), \quad u_t(x,0) = \psi(x), \quad x \in \mathbb{R}, \\
\varphi(x) = -\varphi(-x), \quad \varphi(x+2\pi) = \varphi(x), \\
\psi(x) = -\psi(-x), \quad \psi(x+2\pi) = \psi(x).
\end{gathered} \tag{1.22}
$$

The Cauchy problem (1.22) can be solved with the help of the d'Alembert formula

$$
u(x,t) = \frac{\varphi(x+at) + \varphi(x-at)}{2} + \frac{1}{2a} \int\limits_{x-at}^{x+at} \psi(z)\,dz. \tag{1.23}
$$

We shall show that formula (1.23) can be rewritten as

$$
u(x,t) = \frac{\varphi(x+at) + \varphi(x-at)}{2} + \frac{1}{2} \int\limits_0^t \{\psi(x+a\tau) + \psi(x-a\tau)\} dz. \tag{1.24}
$$

To do this, it suffices to prove the following equality

$$
\frac{1}{2a} \int\limits_{x-at}^{x+at} \psi(z)\,dz = \frac{1}{2} \int\limits_0^t \{\psi(x+a\tau) + \psi(x-a\tau)\}\,d\tau \tag{1.25}
$$

for any odd function $\psi(x)$.

Considering the integral in the left-hand side of (1.25) and changing the corresponding variables gives

$$
\frac{1}{2} \int\limits_0^t \{\psi(x+a\tau) + \psi(x-a\tau)\}\,d\tau =
$$

$$
= \frac{1}{2a} \int\limits_x^{x+at} \psi(z)\,dz - \frac{1}{2a} \int\limits_x^{x-at} \psi(z)\,dz = \frac{1}{2a} \int\limits_{x-at}^{x+at} \psi(z)\,dz.
$$

Thus, the right-hand side of (1.25) turns into its left-hand side. To prove this equality in reverse direction, we write down

$$
\frac{1}{2a} \int\limits_{x-at}^{x+at} \psi(z)\,dz = \frac{1}{2a} \int\limits_{x-at}^{b} \psi(z)\,dz + \frac{1}{2a} \int\limits_b^{x+at} \psi(z)\,dz =
$$

$$
= -\frac{1}{2} \int\limits_t^{\frac{x-b}{a}} \psi(x-a\tau)\,d\tau + \frac{1}{2} \int\limits_{\frac{b-x}{a}}^t \psi(x+a\tau)\,d\tau =
$$

$$
= \frac{1}{2} \int\limits_0^t \psi(x-a\tau)\,d\tau - \frac{1}{2} \int\limits_0^{\frac{x-b}{a}} \psi(x-a\tau)\,d\tau + \frac{1}{2} \int\limits_0^t \psi(x+a\tau)\,d\tau -
$$

$$-\frac{1}{2}\int_0^{\frac{b-x}{a}} \psi(x+a\tau)\,d\tau = \frac{1}{2}\int_0^t [\psi(x+a\tau)+\psi(x-a\tau)]\,d\tau,$$

which completes the proof.

Keeping in mind that $\varphi(x)$ and $\psi(x)$ are the odd 2π-periodic functions, we see now that the function $u(x,t)$ given by (1.23) ((1.24)) satisfies the following boundary condition

$$u(0,t)=0, \quad u(\pi,t)=0 \quad \text{for all } t \geq 0.$$

On the basis of (1.23) ((1.24)) we conclude that

$$\varphi(x) \in \overset{0}{C}{}^2\,[0,\pi], \quad \psi(x) \in \overset{0}{C}{}^1\,[0,\pi], \tag{1.26}$$

is the necessary and sufficient condition for a classical solution of the homogeneous mixed problem to exist, and

$$\varphi(x) \in \overset{0}{C}\,[0,\pi], \quad \psi(x) - (R)\text{– is integrable on } [0,\pi]. \tag{1.27}$$

is that condition for a continuous solution to exist.

Moreover, if the functions $\varphi(x)$ and $\psi(x)$ (the odd 2π-periodic continuations of the functions $\tilde{\varphi}(x)$ and $\tilde{\psi}(x)$ from $[0,\pi]$ to $\mathbb{R}$) can be expanded into the Fourier series

$$\varphi(x) = \sum_{k=1}^{\infty} a_k \sin kx, \quad \psi(x) = \sum_{k=1}^{\infty} b_k \sin kx,$$

with the coefficients computed by (1.21), then a continuous solution of the mixed problem (1.19) can be represented as (1.20).

1.3. Nonhomogeneous mixed problem

This section is conserned with the following nonhomogeneous mixed problem

$$v_{tt} - a^2 v_{xx} = \tilde{f}(x,t), \quad (x,t) \in \overline{Q} = \{0 \leq x \leq \pi,\ 0 \leq t \leq T\}; \tag{1.28}$$

$$v(0,t)=0, \quad v(\pi,t)=0, \quad 0 \leq t \leq T; \tag{1.29}$$

$$v(x,0)=0, \quad v_t(x,0)=0, \quad 0 \leq x \leq \pi. \tag{1.30}$$

We shall show that the function

$$v(x,t) = \frac{1}{2}\int_0^t d\eta \int_0^\eta \{f(x+a(\eta-\tau),\tau) + f(x-a(\eta-\tau),\tau)\}d\tau \equiv$$

$$\equiv \frac{1}{2}\int_0^t d\tau \int_\tau^t \{f(x+a(\eta-\tau),\tau) + f(x-a(\eta-\tau),\tau)\}d\eta, \tag{1.31}$$

satisfies the nonhomogeneous equation (1.28) and conditions (1.29) and (1.30) but if the following condition is fulfilled

$$\tilde{f}(x,t) \in \overset{0}{C}(\overline{Q}), \quad \tilde{f}'_x \in C(\overline{Q}).$$

Here $f(x,t)$ is the odd -periodic continuation of the function $\tilde{f}(x,t)$ with respect to variable x and arbitrary fixed t.

Differentiating (1.31) yields

$$v_t(x,t) = \frac{1}{2}\int_0^t \{f(x+a(t-\tau),\tau) + f(x-a(t-\tau),\tau)\}d\tau; \tag{1.32}$$

$$v_{tt}(x,t) = f(x,t) + \frac{a}{2}\int_0^t \left\{\frac{\partial f(x+a(t-\tau),\tau)}{\partial(x+a(t-\tau))} - \frac{\partial f(x-a(t-\tau),\tau)}{\partial(x-a(t-\tau))}\right\}d\tau; \tag{1.33}$$

$$v_x(x,t) = \frac{1}{2}\int_0^t d\tau \int_\tau^t \{f'_x(x+a(\eta-\tau),\tau) + f'_x(x-a(\eta-\tau),\tau)\}d\eta =$$

$$= \frac{1}{2a}\int_0^t d\tau \int_\tau^t \{f'_\eta(x+a(\eta-\tau),\tau) - f'_\eta(x-a(\eta-\tau),\tau)\}d\eta =$$

$$= \frac{1}{2a}\int_0^t \{f(x+a(t-\tau),\tau) - f(x-a(t-\tau),\tau)\}d\tau;$$

$$v_{xx} = \frac{1}{2a}\int_0^t \left\{\frac{\partial f(x+a(t-\tau),\tau)}{\partial(x+a(t-\tau))} - \frac{\partial f(x-a(t-\tau),\tau)}{\partial(x-a(t-\tau))}\right\}d\tau. \tag{1.34}$$

Combining the expressions (1.33) and (1.34), we obtain $v_{tt} - a^2 v_{xx} = \tilde{f}(x,t)$, i. e. the function $v(x,t)$ given by (1.31) is the solution of the nonhomogeneous equation (1.28); and formulas (1.31) and (1.32) show that $v(x,t)$ satisfies the boundary conditions (1.29) and (1.30).

Thus, if the conditions

$$\tilde{f}(x,t) \in \overset{0}{C}(\overline{Q}) = \{\tilde{f}(x,t) \in \overset{0}{C}(\overline{Q}) \ : \ \tilde{f}(0,t) = \tilde{f}(\pi,t) = 0\}; \tag{1.35}$$

$$\tilde{f}'_x(x,t) \in C(\overline{Q}) \tag{1.36}$$

are fulfilled, then the classical solution of the nonhomogeneous mixed problem (1.28)–(1.30) exists. If only one condition (1.35) is fulfilled, then the continuous solution given by (1.31) exists.

Furthermore, if we assume that the function $f(x,t)$ can be expanded into the Fourier series

$$f(x,t) = \sum_{k=1}^{\infty} f_k(t)\sin kx,$$

then the solution $v(x,t)$ of the nonhomogeneous mixed problem (1.28)–(1.30) can be represented in the similar form [115]. This can be easily checked with the help of (1.31).

Remark 1.1. Consider the following linear nonhomogeneous mixed problem

$$\begin{gathered} u_{tt} - a^2 u_{xx} = \tilde{f}(x,t), \quad (x,t) \in \overline{Q}, \\ u(0,t) = u(\pi,t) = 0, \quad 0 \le t \le T. \\ u(x,0) = \tilde{\varphi}(x), \quad u_t(x,0) = \tilde{\psi}(x), \quad 0 \le x \le \pi. \end{gathered} \tag{1.37}$$

Assuming the conditions (1.26), (1.35) and (1.36) to be satisfied, i. e.

$$\tilde{\varphi}(x) \in \overset{0}{C^2}([0,\pi]), \quad \tilde{\psi}(x) \in \overset{0}{C^1}([0,\pi]);$$

$$\tilde{f}(x,t) \in \overset{0}{C}(\overline{Q}), \quad \tilde{f}'_x(x,t) \in C(\overline{Q}), \tag{1.38}$$

a unique classical solution of this problem exists. And if the condition

$$\tilde{\varphi}(x) \in \overset{0}{C}([0,\pi]), \quad \tilde{\psi}(x) - (R) \text{ –integrable on } [0,\pi],$$

$$\tilde{f}(x,t) \in \overset{0}{C}(\overline{Q}) \tag{1.39}$$

is satisfied, then a unique continuous solution given by

$$u(x,t) = \frac{\varphi(x+at)+\varphi(x-at)}{2} + \frac{1}{2}\int_0^t (\psi(x+a\tau)+\psi(x-a\tau))+$$

$$+\frac{1}{2}\int_0^t d\tau \int_0^\tau \{f(x+a(\eta-\tau),\tau)+f(x-a(\eta-\tau),\tau)\}d\eta \tag{1.40}$$

exists.

1.4. Reduction of the second order quasiwave equation to the first order systems

We consider the following second order quasiwave equation

$$\frac{\partial^2 u}{\partial t^2} - a^2\frac{\partial^2 u}{\partial x^2} = F[u, u_t, u_x, \varepsilon], \quad 0 < \varepsilon \leq 1. \tag{1.41}$$

in a region $Q_\infty = \{-\infty < x < \infty,\ 0 \leq t \leq T\}$.

The (nonlinear) operator F in (1.41) turns each smooth function $u(x,t)$ defined in Q_∞ into a scalar function $F[u, u_t, u_x, \varepsilon]\ (x,t)$ defined in the same region. This operator is of Volterra type, i. e. for $F[u, u_t, u_x, \varepsilon]\ (x, t_0)$, the value $u(x,t)$ depends only on the value $t \leq t_0$ and continuously depends on the parameter ε.

If we introduce the differentiation symbols $\frac{\partial}{\partial t}$ and $\frac{\partial}{\partial x}$ acting according to the rules $\frac{\partial}{\partial t}(u) = \frac{\partial u}{\partial t}$ and $\frac{\partial}{\partial x}(u) = \frac{\partial u}{\partial t}$, equation (1.41) admits the following representations

$$\left(\frac{\partial}{\partial t} - a\frac{\partial}{\partial x}\right)\left(\frac{\partial u}{\partial t} + a\frac{\partial u}{\partial x}\right) = F\left[u, \frac{\partial u}{\partial t}, \frac{\partial u}{\partial x}, \varepsilon\right]; \tag{1.42}$$

$$\left(\frac{\partial}{\partial t} + a\frac{\partial}{\partial x}\right)\left(\frac{\partial u}{\partial t} - a\frac{\partial u}{\partial x}\right) = F\left[u, \frac{\partial u}{\partial t}, \frac{\partial u}{\partial x}, \varepsilon\right]. \tag{1.43}$$

in the class of functions $u \in C^2(Q_\infty)$.

Taking now account of the notation

$$\begin{aligned} u_1 &= \frac{\partial u}{\partial t} + a\frac{\partial u}{\partial x}, \\ u_2 &= \frac{\partial u}{\partial t} - a\frac{\partial u}{\partial x}, \end{aligned} \iff \begin{aligned} \frac{\partial u}{\partial t} &= \frac{u_1+u_2}{2}, \\ \frac{\partial u}{\partial x} &= \frac{u_1-u_2}{2a}, \end{aligned} \tag{1.44}$$

and equalities (1.42) and (1.43), we obtain two system of equations

$$\frac{\partial u_1}{\partial t} - a\frac{\partial u_1}{\partial x} = F\left[u, \frac{u_1+u_2}{2}, \frac{u_1-u_2}{2a}, \varepsilon\right],$$

$$\frac{\partial u_2}{\partial t} + a\frac{\partial u_2}{\partial x} = F\left[u, \frac{u_1+u_2}{2}, \frac{u_1-u_2}{2a}, \varepsilon\right], \quad \frac{\partial u}{\partial t} = \frac{u_1+u_2}{2}; \tag{1.45}$$

$$\frac{\partial u_1}{\partial t} - a\frac{\partial u_1}{\partial x} = F\left[u, \frac{u_1+u_2}{2}, \frac{u_1-u_2}{2a}, \varepsilon\right],$$

$$\frac{\partial u_2}{\partial t} + a\frac{\partial u_2}{\partial x} = F\left[u, \frac{u_1+u_2}{2}, \frac{u_1-u_2}{2a}, \varepsilon\right], \quad \frac{\partial u}{\partial x} = \frac{u_1+u_2}{2a}; \tag{1.46}$$

Each solution $u \in C^2(Q_\infty)$ of equation (1.41) is thus a solution (u, u_1, u_2) of the systems (1.45) and (1.46) provided the functions u, u_1, u_2 are related by equality (1.44).

We shall now prove that each smooth solution of, say, system (1.45) is the solution of equation (1.41). Indeed, adding the first two equations in (1.45) yields

$$\frac{\partial}{\partial t}\left(\frac{u_1+u_2}{2}\right) - a^2\frac{\partial}{\partial x}\left(\frac{u_1-u_2}{2a}\right) = F\left[u, \frac{u_1+u_2}{2}, \frac{u_1-u_2}{2a}, \varepsilon\right]. \tag{1.47}$$

Combining equation (1.47) with the third equation in (1.45) and the notation (1.44) we conclude that the function $u(x,t)$ satisfies equation (1.41). Hence, each smooth solution of system (1.45) is the solution of equation (1.41).

1.5. Reduction of the quasiwave equation to a system of integral equations

Let the axis $x(t=0)$ be the initial line for equation (1.41) and let the initial condition

$$u(x,0) = \varphi(x), \quad u_t(x,0) = \psi(x), \quad -\infty < x < \infty \tag{1.48}$$

be given. Using equalities in (1.44) we then obtain the following initial condition for system (1.45)

$$\begin{aligned} u_1(x,0) &= \psi(x) + a\varphi'(x) \equiv \varphi_1(x), \\ u_2(x,0) &= \psi(x) - a\varphi'(x) \equiv \varphi_2(x), \\ u(x,0) &= \varphi(x). \end{aligned} \tag{1.49}$$

Going over from the Cauchy problem (1.45), (1.49) to the system of integral equations we find

$$u_1(x,t) = \varphi_1(x+at) + \int_0^t \Phi[u, u_1, u_2, \varepsilon](x - a(\tau - t), \tau)\, d\tau,$$

$$u_2(x,t) = \varphi_2(x-at) + \int_0^t \Phi[u, u_1, u_2, \varepsilon](x + a(\tau - t), \tau)\, d\tau,$$

$$u(x,t) = \varphi(x) + \frac{1}{2}\int_0^t \{u_1, (x,\theta) + u_2(x,\theta)\}\, d\theta, \tag{1.50}$$

where $\Phi[u, u_1, u_2, \varepsilon] = F[u,\ (u_1+u_2)/2,\ (u_1-u_2)/2a,\ \varepsilon]$.

Finding a solution of the Cauchy problem (1.41), (1.48) is thus equivalent to finding a solution of the system of integral equations (1.50). Moreover, this Cauchy problem is equivalent to the following system of integral equations (see 1.7.):

$$u(x,t) = \varphi(x) + \frac{1}{2}\int_0^t (\varphi_1(x+a\theta) + \varphi_2(x-a\theta))\,d\theta +$$

$$+\frac{1}{2}\int_0^t d\theta \int_0^\theta \{F[u,u_t,u_x,\varepsilon](x-a(\tau-\theta),\tau) + F[u,u_t,u_x,\varepsilon](x+a(\tau-t),\tau)\}\,d\tau \equiv$$

$$\equiv \frac{\varphi(x+at)+\varphi(x-at)}{2} + \frac{1}{2}\int_{x-at}^{x+at} \psi(\alpha)\,d\alpha + \frac{1}{2}\int_0^t d\tau \int_\tau^t \{F[u,u_t,u_x,\varepsilon](x-a(\tau-\theta),\tau)+$$

$$+F[u,u_t,u_x,\varepsilon](x+a(\tau-\theta),\tau)\}\,d\theta, \tag{1.51}$$

$$u_t(x,t) = \frac{\partial}{\partial t}(u(x,t)), \quad u_x(x,t) = \frac{\partial}{\partial x}(u(x,t)).$$

In system (1.51), the sum of the first two terms of the first equation is the solution of the homogeneous equation $u_{tt} - a^2 u_{xx} = 0$ (according to the d'Alembert formula) and the third term is the solution of the nonhomogeneous equation (1.41).

Definition 1.1. The solution of the system of intergal equations (1.51) is referred to as a smooth generalized solution or simply smooth solution of the problem (1.41), (1.48); and the solution of the system of integral equations (1.50) is referred to as a continuously generalized or simply continuous solution of the Cauchy problem (1.45), (1.49) [1, 2, 61].

1.6. Quasilinear mixed problem

The present section deals with the existence of a solution of the mixed problem

$$u_{tt} - a^2 u_{xx} = \tilde{F}[u,u_t,u_x,\varepsilon], \quad 0 < \varepsilon \le 1; \tag{1.52}$$

$$u(0,t) = u(l,t) = 0, \quad 0 \le t \le T; \tag{1.53}$$

$$u(x,0) = \tilde{\varphi}(x), \quad u_t(x,0) = \tilde{\psi}(x), \quad 0 \le t \le l. \tag{1.54}$$

The operator $\tilde{F}$ turns each smooth function $u(x,t)$ defined in $\Pi_T = \{0 \le x \le l,\ 0 \le t \le T;\}$ into a scalar function $\tilde{F}[u,u_t,u_x,\varepsilon]\ (x,t)$ defined in $\Pi_T \times (0,1]$ and its values continuously depend on the parameter ε.

As shown above, the passage from equation (1.52) to the first order system (1.45)

$$u_{it} + (-1)^i a u_{ix} = \tilde{F}[u,(u_1+u_2)/2,\ (u_1+u_2)/2a,\varepsilon], \quad i=1,2,$$

$$u_t = (u_1+u_2)/2 \tag{1.55}$$

makes use of the equalities in (1.44): $u_1 = u_t + au_x$, $u_2 = u_t - au_x$. Taking into account these equalities and the conditions (1.53) and (1.54) we obtain the following boundary and initial conditions for system (1.55)

$$u_1(0,t) + u_2(0,t) = 0, \quad u_1(l,t) + u_2(l,t) = 0; \tag{1.56}$$

$$u_1(x,0) = \tilde{\psi}(x) + a\tilde{\varphi}'(x) \equiv \tilde{\varphi}_1(x),$$

$$u_2(x,0) = \tilde{\psi}(x) - a\tilde{\varphi}'(x) \equiv \tilde{\varphi}_2(x), \quad u(x,0) = \tilde{\varphi}(x). \tag{1.57}$$

It follows that the mixed problem (1.52)–(1.54) is equivalent, in the region Π_T, to the corresponding mixed problem (1.55)–(1.57) for the first order hyperbolic system.

Theorem 1.2. *Let the following conditions be satisfied:*

1) $\tilde{\varphi}(x) \in \overset{0}{C}{}^1([0,l]) = \{\tilde{\varphi}(x) \in C^1 : \tilde{\varphi}(0) = \tilde{\varphi}(l) = 0\}$, $\tilde{\psi}(x) \in \overset{0}{C}([0,l])$;

2) the function $\tilde{F}[u, u_t, u_x, \varepsilon]\ (x,t)$ *is determined for any smooth function* $u(x,t)$*, is continuous in* $\Pi_T \times (0,1]$ *and satisfies the condition: if the function* $u \in C^1(\Pi_T)$ *is replaced by any other function* $\tilde{u} \in C^1(\Pi_T)$*, then*

$$\int_0^t \max_x \left|\Delta\tilde{F}[u, u_t, u_x, \varepsilon](x,t)\right| d\tau \le K \int_0^l \{\max_{x;\theta\le\tau} |\Delta u\,(x,\theta)| +$$

$$+ \max_{x;\theta\le\tau} |u_t\,(x,\theta)| + \max_{x;\theta\le\tau} |u_x\,(x,\theta)|\} d\tau,$$

where $\Delta u = \tilde{u} - u$ *and so on, and the constant* K *is determined by the operator* $\tilde{F}$ *(by the values of* T *and* ε *in particular).*

Then the mixed problem (1.52)–(1.54) has one and only one smooth solution in the rectangular Π_T.

Proof. Let the rectangular $\Pi_\infty = \{0 \le x \le l,\ 0 \le t < \infty\}$ be partitioned into the regions $\Delta_1, \Delta_2, \Delta_3, \ldots$ by the lines that are given by the equations $x = at - kl$, $x = -at + (k+1)l$, $k = 0, 1, 2, \ldots$ and start at the points $(0, kl/a)$, $(l, kl/a)$ inside Π_∞ (Fig. 1).

Let $(x,t) \in \Delta_1$. For that set of points, we get the Cauchy problem (1.55) and (1.57), which is equivalent to the system of intergal equations

$$u_1(x,t) = \tilde{\varphi}_1(x + at) + \int_0^t \tilde{\Phi}[u, u_1, u_2, \varepsilon](x - a(\tau - t), \tau)\, d\tau,$$

$$u_2(x,t) = \tilde{\varphi}_2(x - at) + \int_0^t \tilde{\Phi}[u, u_1, u_2, \varepsilon](x + a(\tau - t), \tau)\, d\tau,$$

$$u(x,t) = \tilde{\varphi}(x) + \int_0^t [u_1, (x,\theta) + u_2(x,\theta)]/2 d\theta, \tag{1.58}$$

where $\tilde{\Phi}[u, u_1, u_2, \varepsilon] = \tilde{F}[u,\ (u_1 + u_2)/2,\ (u_1 - u_2)/2a,\ \varepsilon]$.

If we now take the functions $u_1^0 = \tilde{\varphi}_1(x + at)$, $u_2^0 = \tilde{\varphi}_1(x - at)$, $u^0 = \tilde{\varphi}(x)$ as the zero approximation and use the method of successive approximations, then, keeping in mind the condition in Theorem 1.2, we see that the system of Volterra type integral equations (1.58) has one and only one continuous solution in the region Δ_1.

Our next claim is that the mixed problem (1.55)–(1.57) is equivalent, in the other regions $\Delta_2, \Delta_3, \Delta_4, \ldots$ to the corresponding systems of Volterra type integral equations.

Indeed, let $(x,t) \in \Delta_2$ (see Fig. 1). For that set of points, the mixed problem (1.55)–(1.57) is equivalent to the system of integral equations

$$u_1(x,t) = \tilde{\varphi}_1(x + at) + \int_0^t \tilde{\Phi}[u, u_1, u_2, \varepsilon](x - a(\tau - t), \tau)\, d\tau,$$

$$u_2(x,t) = u_2(0, t_0) + \int_0^t \tilde{\Phi}[u, u_1, u_2, \varepsilon](x + a(\tau - t), \tau)\, d\tau,$$

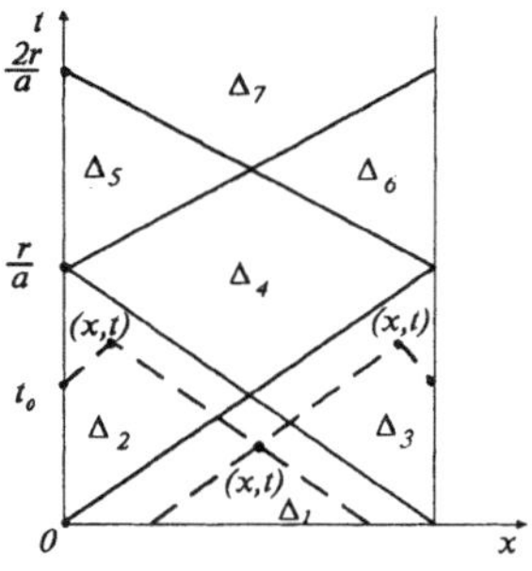

Fig. 1.

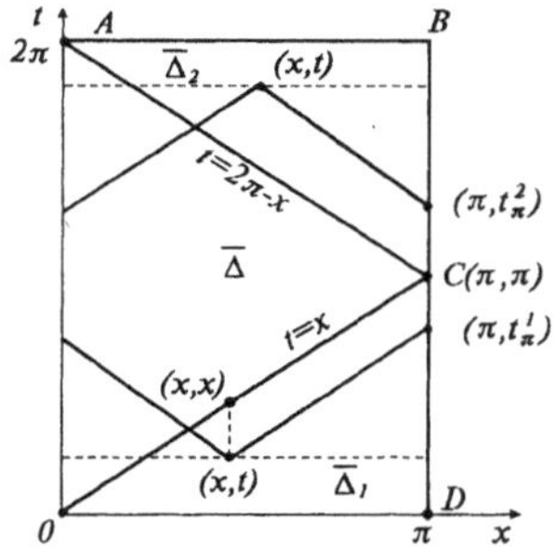

Fig. 2.

$$u(x,t) = \tilde{\varphi}(x) + \frac{1}{2}\int_0^t [u_1(x,\theta) + u_2(x,\theta)]\, d\theta, \tag{1.59}$$

where $t_0 = t - x/a$; $0 < t_0 \leq t$. The boundary condition $u_1(0,t) + u_2(0,t) = 0$ now gives $u_2(0,t_0) = -u_1(0,t_0)$. Since the value of $u_1(0,t_0)$ is always defined according to the first equation in system (1.59), we obtain finally the following system of Volterra type integral equations

$$u_1(x,t) = \tilde{\varphi}_1(x+at) + \int_0^t \tilde{\Phi}[u,u_1,u_2,\varepsilon](x - a(\tau - t),\tau)\, d\tau,$$

$$u_2(x,t) = -\tilde{\varphi}_1(at_0) - \int_0^{t_0} \tilde{\Phi}[u,u_1,u_2,\varepsilon](-a(\tau - t_0),\tau)\, d\tau +$$

$$+ \int_{t_0}^t \tilde{\Phi}[u,u_1,u_2,\varepsilon](x + a(\tau - t),\tau)\, d\tau,$$

$$u(x,t) = \tilde{\varphi}(x) + \frac{1}{2}\int_0^t [u_1,(x,\theta) + u_2(x,\theta)]\, d\theta. \tag{1.60}$$

Let $(x,t) \in \Delta_3$ (see Fig. 1). The mixed problem id then equivalent to the following system of intergal equations

$$u_1(x,t) = u_1(l,t_1) + \int_0^t \tilde{\Phi}[u,u_1,u_2,\varepsilon](x - a(\tau - t),\tau)\, d\tau,$$

$$u_2(x,t) = \tilde{\varphi}_2(x - at) + \int_0^t \tilde{\Phi}[u,u_1,u_2,\varepsilon](x + a(\tau - t),\tau)\, d\tau,$$

$$u(x,t) = \tilde{\varphi}(x) + \frac{1}{2}\int_0^t \{u_1,(x,\theta) + u_2(x,\theta)\}\, d\theta.$$

By using the boundary condition $u_1(l,t) + u_2(l,t) = 0$, we can rewrite this system as

$$u_1(x,t) = -\tilde{\varphi}_2(l - at_l) - \int_0^{t_l} \tilde{\Phi}[u, u_1, u_2, \varepsilon](l + a(\tau - t_l), \tau)\, d\tau +$$

$$+ \int_{t_l}^{t} \tilde{\Phi}[u, u_1, u_2, \varepsilon](x - a(\tau - t_l), \tau)\, d\tau,$$

$$u_2(x,t) = \tilde{\varphi}_2(x - at) + \int_0^t \tilde{\Phi}[u, u_1, u_2, \varepsilon](x + a(\tau - t), \tau)\, d\tau,$$

$$u(x,t) = \tilde{\varphi}(x) + \frac{1}{2}\int_0^t \tilde{\Phi}\{u_1(x,\theta) + u_2(x,\theta\}\, d\theta, \tag{1.61}$$

where $t_2 = t - (l - x)/a,\ 0 < t_l \le t$.

Now let $(x,t) \in \Delta_4$. For that set of points, the mixed problem (1.55)–(1.57) is equivalent to the following system of integral equations

$$u_1(x,t) = u_1(0, t_0) + \int_{t_0}^{t} \tilde{\Phi}[u, u_1, u_2, \varepsilon](x - a(\tau - t), \tau)\, d\tau,$$

$$u_2(x,t) = u_2(l, t_l) + \int_{t_l}^{t} \tilde{\Phi}[u, u_1, u_2, \varepsilon](x + a(\tau - t), \tau)\, d\tau,$$

$$u(x,t) = \tilde{\varphi}(x) + \frac{1}{2}\int_0^t \{u_1(x,\theta) + u_2(x,\theta)\}\, d\theta, \tag{1.62}$$

where $t_0 = t - x/a$; $t_l = t - (l - x)/a$; $0 < t_0 \le l/a$; $0 < t_l \le l/a$.

Since the functions $u_1(0, t_0)$ and $u_2(l, t_l)$, $0 < t_0 < l/a$, $0 < t_l < l/a$ are given by the systems of integral equations (1.60) and (1.61), the system of integral equations can always be solved with the method of successive approximations provided the condition in Theorem 1.2 is fulfilled.

Proceeding analogously with the regions Δ_k, $k = 5, 6, \ldots$, covering $\Pi_T = \{0 \le x \le l, 0 \le t \le T\}$ we deduce by induction that the assertion in Theorem 1.2 holds for all $(x,t) \in \Pi_T$.

We have thus shown that the mixed problem (1.55)–(1.57) for the first order hyperbolic system has one and only one solution continuous in the rectangular Π_T. According to the above definition, this implies the existence of a smooth solution of the mixed problem (1.52)–(1.54) for the second order quasiwave equation, which proves the theorem.

Remark 1.2. It should be noted that the classical solution of the mixed problem (1.52)–(1.54) might not exist even if the function $\tilde{F}[u, u_t, u_x, \varepsilon]\ (x,t)$ is smooth enough. This relates to the fact that a solution of the first order hyperbolic system can have discontinuities along characteristics. The additional condition for the classical solution to exist is the condition of agreement at the points $(0,0)$ and $(l,0)$ imposed on the initial and boundary functions and on the operator in the right-hand side of (1.52).

1.7. A property of solutions of quasilinear mixed problem

Let $\tilde{A}$ and A^0 be the function spaces

$$\tilde{A} = \{u(x,t) \in C^1(\Pi_T)\}, \quad \Pi_T = [l;0,T];$$

$$A^0 = \{u(x,t) \in C^1(Q_\infty) : u(x+2l,t) = u(x,t) = -u(-x,t)\},$$
$$Q_\infty = \mathbb{R} \times [0,T].$$

Definition 1.2. If the operator $\tilde{F}[u,u_t,u_x,\varepsilon]$ defined on the set $\tilde{A}$ is also defined on the set A^0 and its values $\tilde{F}[u,u_t,u_x,\varepsilon]$ on this set satisfy the conditions

$$1) \quad F[u,u_t,u_x,\varepsilon](x+2l,t) = F[u,u_t,u_x,\varepsilon](x,t);$$

$$2) \quad F[u,u_t,u_x,\varepsilon](-x,t) = -F[u,u_t,u_x,\varepsilon](x,t),$$

then we say that $\tilde{F}$ admits an odd $2l$-periodic continuation from the set $\tilde{A}$ to the set A^0 with respect to variable x. Its values are denoted as $F[u,u_t,u_x,\varepsilon](x,t)$.

Theorem 1.3. *Let the conditions 1) and 2) in Theorem 1.2 be fulfilled; and let the following condition be also fulfilled: the operator $\tilde{F}$ admits an odd $2l$-periodic continuation from the set $\tilde{A} = \{u(x,t) \in C^1(\Pi_T)\}$ to the set $A^0 = \{u(x,t) \in C^1(Q_\infty) : u(x+2l,t) = u(x,t) = -u(-x,t)\}$ with respect to variable x.*

The mixed problem (1.52)-(1.54) has then one and only one smooth solution $u \in A^0$, which satisfies the system of integral equation (1.51).

As in the proof of Theorem 1.2, we partition the rectangular $\Pi_\infty = \{0 \le x \le l,\ 0 \le t < \infty\}$ into the regions $\Delta_1, \Delta_2, \ldots$ by the lines $x = at - kl$, $x = -at + (k+1)\,l$, $k = 0,1,2,\ldots$ (see Fig. 1).

Let $(x,t) \in \Delta_1$. The functions $\varphi(x)$ and $\psi(x)$ are the odd continuations of $\tilde{\varphi}(x)$ and $\tilde{\psi}(x)$ from $[0,l]$ to the real line $\mathbb{R}$. We set $\varphi_1(x) = \psi(x) + a\varphi'(x)$ and $\varphi_2(x) = \psi(x) - a\varphi'(x)$. On the set A^0, the system of integral equations (1.58) turns then into system (1.50). Combining the equations in this system with the notation $(u_1+u_2)/2 = u_t$, $(u_1-u_2)/2 = u_x$ we arrive at different expressions of the first equation in system (1.51). Indeed, substituting $u_1(x,t)$ and $u_2(x,t)$ into the third equation in (1.50) yields

$$u(x,t) = \varphi(x) + \frac{1}{2}\int_0^t \{\varphi_1(x+a\theta) + \varphi_2(x+a\theta)\}\,d\theta +$$
$$+\frac{1}{2}\int_0^t d\theta \int_0^\theta \{F[u,u_t,u_x,\varepsilon](x-a(\tau-\theta),\tau) +$$
$$+F[u,u_t,u_x,\varepsilon](x+a(\tau-\theta),\tau)\}\,d\tau. \tag{1.63}$$

Taking account of the equalities $\varphi_1(x) = \psi(x) + a\varphi'(x)$ and $\varphi_2(x) = \psi(x) - a\varphi'(x)$ and reversing the order of integration in the last term of equality (1.63), we obtain

$$u(x,t) = \frac{\varphi(x+at) + \varphi(x-at)}{2} + \frac{1}{2}\int_0^t \{\psi(x+a\theta) + \psi(x-a\Theta)\}\,d\theta +$$

$$+\frac{1}{2}\int_0^t d\tau \int_0^t \{F[u,u_t,u_x,\varepsilon](x-a(\tau-\theta),\tau) + F[u,u_t,u_x,\varepsilon](x+a(\tau-\theta),\tau)\}\,d\theta. \tag{1.64}$$

Changing variables in (1.64) allows us to rewrite it in the following form suitable for differentiation

$$u(x,t) = \frac{\varphi(x+at)+\varphi(x-at)}{2} + \frac{1}{2a}\int\limits_{x-at}^{x+at} \psi(\alpha)\,d\alpha +$$

$$+\frac{1}{2a}\int\limits_0^t d\tau \int\limits_{x-a(t-\tau)}^{x+a(t-\tau)} F[u,u_t,u_x,\varepsilon](\eta,\tau)\,d\eta. \qquad (1.65)$$

Combining equalities (1.63)–(1.65) we thus arrive at the system of integral equations (1.51).

Let $(x,t) \in \Delta_2$ (see Fig. 1) and let the conditions 1 and 2 in Theorem 1.2 be fulfilled. A smooth solution of the mixed problem (1.52)–(1.54) then satisfies, on the set Δ_2, the system of integral equations (1.60). Using the equality

$$\varphi_1(-x) = -\varphi_2(x) \qquad (1.66)$$

we shall prove that system (1.60) coincides with system (1.50) on the set of functions A^0. To do this, it suffices to show that the second equation in (1.60) can be transformed, on the set A^0, into the second equation in (1.50). Writing $t_0 = t - x/a$ we have

$$u_2(x,t) = -\varphi_2(at-x) - \int\limits_0^{t_0} \Phi[u,u_1,u_2,\varepsilon](-a(\tau-(t-x/a)),\tau)\,d\tau +$$

$$+\int\limits_0^{t_0} \Phi[u,u_1,u_2,\varepsilon](x+a(\tau-t),\tau)\,d\tau =$$

$$= \varphi_2(x-at) + \int\limits_0^{t_0} \Phi[u,u_1,u_2,\varepsilon](x+a(\tau-t),\tau)\,d\tau +$$

$$+\int\limits_0^{t} \Phi[u,u_1,u_2,\varepsilon](x+a(\tau-t),\tau)\,d\tau \equiv$$

$$\equiv \varphi_2(x-at) + \int\limits_0^{t} \Phi[u,u_1,u_2,\varepsilon](x+a(\tau-t),\tau)\,d\tau$$

by virtue of the condition 1 in Theorem 1.3 and equality (1.66).

Hence, system (1.60) has, on the set A^0, the same form as system (1.50). The smooth solution $u(x,t)$ of the quasilinear mixed problem (1.52)–(1.54) is therefore determined, on the set Δ_2, by the solution of the system of integral equations (1.51).

Proceeding by induction we see that the assertion in Theorem 1.3 holds for all the points of Π_T. This completes the proof.

1.8. Justification of the asymptotic methods to be applied to the investigation of quasilinear mixed problems

In this section we formulate one another statement on the representation of a solution of the mixed problem (1.52)–(1.54), which was formally used in building up the asymptotic methods for the investigation of solutions of quasiwave mixed problems [19, 26, 28, 30, 32, 39, 49, 50, 140–144, 148].

Theorem 1.4. *Let the condition in Theorem 1.3 and the following condition be satisfied: for any smooth function* $u(x,t) = \sum_{k=1}^{\infty} u_k(t)\sin\frac{k\pi x}{l} \in A^0$, *the function* $F[u, u_t, u_x, \varepsilon](x,t) \in A^0$ *belongs to* $\sin\frac{k\pi x}{l}$ *and can be expanded into the Fourier series in* $\sin\frac{k\pi x}{l}$, *i. e.* $F[u, u_t, u_x, \varepsilon](x,t) = \sum_{k=1}^{\infty} F_k(t,\varepsilon)\sin\frac{k\pi x}{l}$.

The solution $u(x,t) \in A^0$ *of the mixed problem (1.52)–(1.54) then admits the representation*

$$u(x,t) = \sum_{k=1}^{\infty} (z_k(t,\varepsilon)\cos\omega_k t + \omega_k(t,\varepsilon)\sin\omega_k t)\sin\frac{k\pi x}{l}, \quad \omega_k = \frac{k\pi a}{l}.$$

Proof. Indeed, let us take the function

$$u_0(x,t) = \frac{\varphi(x+at)+\varphi(x-at)}{2} + \frac{1}{2}\int_0^t (\psi(x+a\theta)+\psi(x-a\theta))d\theta \in A^0,$$

as the zero approximation. The system of integral equations (1.51) and the condition in Theorem 1.4 then allow us to construct a uniformly converging sequence of functions $\{u_n(x,t)\}$, $\{u_{nt}(x,t)\}$, $\{u_{nx}(x,t)\}$ with the $(n+1)$-th term of the sequence $\{u_n(x,t)\}$ being of the form

$$u_{n+1}(x,t) = \frac{\varphi(x+at)+\varphi(x-at)}{2} + \frac{1}{2}\int_0^t (\psi(x+a\theta)+\psi(x-a\theta))d\theta +$$

$$+\frac{1}{2}\int_0^t d\tau \int_\tau^t \{F[u_n, u_{nt}, u_{nx}, \varepsilon](x - a(\tau-\theta), \tau) +$$

$$+F[u_n, u_{nt}, u_{nx}\varepsilon](x + a(\tau-\theta), \tau)\}\,d\theta, \quad n = 0, 1, 2, \ldots. \tag{1.67}$$

Let $\varphi(x) = \sum_{k=1}^{\infty} \varphi_k \sin\frac{k\pi x}{l}$ and $\psi(x) = \sum_{k=1}^{\infty} \psi_k \sin\frac{k\pi x}{l}$ be the expansions of the odd $2l$-periodic initial functions $\varphi(x)$ and $\psi(x)$, which are odd $2l$-periodic continuations of the initial functions $\tilde{\varphi}(x)$ and $\tilde{\psi}(x)$ from $[0,l]$ to $\mathbb{R}$.

Equality (1.67) and the condition in Theorem 1.4 now give

$$u_{n+1}(x,t) = \sum_{k=1}^{\infty} \sin\frac{k\pi x}{l}\left(\varphi_k\cos\omega_k t + \frac{1}{\omega_k}\psi_k\sin\omega_k t\right) +$$

$$+\sum_{k=1}^{\infty}\sin\frac{k\pi x}{l}\left(\frac{\sin\omega_k t}{\omega_k}\int_0^t F_{kn}(\tau,\varepsilon)\cos\omega_k\tau d\tau - \right.$$

$$\left. - \frac{\cos\omega_k t}{\omega_k}\int_0^t F_{kn}(\tau,\varepsilon)\sin\omega_k\tau d\tau\right), \quad n = 0, 1, 2, \ldots, \tag{1.68}$$

where $\omega_k = k\pi a/l$, and F_{kn} are the Fourier coefficients in the expansion of the function $F[u_n, u_{nt}, u_{nx}, \varepsilon](x,t)$ into series in $\sin \dfrac{k\pi x}{l}$.

By virtue of the condition in Theorem 1.4, the sequence $\{u_n(x,t)\}$ converges uniformly. If we introduce the notation $\lim_{n\to\infty} u_n(x,t) = u(x,t)$, then from (1.68) we obtain

$$u(x,t) = \sum_{k=1}^{\infty} (z_k(t,\varepsilon) \cos \omega_k t + \omega_k(t,\varepsilon) \sin \omega_k t) \sin \frac{k\pi x}{l},$$

which is our assertion.

Remark 1.3. Let

$$F[u, u_t, u_x, \varepsilon](x,t) = \varepsilon f(x, t, u, u_t, u_x) \tag{1.69}$$

and let the condition in Theorem 1.4 be satisfied. The averaging method can then be used to investigate the quasilinear mixed problem (1.52)–(1.54), and if the condition in Theorem 1.4 is satisfied, the Krylov-Bogolyubov asymptotic expansions hold [66]. In the last case, the interval $[0,T]$ where a solution exists widens, and if $f(x,t,u,u_t,u_x)$ is an analytic function of variables u, u_t, u_x in the region $\Pi_T \times \mathbb{R}^3$, then the solution of the mixed problem (1.52)–(1.54) is the analytic function of the parameter ε, $0 < \varepsilon \ll 1$ [145].

1.9. A periodic boundary value problem

Given the quasiwave equation

$$u_{tt} - a^2 u_{xx} = \tilde{F}[u, u_t, u_x, \varepsilon], \quad 0 < \varepsilon \ll 1, \tag{1.70}$$

we are interested in additional conditions ensuring the existence of a solution of (1.70).

If the axis $x(t=0)$ is the initial line and the initial and boundary conditions are given as $u(x,0) = \tilde{\varphi}(x)$, $u_t(x,0) = \tilde{\psi}(x)$, $u(0,t) = u(l,t) = 0$ then the above mixed problem (1.52)–(1.54) is solvable under certain condition (see §§1.6–1.8).

Let the axis $t(x=0)$ be the initial line for equation (1.70) and let the initial condition

$$u(x,0) = \nu(t), \quad u_x(0,t) = \mu(t), \quad t \in \mathbb{R} \tag{1.71}$$

be given.

Since equation (1.70) is equivalent, in the class of smooth functions, to the following first order system of the form (1.45)

$$u_{it} + (-1)^i a u_{ix} = \tilde{F}[u, (u_1+u_2)/2, (u_1-u_2)/2a, \varepsilon], \quad i = 1,2,$$

$$u_x = (u_1 - u_2)/2a, \tag{1.72}$$

the initial condition (1.71) determines the initial condition for system (1.72). Indeed, according to the equalities $u_1 = u_1 + au_2$, $u_2 = u_t - au_x$ we have

$$u_1(0,t) = \nu'(t) + a\mu(t) \equiv \mu_1(t),$$

$$u_2(0,t) = \nu'(t) - a\mu(t) \equiv \mu_2(t), \quad u(0,t) = \nu(t). \tag{1.73}$$

Consider the characteristic triangle $\Delta = \{(x,t) \in \mathbb{R}^2 : 0 \le x \le l, x/a \le t \le 2la - x/a\}$ (see Fig. 1). Let $(x,t) \in \Delta$. The characteristics originating at this point then meet the axis

t at the points $(0, t_0^1 = t + x/a)$ and $(0, t_0^2 = t - x/a)$. Therefore, this enables the Cauchy problem (1.72), (1.73) to be replaced by the system of integral equations

$$u_1(x,t) = \mu_1\left(t + \frac{x}{a}\right) + \int\limits_{t_0^{(1)}}^{t} \tilde{\Phi}[u, u_1, u_2, \varepsilon](x - a(\tau - t), \tau)\, d\tau \equiv$$

$$\equiv \mu_1\left(t + \frac{x}{a}\right) + \frac{1}{a}\int\limits_0^x \tilde{\Phi}[u, u_1, u_2, \varepsilon]\left(\eta, t + \frac{x-\eta}{a}\right) d\eta,$$

$$u_2(x,t) = \mu_2\left(t - \frac{x}{a}\right) + \int\limits_{t_0^{(2)}}^{t} \tilde{\Phi}[u, u_1, u_2, \varepsilon](x + a(\tau - t), \tau)\, d\tau \equiv$$

$$\equiv \mu_2\left(t - \frac{x}{a}\right) + \frac{1}{a}\int\limits_0^x \tilde{\Phi}[u, u_1, u_2, \varepsilon]\left(\eta, t + \frac{\eta - x}{a}\right) d\eta,$$

$$u(x,t) = \nu(t) + \frac{1}{2a}\int\limits_0^x (u_1(\xi,t) - u_2(\xi,t))\, d\xi,$$

$$\tilde{\Phi}[u, u_1, u_2, \varepsilon] = \tilde{F}[u, , (u_1 + u_2)/2, (u_1 - u_2)/2a, \varepsilon]. \tag{1.74}$$

Solving the Cauchy problem (1.70), (1.71) is thus equivalent to solving the system of integral equations (1.74). Moreover, the above Cauchy problem is equivalent to the following system of integral equations

$$u(x,t) = \nu(t) + \frac{1}{2a}\int\limits_0^x (\mu_1(t + \xi/a) - \mu_2(t - \xi/a))\, d\xi -$$

$$-\frac{1}{2a^2}\int\limits_0^x d\xi \int\limits_0^\xi \left\{\tilde{F}[u, u_t, u_x, \varepsilon]\left(\eta, t + \frac{\xi - \eta}{a}\right) + \tilde{F}[u, u_t, u_x, \varepsilon]\left(\eta, t + \frac{\eta - \xi}{a}\right)\right\} d\eta \equiv$$

$$\equiv \frac{\nu(t + x/a) + \nu(t - x/a)}{2} + \frac{a}{2}\int\limits_{t-x/a}^{t+x/a} \mu(\alpha)\, d\alpha - \frac{1}{2a}\int\limits_0^x d\eta \int\limits_{t+\frac{\eta - x}{a}}^{t+\frac{x-\eta}{a}} \tilde{F}[u, u_t, u_x, \varepsilon](\eta, \tau)\, d\tau, \tag{1.75}$$

$$u_t = \frac{\partial}{\partial t}(u), \quad u_x = \frac{\partial}{\partial x}(u).$$

If we now assume that the initial condition

$$u(l,t) = \nu(t), \quad u_x(l,t) = \mu(t), \quad t \in \mathbb{R}, \tag{1.76}$$

is defined at the side $x = l$, then the Cauchy problem (1.70), (1.76) is equivalent, in the triangle $\tilde{\Delta} = \{(x,t) \in \mathbb{R}^2 : 0 \le x \le l,\ (l - x)/a \le t \le (l + x)/a$ (see Fig. 1), to the following system of integral equations

$$u(x,t) = \nu(t) - \frac{1}{2a}\int\limits_0^t \left(\mu_1\left(t - \frac{l-\xi}{a}\right) - \mu_2\left(t + \frac{l-\xi}{a}\right)\right) d\xi -$$

$$-\frac{1}{2a^2}\int_x^l d\xi\int_\xi^l\left\{\tilde F[u,u_t,u_x,\varepsilon]\left(\eta,t+\frac{\eta-\xi}{a}\right)+\right.$$

$$\left.+\tilde F[u,u_t,u_x,\varepsilon]\left(\eta,t+\frac{\xi-\eta}{a}\right)\right\}d\eta\equiv\frac{\nu\left(t+\frac{l-x}{a}\right)+\nu\left(t-\frac{l-x}{a}\right)}{2}-$$

$$-\frac{a}{2}\int_{t-\frac{l-x}{a}}^{t+\frac{l-x}{a}}\mu(\alpha)d\alpha-\frac{1}{2a}\int_x^l d\eta\int_{t-\frac{l-x}{a}}^{t+\frac{l-x}{a}}\tilde F[u,u_t,u_x,\varepsilon](\eta,\tau)d\tau, \tag{1.77}$$

$$u_t=\frac{\partial}{\partial t}(u),\quad u_x=\frac{\partial}{\partial x}(u).$$

Remark 1.4. Since the function ε is the parameter in system (1.75), the following conclusion can be derived. If the operator $\tilde F[u,u_t,u_x,\varepsilon]$ admits a T-periodic continuation on the set of functions $A=\{u(x,t)\in C^1([0,l\}\times\mathbb{R}) : u(x,t+T)=u(x,t)]$ with respect to variable t, then the solution of the Cauchy problem is the T-periodic function of variable t.

Remark 1.5. The first equations in (1.75) and (1.77) contain both the solution of homogeneous equation $u_{tt}-a^2u_{xx}=0$ and the solution of nonhomogeneous equation (1.70). At the same time, the operators of the following form

$$(S_1g)(x,t)=-\frac{1}{2}\int_0^x d\xi\int_{t-x+\varepsilon}^{t+x-\varepsilon}g(\xi,\tau)\,d\tau,$$

$$(S_2g)(x,t)=-\frac{1}{2}\int_x^\pi d\xi\int_{t+x-\varepsilon}^{t-x+\varepsilon}g(\xi,\tau)\,d\tau,$$

which occur in a more general form in the first equations of the systems in question, enabled [21] to investigate analytically the periodic boundary value problem

$$u_{tt}-u_{xx}=g\,(x,t),$$

$$u(0,t)=u(\pi,t)=0,\quad u(x,t+T)=u\,(x,t),$$

for the first time. This boundary value problem is studied in detail in Chapter 3 for both linear and nonlinear cases, i. e. where the quasiwave equation is of the form $u_{tt}-u_{xx}=F[u,u_t,u_x,\varepsilon]$ provided the homogeneous equation $u_{tt}-u_{xx}=0$ has only the zero solution.

The difficulty caused by small denominators is one of the reasons that periodic boundary value problems for hyperbolic equations (both linear and nonlinear) have begun to be studied only recently. In the last decade, the periodic boundary value problem

$$u_{tt}-a^2u_{xx}=F[u,u_t,u_x,\varepsilon],\quad 0<\varepsilon\ll 1,$$

$$u(0,t)=u(\pi,t)=0,\quad t\in\mathbb{R},$$

$$u(x,t+T)=u(x,t),\quad 0\le x\le\pi,\quad t\in\mathbb{R} \tag{1.78}$$

was studied in many works. The first of them was [6], which deals with the following problem

$$z_{tt}-a^2z_{xx}=\Phi(x,t)+\mu f(z),$$

$$z(0,t)=z(1,t)=0,\quad z(x,0)=z(x,1),\quad z_t(x,0)=z_t(x,1) \tag{1.79}$$

in a region $\{0 \leq x \leq 1,\ 0 \leq t \leq 1\}$. It proves that if $a = (2m+1)/q$ (m, q are positive integers) and certain restrictions are imposed on the functions $\Phi(x,t)$ and $f(z)$ $\left(f(z) = \sum_{k=1}^{\infty} a_{2k-1} z^{2k-1}\right)$, then problem (1.79) has a unique classical solution in the class of functions represented by the series

$$z(x,t) = \sum_{k=0}^{\infty} z_{2k+1}(t) \sin(2k+1)\pi x$$

provided $|\mu|$ is small enough.

Consider the following boundary value problem

$$u_{tt} - u_{xx} = \varepsilon f(x,t,u); , \tag{1.80}$$

$$u(0,t) = u(\pi,t) = 0, \tag{1.81}$$

where ε is a small parameter and $f(x, t+2\pi, u) = f(x,t,u)$.

Let C^{∞} denote the space of real functions that are infinitely differentiable in x and t, 2π-periodic in variable t, and defined on the interval $0 \leq x \leq \pi$. Let C_0^{∞} denote a subspace of C^{∞} that includes functions whose support with respect to x is contained in $(0,\pi)$. Let H_r (H_r^0) be a completion of C^{∞} (C_0^{∞}) with respect to the norm

$$\|u\|_r^2 = \int_0^T \int_0^{\pi} u(1-\Delta)^r u\,dx\,dt,$$

where $\Delta = \partial^2/\partial t^2 + \partial^2/\partial x^2$.

The following statement concerning the boundary value problem (1.80) was established first.

Theorem 1.5. [175]. *If $f(x,t,u) \in C^r$ in all the variables, $f(x,t+2\pi,u) = f(x,t,u)$ and $T_u \geq \beta > 0$, then for any sufficiently small $|\varepsilon| < \varepsilon_0$ equation (1.80) has a solution $u \in H_r(u(x,t,\varphi) = v(x,t) + \varepsilon w(x,t,\varepsilon)$, $v \in H_r$, $w \in H_{r+1})$ periodic in t and such that $u(0,t) = u(\pi,t) = 0$, where $v(x,t)$ is the solution of the homogeneous equation $u_{tt} - u_{xx} = 0$ and $\varepsilon w(x,t,\varepsilon)$ is the partial solution of the nonhomogeneous equation (1.80).*

Example. Consider the boundary value problem

$$u_{tt} - u_{xx} = \varepsilon u, \quad u(0,t) = u(\pi,t) = 0. \tag{1.82}$$

Here $f(x,t,u) = u$ and the condition in Theorem 1.5 is fulfilled. Let the axis Ox $(t=0)$ be the initial line for equation (1.82) and let the initial condition $u(x,0) = \varphi(x) \in \overset{0}{C}{}^2$, $u_t(x,0) = \psi(x) \in \overset{0}{C}{}^1$ be given. It is easily seen that the boundary value problem (1.82) does not have any nontrivial solution 2π-periodic in t.

Let the axis Ot $(x=0)$ be now the initial line for equation (1.82) and let $u(0,t) = 0$, $u_x(0,t) = \mu(t)$ be the initial condition. The Cauchy problem

$$u_{tt} - u_{xx} = \varepsilon u, \quad u(0,t) = 0, \quad u_x(0,t) = \mu(t) \tag{1.83}$$

is then equivalent, in the triangle $\overline{\Delta} = \{0 \leq x \leq \pi,\ x \leq t \leq 2\pi - x\} \subset \Pi_{2\pi} = \{0 \leq x \leq \pi,\ 0 \leq t \leq 2\pi\}$ to the integral equation of the form (1.75)

$$u(x,t) = \frac{1}{2}\int_0^x (\mu(t+\xi) + \mu(t-\xi))\,d\xi - \frac{\varepsilon}{2}\int_0^x d\xi \int_0^{\xi} \{u(\eta, t+\xi-\eta) + u(\eta, t-\xi+\eta)\}\,d\eta,$$

which has 2π-periodic solution $u_{\overline{\Delta}}(x,t)$ satisfying the first boundary condition $u(0,t)=0$ provided $0<|\varepsilon|<\varepsilon_0$, $\mu(t+2\pi)=\mu(t)$. The second boundary condition t $u(\pi,\pi)=0$ is obviously satisfied not with any initial function $\mu(t)$. Hence, a continuous and 2π-periodic solution $u(x,t)$ of the boundary problem

$$u_{tt}-u_{xx}=\tilde{F}[u,u_t,u_x,\varepsilon]\equiv\varepsilon\tilde{F}[u,u_t,u_x]; \tag{1.84}$$

$$u(0,t)=u(\pi,t)=0, \tag{1.85}$$

is to be sought in the set of functions satisfying the system of integral equations (1.75) $(a=1)$ and the condition $u(\pi,\pi)=0$. Indeed, the following statement holds.

Theorem 1.6. [149]. *Let the function $\tilde{F}[u,u_t,u_x]\,(x,t)$ be defined in a region $\Omega_{2\pi}=\{0\le x\le\pi,\ 0\le t\le 2\pi,\ |u|\le a,\ |u_t|\le a,\ |u_x|\le a\}$ and be continuous in variables x,t and satisfy a Lipschitz condition with respect to u, u_t, u_x with a constant K independent of x, t. For the boundary value problem (1.84), (1.85) to have, in the rectangular $\Pi_{2\pi}=\{0\le x\le\pi,\ 0\le t\le 2\pi\}$ a unique smooth nontrivial solution it is necessary and sufficient that the condition*

$$\int_0^\pi (u_x(0,\pi+\xi)+u_x(0,\pi-\xi))d\xi-\varepsilon\int_0^\pi d\xi\int_0^\xi\Big\{\tilde{F}_1[u,u_t,u_x](\eta,\pi+\xi-\eta)+$$
$$+\tilde{F}_1[u,u_t,u_x](\eta,\pi-\xi+\eta)\Big\}d\eta=0 \tag{1.86}$$

be satisfied.

Proof. Let the rectangular $\Pi_{2\pi}$ be partitioned into the three pieces

$$\overline{\Delta}=\{(x,t)\in\Pi_{2\pi}\ :0\le x\le\pi,\quad x\le t\le 2\pi-x\};$$
$$\overline{\Delta}_1=\{(x,t)\in\Pi_{2\pi}\ :0\le x\le\pi,\quad 0\le t\le x\};$$
$$\overline{\Delta}_2=\{(x,t)\in\Pi_{2\pi}\ :0\le x\le\pi,\quad 2\pi-x\le t\le 2\pi\}$$

by the lines given by the equations $t=x$ and $t=2\pi-x$ (Fig. 2). Let the boundary problem (1.84), (1.85) have a smooth solution in $\Pi_{2\pi}$ such that $|u_x(0,t)|=|\mu(t)|<a/2\pi$. Clearly, $\mu(t)\not\equiv 0$ and the above solution is defined in every triangle $\overline{\Delta}$, $\overline{\Delta}_1$, $\overline{\Delta}_2$. Let $(x,t)\in\overline{\Delta}$. The Cauchy problem

$$u_{tt}-u_{xx}=\varepsilon\tilde{F}_1[u,u_t,u_x],\quad u(0,t)=0,\quad u_x(0,t)=\mu(t) \tag{1.87}$$

considered in the class of smooth functions is then equivalent, in the triangle $\overline{\Delta}$, to the system of integral equations of the form (1.75) with $a=1$, $\nu(t)=0$, $\mu_1(t)=\mu(t)$, $\mu_2(t)=-\mu(t)$, i. e.

$$u(x,t)=\frac{1}{2}\int_0^x(\mu(t+\xi)+\mu_2(t-\xi))\,d\xi-\frac{\varepsilon}{2}\int_0^x d\xi\int_0^\xi\{\tilde{F}_1[u,u_t,u_x]\,(\eta,t+\xi-\eta)+$$
$$+\tilde{F}[u,u_t,u_x]\,(\eta,t-\xi+\eta)\}\,d\eta\equiv$$
$$\equiv\frac{1}{2}\int_{t-x}^{t+x}\mu(\alpha)\,d\alpha-\frac{\varepsilon}{2}\int_0^x d\eta\int_{t-x+\eta}^{t+x-\eta}\tilde{F}[u,u_t,u_x](\eta,\tau)\,d\tau, \tag{1.88}$$

$$u_t = \frac{\partial}{\partial t}(u), \quad u_x = \frac{\partial}{\partial x}(u).$$

It follows that if $|\varepsilon| < \varepsilon_0$, then the Cauchy problem (1.87) has a unique smooth nontrivial solution $u_{\overline{\Delta}}(x,t)$ in the triangle $\overline{\Delta} \subset \Pi_{2\pi}$. According to the assumption, the smooth solution $u(x,t)$ of the boundary value problem (1.84), (1.85) satisfies the condition $u(\pi,\pi) = 0$. So, it is obvious that $u_{\overline{\Delta}}(\pi,\pi)$. Combining it with (1.88) we see that the condition (1.86) in Theorem 1.6 is fulfilled.

Let the condition (1.86) be now fulfilled with a solution $u(x,t)$ of the boundary value problem (1.84), (1.85) such that $|u_x(0,t)| = |\mu(t)| < a/2\pi$. We shall prove that this solution is smooth in $\Pi_{2\pi}$ and can be represented as

$$u(x,t) = \begin{cases} u_{\overline{\Delta}}(x,t), & \text{если} \quad (x,t) \in \overline{\Delta}, \\ u_{\overline{\Delta}_1}(x,t), & \text{если} \quad (x,t) \in \overline{\Delta}_1, \\ u_{\overline{\Delta}_2}(x,t), & \text{если} \quad (x,t) \in \overline{\Delta}_2. \end{cases} \tag{1.89}$$

Indeed, let $(x,t) \in \overline{\Delta}$. According to the above, for any continuous initial function $|u_x(0,t)| = |\mu(t)| < a/2\pi$ there is a unique smooth solution $u_{\overline{\Delta}}(x,t)$ of the Cauchy problem (1.87). If the initial function $\mu(t)$ and solution $u_{\overline{\Delta}}(x,t)$ satisfy (1.86), then the boundary condition $u_{\overline{\Delta}}(x,t)$, $0 \le t \le 2\pi$, $u_{\overline{\Delta}}(\pi,\pi) = 0$ is fulfilled.

Since the values of the solution $u_{\overline{\Delta}}(x,t)$ and its derivatives are known at the sides OC and AC of the triangle $\overline{\Delta}$ (see Fig. 2), the existence of the solution of the problem (1.84) on the sets $\overline{\Delta}_1$ and $\overline{\Delta}_2$, (1.85) can be proven as follows.

Let $(x,t) \in \overline{\Delta}_1$. Considered in the class of smooth functions, the boundary value problem (1.84), (1.85) is then equivalent, in $\overline{\Delta}_1$, to the system of integral equations

$$u_1(x,t) = u_1(0,t+x) - \varepsilon \int_0^x \tilde{\Phi}_1[u,u_1,u_2]\,(\eta, t+x-\eta)\,d\eta,$$

$$u_2(x,t) = u_2(\pi, t_\pi^1) - \varepsilon \int_x^\pi \tilde{\Phi}_1[u,u_1,u_2]\,(\eta, t-x+\eta)\,d\eta,$$

$$u(x,t) = u(x,\tau) + \frac{1}{2}\int_x^t [u_1(x,\tau) + u_2(x,\tau)]\,d\tau, \tag{1.90}$$

Here $u_1 = u_1 + u_x$; $u_2 = u_t - u_x$; $(\pi, t_\pi^1) = (\pi, \pi + t - x)$ is the point the characteristic $\xi = \tau - t + x$ meets the line $\xi = \pi$ at and

$$\tilde{\Phi}_1[u,u_1,u_2] = \tilde{F}_1[u, (u_1+u_2)/2, \ (u_1-u_2)/2].$$

Taking account of the condition $u(\pi,t) = 0 \forall t \in [0,2\pi]$ we obtain $u_1 \equiv (u_t + u_2)/2 = 0$ for $x = \pi$.

Thus, $u_2(\pi, t_\pi^1) = -u_t(\pi, t_\pi^1)$ and (1.88) allows us to rewrite (1.90) as

$$u_1(x,t) = u_1(0,t+x) - \varepsilon \int_0^x \tilde{\Phi}_1[u,u_1,u_2]\,(\eta, t+x-\eta)\,d\eta,$$

$$u_2(x,t) = -u_2(0, 2\pi + t - x) + \varepsilon \int_0^x \tilde{\Phi}_1[u,u_1,u_2]\,(\eta, 2\pi + t - x - \eta)\,d\eta-$$

$$-\varepsilon \int_{x}^{\pi} \tilde{\Phi}_1[u, u_1, u_2](\eta, t - x + \eta)\, d\eta, \tag{1.91}$$

$$u(x,t) = \frac{1}{2}\int_{0}^{2x} u_x(0,\alpha)\, d\alpha - \frac{\varepsilon}{2}\int_{0}^{x} d\eta \int_{\eta}^{2x-\eta} \tilde{\Phi}_1[u, u_1, u_2]\,(\eta, \tau) +$$

$$+\frac{1}{2}\int_{x}^{t} \{u_1(x,\tau) + u_2(x,\tau)\}\, d\tau.$$

By virtue of the notation $u_1 = u_t + u_x$ and condition $u_t(0,t) = 0$, the value $u_1(0,t) = u_t(0,t) + u_x(0,t) \equiv u_x(0,t) = \mu(t)$ is the known function. Hence, for a sufficiently small ε, the system of Fredholm integral equations (1.91) has a unique continuous solution $u_{\overline{\Delta}_1}(x,t)$ in the triangle Δ_1, which corresponds to the existence of a smooth solution of equation (1.84). This solution satisfies, in the triangle $\overline{\Delta}_1$, the boundary conditions Δ_1 and $\Delta_{\overline{\Delta}_1}(0,0) = 0$ and $u_{\overline{\Delta}_1}(\pi,t) = 0 \forall t : 0 \le t \le \pi$.

Let now $(x,t) \in \overline{\Delta}_2$ (see Fig. 2). Considered in the class of smooth functions on this set, the boundary problem (1.84), (1.85) is then equivalent to the system of integral equations

$$u_1(x,t) = -u_2(0, t + x - 2\pi) - \varepsilon \int_{0}^{x} \tilde{\Phi}_1[u, u_1, u_2]\,(\eta, t + x - 2\pi + \eta)\, d\eta +$$

$$+\varepsilon \int_{x}^{\pi} \tilde{\Phi}_1[u, u_1, u_2]\,(\eta, t + x - \eta)\, d\eta,$$

$$u_2(x,t) = u_2(0, t - x) + \varepsilon \int_{0}^{x} \tilde{\Phi}_1[u, u_1, u_2]\,(\eta, t - x + \eta)\, d\eta,$$

$$u(x,t) = u(x, 2\pi - x) + \frac{1}{2}\int_{2\pi - x}^{t} \{u_1(x,\tau) + u_2(x,\tau)\}\, d\tau. \tag{1.92}$$

By virtue of the notation $u_2 = u_t - u_x$ and condition $u_t(0,t) = 0$, the function $u_2(0,t) \equiv -u_x(0,t) = -\mu(t)$ is known and the value of $u(x, 2\pi - x)$ is defined by the function $u_{\overline{\Delta}}(x, 2\pi - x)$. For a sufficiently small ε, the system of Fredholm integral equations (1.92) has therefore a unique continuous solution $u_{\overline{\Delta}_2(x,t)}$ in the triangle Δ_2, which corresponds to the existence of a smooth solution of the boundary problem (1.84), (1.85).

Thus, each smooth solution of the boundary problem (1.84), (1.85) satisfying (1.86) exists and can be represented as (1.89). This completes the proof.

Remark 1.6. The proof of Theorem 1.6 enables us to show easily that the third equations in (1.91) and (1.92) can be replaced, in the class of smooth functions, by the equation

$$u(x,t) = \frac{1}{2}\int_{\pi}^{x} \{u_1(\xi,t) - u_2(\xi,t)\}\, d\xi, \tag{1.93}$$

Furthermore, by the known initial value $u_x(0,t) = \mu(t)$ it is possible to construct solutions in the triangles $\overline{\Delta}$, $\overline{\Delta}_1$, $\overline{\Delta}_2$ independently of each other. These solutions are obviously discontinuous. If we assume that the functions $u_1(x,t)$ and $u_2(x,t)$ are 2π-periodic in variable t, then the function $u(x,t)$ given by (1.93) is also 2π-periodic in variable t.

Remark 1.7. Generally speaking, the continuous solution $u(x,t)$ of the boundary value problem (1.84), (1.85) constructed in $\Pi_{2\pi}$ according to Theorem 1.6 does not satisfy the periodicity condition, i. e. $u(x,0) \neq u(x,0+2\pi)$ for all $x \in [0,\pi]$. We study this question in detail in Chapter 4.

CHAPTER 2

PERIODIC SOLUTIONS OF THE WAVE ORDINARY DIFERENTIAL EQUATIONS OF SECOND ORDER

In the present chapter we prove several lemmas and theorems related to the construction of algorithms for finding periodic solutions of the nonlinear wave ordinary differential equations of second order [77, 148–155]. These lemmas and theorems will be used to study partial differential equations (of hyperbolic type) in subsequent chapters.

2.1. Preliminary remarks

In the subsequent account we use the expression $(r, q) = 1$ to denote that the integers r and q are coprime.

Lemma 2.1. *Let $f(t)$ be a continuous T-periodic function. If there are positive integers p and q such that the condition*

$$\omega T q = (2p-1)\pi, (2p-1, \omega q) = 1, \tag{2.1}$$

is fulfilled for the period T, then the operator

$$(Pf)(t) = \frac{1}{\omega}\int_0^t f(\tau)\sin\omega(t-\tau)d\tau - \frac{1}{2\omega}\int_0^{Tq} f(\tau)\sin\omega(t-\tau)d\tau \equiv$$

$$\equiv \frac{1}{2\omega}\int_0^{Tq} Q(\tau)f(\tau)\sin\omega(t-\tau)d\tau, \tag{2.2}$$

where

$$Q(\tau) = \begin{cases} 1, & 0 \le \tau \le t, \\ -1, & t < \tau \le Tq, \end{cases} \tag{2.3}$$

sends the T-periodic function $f(t)$ into the T-periodic function $(Pf)(t)$.

Proof. Replacing t by $t + T$ in equality (2.2) yields

$$(Pf)(t+T) = \frac{1}{\omega}\int_0^{t+T} f(\tau)\sin\omega(t+T-\tau)d\tau -$$

$$-\frac{1}{2\omega}\int_0^{Tq} f(\tau)\sin\omega(t+T-\tau)d\tau = \frac{1}{\omega}\int_{-T}^{t} f(\theta)\sin\omega(t-\theta)d\theta -$$

$$-\frac{1}{2\omega}\int_{-T}^{Tq-T} f(\theta)\sin\omega(t-\theta)d\theta = (Pf)(t) +$$

$$+\frac{1}{2\omega}\int_{-T}^{0} f(\theta)\sin\omega(t-\theta)d\theta+\frac{1}{2\omega}\int_{Tq-T}^{Tq} f(\theta)\sin\omega(t-\theta)d\theta.$$

Changing the variable $\theta = Tq+\tau$ in the last integral and taking account of (2.1) we obtain $(Pf)(t+T) = (Pf)(t)$, and the lemma follows.

Lemma 2.2. *Let $t(f)$ be a continuous $2T$-periodic function. If there are positive integers $r = 2k$ and $q = 2s-1$ such that the condition*

$$\omega Tq = r\pi,\ r = 2k,\ q = 2s-1,\ (r,\omega q) = 1,\ f(t+T) = -f(t), \tag{2.4}$$

is fulfilled for T, then the operator P given by (2.2) sends the $2T$-periodic function $f(t)$ into $2T$-periodic function $(Pf)(t)$.

The proof of Lemma 2.2 is similar to that of Lemma 2.1.

It is now a simple matter to see that the following statement is true.

Lemma 2.3. *Let $f(t)$ be a function continuous on $[0,Tq]$. Then the function $x(t) = (Pf)(t)$ is a solution of the nonhomogeneous equation $\ddot{x}+\omega^2x = f(t)$.*

Remark 2.1. Let $f(x)-$ be a T-periodic function. The condition (2.1) makes it impossible for the function $f(\tau)\sin\omega(t-\tau)$ to be T-periodic in variables t and τ. A question then arises about a method for constructing an algorithm for finding periodic solutions of the wave equation $\ddot{x}+\omega^2x = f(t)$ in the case where the function $f(\tau)\sin\omega(t-\tau)$ has the same period T as the function $f(t)$. Any number of the form $T = 2\pi n_0/\omega$, n_o is a fixed positive integer, can be this period.

Lemma 2.4. *If a function $f(\tau)\sin\omega(t-\tau)$ is $2\pi n_0/\omega$-periodic in t and τ, then the operator*

$$(P_1f)(t) = \frac{1}{\omega}\int_0^t\left(f(\tau)\sin\omega(t-\tau)-\frac{1}{T}\int_0^T f(s)\sin\omega(t-s)ds\right)d\tau \equiv$$

$$\equiv \frac{1}{\omega}\int_0^T Q_1(\tau)f(\tau)\sin\omega(t-\tau)d\tau \tag{2.5}$$

where

$$Q_1(\tau) = \begin{cases} 1-\frac{t}{T}, & 0\le\tau\le t,\\ -\frac{t}{T}, & t<\tau\le T\end{cases} \tag{2.6}$$

sends each $2\pi n_0/\omega$-periodic function $f(t)$ into $2\pi n_0/\omega$-periodic function $(P_1f)(t)$. Furthermore, if the condition

$$\frac{1}{T}\int_0^T f(s)\sin\omega(t-s)ds = 0, \tag{2.7}$$

is fulfilled, then the function $x(t) = (P_1f)(t)$ is a periodic solution of the second order nonhomogeneous equation $\ddot{x}+\omega^2x = f(t)$.

Note that the condition (2.7) coincides entirely with the common condition for the right-hand side $f(t)$ of the linear nonhomogeneous equation $\ddot{x}+\omega^2x = f(t)$ to be orthogonal to the fundamental functions $\cos\omega t$and $\sin\omega t$ of the corresponding homogeneous equation $\ddot{x}+\omega^2x = 0$.

At last, it should be noted that periodic solutions of the second order wave equations can be investigated with the help of the numerical-analytical method due to Samoilenko

[103]. We shall exploit the assertions of this method to study the second order wave equations. So, we cite the basic results.

Let $x = (x_1, x_2, \ldots, x_n)$ be a point in the n-dimensional Euclidean space $\mathbb{R}^n$, D a closed bounded region of $\mathbb{R}^n$, Γ its boundary. For the point x, let $|x|$ denote a vector with coordinates $|x_1|, |x_2|, \ldots, |x_n|$. In the present section, let us agree to denote vectors and matrices with nonnegative numerical componenets by a single letter with the superscript $+$ and to interpret inequalities between them component-wise.

By the d^+-neighbourhood of a point $x^0 \in \mathbb{R}^n$ we mean the set of points $x \in \mathbb{R}^n$ such that $|x - x^0| < d^+$.

Let $f(t,x) = (f_1(t,x), f_2(t,x), \ldots, f_n(t,x))$ be a function defined and continuous in the region

$$(t,x) \in \mathbb{R} \times D = (-\infty, +\infty) \times D, \tag{2.8}$$

periodic in t with the period T and assuming values in $\mathbb{R}^n$. The function $|f(t,x)| = (|f_1(t,x)|, \ldots, |f_n(t,x)|)$ is supposed to be bounded by a vector M^+,

$$|f(t,x)| \leq M^+ \tag{2.9}$$

and satisfy a Lipschitz condition with a matrix K^+,

$$|f(t,x') - f(t,x")| \leq K^+|x' - x''| \tag{2.10}$$

for all $t \in \mathbb{R}$ and $x, x', x" \in D$.

We relate the region D to the function $f(t,x)$ and denote by D_f a set of points of $\mathbb{R}^n$ such that each point is contained in D together with its $\frac{T}{2}M^+$-neighbourhood.

Definition 2.1. A system of differential equations

$$\frac{dx}{dt} = f(t,x), \tag{2.11}$$

the left-hand side of which is defined and continuous and T-periodic in t in the region (2.8) is said to be a T-system in the region D provided (1) the conditions (2.9) and (2.10) are fulfilled, (2) the set D_f is nonempty,

$$D_f \neq \emptyset; \tag{2.12}$$

and (3) the largest eigenvalue $\lambda_{\max}$ of the matrix $Q^+ = \frac{T}{\pi}K^+$ does not exceed one,

$$\lambda_{\max} < 1 \tag{2.13}$$

Along with system (2.11) we shall consider the following system

$$\frac{dx}{dt} + f(t,x) - \mu, \tag{2.14}$$

where $\mu = (\mu_1, \mu_2, \ldots, \mu_n)$ is a parameter.

Definition 2.2. A value of the parameter μ for which the solution of (2.14) assuming the value $t = \tau$ at $x = x_0$ is T-periodic is called a Δ-constant at the point (τ, x_0) with respect to the system of equations (2.11) provided this value of μ is unique.

Obviously, the points (τ, x_0) for which the Δ-constant equals zero define the initial values of the periodic solutions of (2.11).

Let $\overline{f(t,x)}$ denote the integral mean value with respect to time on the interval $[0,T]$ $\overline{f(t,x)} = \frac{1}{T}\int_0^T f(s, x(s))ds$.

The investigation of periodic solutions of T-systems essentially uses the statements and estimates given below.

Lemma 2.5 *(basic lemma [103]). Let the system of equations (2.11) be a T-system in the region D. Then for the sequence of functions T-periodic in t*

$$x_m(t,\tau,x_0) = x_0 + \int_\tau^t \left(f(t, x_{m-1}(t,\tau,x_0)) - \overline{f(t, x_{m-1}(t,\tau,x_0))} \right) dt \tag{2.15}$$

converges uniformly with respect to $m \to \infty$

$$(t,\tau,x_0) \in \mathbb{R} \times \mathbb{R} \times D_f \tag{2.16}$$

to a function $x^0(t,\tau,x_0)$ defined in the region (2.16), T-periodic in t, τ and satisfying the following system of equations

$$x(t,\tau,x_0) = x_0 + \int_\tau^t \left(f(t, x(t,\tau,x_0)) - \overline{f(t, x(t,\tau,x_0))} \right) dt. \tag{2.17}$$

Moreover, for all $m \geq 1$ and $t \in \mathbb{R}$ the inequality

$$|x^0(t,\tau,x_0) - x_m(t,\tau,x_0)| \leq$$

$$\leq \alpha_1(t) Q^{+m}(E^+ - Q^+)^{-1} + \sum_{j=1}^{3} \delta_{jm} B_j^+ M^+, \tag{2.18}$$

holds, where $\alpha_1(t) = 2(t-\tau)\left(1 - \frac{t-\tau}{T}\right)$;

$$\delta_{jm} = \begin{cases} 0, & j \neq m, \\ 1, & j = m; \end{cases} \tag{2.19}$$

$$B_1^+ = \frac{1}{30\pi^2}((10\pi^3 - 30\pi^2)E^+ + (3\pi^3 - 30\pi)TK^+ + (\pi^3 - 30)T^2K^{+2});$$

$$B_2^+ = \frac{1}{30\pi}((3\pi^3 - 30\pi)E^+ + (\pi^3 - 30)TK^+); \; B_3^+ = \frac{\pi^3 - 30}{30}E^+. \tag{2.20}$$

Theorem 2.1 [103].*For any point $(\tau, x_0) \in \mathbb{R} \times D_f$ there exists a Δ-constant with respect to a system given in the region (2.8), which is defined as*

$$\Delta(\tau,x_0) \equiv \overline{f(t, x_0(t,\tau,x_0))} = \lim_{m\to\infty} \bar{\Delta}_m(\tau,x_0) \equiv \lim_{m\to\infty} \overline{f(t, x_m(t,\tau,x_0))}.$$

Theorem 2.2 [103].*Let the system of equations (2.11) be the T-system in the region (2.8). Then*

(1) the solution $x(t,\tau,x_0)$ of (2.11) assuming value $t = \tau$ at $x_0 \in D_f$ is T-periodic if and only if the Δ-constant at the point τ, x_0 equals zero and then

$$x(t,\tau,x_0) = x^0(t,\tau,x_0);$$

(2) if $f(t,x) = -f(-t,x)$ for $t \in \mathbb{R}$, $x \in \mathbb{R}$, then $\Delta_m(\tau,x_0) \equiv 0$ for all $m = 0,1,2,\ldots,\tau \in \mathbb{R}, x_0 \in D_f$..

Theorem 2.3 [103].*Let the following conditions be fulfilled for the T-system given in the region (2.8)*

(1) the mapping

$$\Delta_m : D_f \to \mathbb{R}^n,\ \Delta_m(x_0) = \overline{f(t, x_m(t, \tau, x_0))}$$

has an isolated singular point $\Delta_m(x_0) = 0$; *for certain real* τ *and integer* m,

(2) the index of this point does not vanish,

(3) there exists a closed convex region $D_1 \subset D_f$ *with* x_0 *its only singular point such that the following inequality is fulfilled at its boundary* Γ_{D_1}

$$\inf_{x\in\Gamma_{D_1}} |\Delta_m(x)| \geq \frac{\pi}{3} \left| Q^{+(m+1)} \left((E^+ - Q^+)^{-1} + \sum_{j=1}^{3} \delta_{jm} B_j^+ \right) M^+ \right|, \tag{2.21}$$

here $Q^+ = \frac{TK^+}{\pi}$; $m \geq 1$..

Then system (2.11) has a periodic solution $x = x(t)$ *such that* $x(\tau) \in D_1$.

The assertion in Theorem 2.3 also holds for $m = 0$ with the change that inequality (2.21) is replaced by the inequality

$$\inf_{x\in\Gamma_{D_1}} |\overline{f(t,x)}| \geq \frac{\pi}{3} |Q^+ M^+|. \tag{2.22}$$

To check the condition (2) in Theorem 2.3 it is necessary to compute the index of an isolated singular point. For a plane, this computation can be easily carried out [103]. Furthermore, the index of a singular point $x = x_0$ of the mapping Δ_m does not vanish whenever the Jacobian of the function $\Delta_m(x)$ at the point x_0 does not vanish

$$\det \left(\frac{\partial \Delta_m(x_0)}{\partial x} \right) \neq 0.$$

provided Δ_m is continuously differentiable.

To check the condition (3) in Theorem 3.3 it is necessary to choose an appropriate region D_1. However, for a number of systems, for example, the system of standard form given below, this condition is fulfilled with a region chosen more or less arbitrarily.

Consider a system of standard form T-periodic in time

$$\frac{dx}{dt} = \varepsilon f(t, x), \tag{2.23}$$

where ε is a small parameter, $f(t, x)$ a function continuous in t and satisfying a Lipschitz condition with respect to x in a closed bounded region $D \subset \mathbb{R}^n$.

Theorem 2.3 enables us to prove the following statement.

Theorem 2.4 [103].*For the system of equations (2.23) to possess a* T*-periodic solution continuously depending on* ε *in a neighbourhood of the point* $\varepsilon = 0$ *for all sufficiently small values of* ε *it is necessary that the averaged system of equations*

$$\frac{d\xi}{dt} = \varepsilon f_0(\xi) \equiv \varepsilon \frac{1}{T} \int_0^T f(s, \xi) ds \tag{2.24}$$

have an equilibrium state $\xi = \xi_0 \in D$ *and sufficient that this state be isolated of zero index.*

Indeed, the following argument shows that the condition in Theorem 2.4 is sufficient for a T-periodic solution of (2.23) to exist. The system (2.23) is the T-system in the region (2.8) for small values of ε. If a ball of sufficiently small radius ρ and centre at the isolated singular point $\xi = \xi_0$ of the mapping $f_0(\xi)$ is assumed as the region D_1 in Theorem 2.3, then inequality (2.22) turns into

$$\inf_{|\xi-\xi_0|=\rho} \varepsilon|f_0(\xi)| \geq \frac{\pi}{3}Q^+ \left|\varepsilon^2 \frac{T}{\pi} K^+ M^+\right|, \tag{2.25}$$

where $M^+ = \max_{[0,T]\times D} |f(t,x)|$ and K^+ is the matrix in the Lipschitz condition for the function $f(t,x)$ with respect to $x \in D$.

Since the singular point is isolated, we have

$$\inf_{|\xi-\xi_0|=\rho} |f_0(\xi)| = g > 0, \tag{2.26}$$

for sufficiently small ρ. Hence, inequality (2.25) takes the form

$$g \geq \varepsilon \frac{T}{3} |K^+ M^+|$$

and holds for all the values of ε satisfying

$$\varepsilon \leq \varepsilon_0 = \frac{3g}{T|K^+M^+|}. \tag{2.27}$$

Therefore, all the conditions in Theorem 2.3 are fulfilled for $\varepsilon \leq \varepsilon_0$, i. e. system (2.23) has a T-periodic solution $x(t,\varepsilon)$ satisfying

$$|x(0,\varepsilon) - \xi_0| < \rho; \tag{2.28}$$

$$|x(t,\varepsilon) - x(0,\varepsilon)| \leq \varepsilon \frac{T}{2} M^+. \tag{2.29}$$

The argument below shows that the conditions in Theorem 2.4 are necessary for a T-periodic solution of (2.23) continuously depending on ε at the point $\varepsilon = 0$ to exist. Since system (2.23) is the T-system and has a T-periodic solution for all sufficiently small values of ε, the Δ-constant has to be zero at the initial values $t = 0$, $x = x_0$ of the periodic solution. Obviously, we have

$$\Delta(x_0) = \varepsilon f(x_0) + \varepsilon^2 f_1(x_0, \varepsilon), \tag{2.30}$$

where $f_1(x_0,\varepsilon)$ is a bounded function continuous for all sufficiently small values of ε. Therefore, $\Delta(x_0) = 0$ holds for all sufficiently small ε only if $f(x_0) = 0$, which is the desired conclusion.

As can be seen from Definition 2.1, the condition for the system of differential equations (2.11) to be a T-system does not involve the number of equations. This allows the abstract scheme of the numerical-analytical method to be easily extended over the countable systems of differential equations. So, this method enables us to obtain, for these systems, the statements and estimates similar to that established above for equations in finite-dimensional space.

Indeed, let $x = (x_1, x_2, \ldots, x_n, \ldots)$ be a point of the space m of bounded number sequences with the norm

$$\|x\| = \sup_n |x_n|, \tag{2.31}$$

and let $f(t,x) = (f_1(t,x), \ldots, f_n(t,x), \ldots)$ be a continuous function of variables t, x ranging in the domain

$$t \in \mathbb{R},\ x \in D \tag{2.32}$$

Here D is a bounded closed region of the space m.

Consider a countable system of differential equations

$$\frac{dx}{dt} = f(t, x), \tag{2.33}$$

The function $f(t, x)$ is regarded as the T-periodic function of variable t satisfying the following inequalities

$$|f(t, x)| \leq M^+; \tag{2.34}$$

$$|f(t, x') - f(t, x'')| \leq K^+|x' - x''| \tag{2.35}$$

for all $t,\ x,\ x',\ x''$ in the region

$$t \in \mathbb{R},\ x, x',\ x'' \in D. \tag{2.36}$$

Here $M^+ = (M_1, \ldots, M_n, \ldots)$; $K^+ = (K_{ij}, i, j = 1, 2, 3, \ldots)$ and are an infinite-dimensional vector and matrix with nonnegative elements, respectively.

By a solution of the system of equations (2.33) we mean a function $x(t) = (x_1(t), \ldots, x_n(t), \ldots)$ defined for t from an interval a, b and assuming values in the space m which is continuously differentiable and satisfies system (2.33). There are natural conditions ensuring the existence of a solution of (2.33) (see [89, 103]).

The matrix K^+ generates a linear operator K^+ acting in the space m provided

$$\sup_i \sum_{j=1}^{\infty} K_{ij} < \infty, \tag{2.37}$$

The norm of the operator K^+ in the space m is determined by the left-hand side in inequality (2.37)

$$\|K^+\| = \sup_i \sum_{j=1}^{\infty} K_{ij}. \tag{2.38}$$

he operator K^+ is said to be totally regular if

$$\|K^+\| < 1. \tag{2.39}$$

The assumptions ensuring that the system of equations (2.33) is the T-system are as follows: the conditions (2.34) and (2.35) hold, the region D_f is nonempty,

$$D_f \neq \emptyset\cdot \tag{2.40}$$

and the operator $\frac{T}{\pi}K^+$ is totally regular,

$$\frac{T}{\pi}\|K^+\| \leq q < 1. \tag{2.41}$$

Theorem 2.5 [103]. *Let the system of equations (2.33) be the T-system in the region (2.32). Then*

(1) the sequence of functions T-periodic in t, τ

$$x_m(t, \tau, x_0) = x_0 + \int_{\tau}^{t} \left(f(s, x_{m-1}(s, \tau, x_0)) - \overline{f(s, x_{m-1}(s, \tau, x_0))} \right) ds,$$

$$m = 1, 2, \ldots, \tag{2.42}$$

converges uniformly with respect to

$$(t,\tau,x_0)\in\mathbb{R}\times\mathbb{R}\times D_f \tag{2.43}$$

to a function $x^0(t,\tau,x_0)$ *defined in the region (2.43) and satisfying the equation*

$$x(t,\tau,x_0)=x_0+\int_\tau^t\Big(f(s,x(s,\tau,x_0))-\overline{f(s,x(s,\tau,x_0))}\Big)\,ds, \tag{2.44}$$

Furthermore, the inequalities

$$\|x^0(t,\tau,x_0)-x_m(t,\tau,x_0)\|\le$$

$$\le\frac{\|M^+\|T}{2}q^m\left(\frac{1}{1-q}+\sum_{j=1}^{3}\delta_{jm}\beta_j\right)=\frac{\|M^+\|}{2}\rho_m,\ m=1,2,\ldots, \tag{2.45}$$

hold, β_j *being constants given by (2.20) where the matrices* E^+ *and* K^+ *are replaced by the numbers* 1 *and* $\|K^+\|$*, respectively.*

(2) any point $(\tau,x_0)\in R\times D_f$ *has a Delta-constant given by the equality*

$$\Delta(\tau,x_0)\equiv\overline{f(t,x^0(t,\tau,x_0))}=\lim_{m\to\infty}\Delta_m(\tau,x_0)\equiv$$

$$\equiv\lim_{m\to\infty}\overline{f(t,x_m(t,\tau,x_0))}; \tag{2.46}$$

(3) the solution $x=x(t,\tau,x_0)$ *of (2.33) assuming the value* $t=\tau$ *at* $x_0\in D_f$ *is* T*-periodic if and only if the* Δ*-constant at the point* (τ,x_0) *equals zero. In this case*

$$x(t,\tau,x_0)=x^0(t,\tau,x_0); \tag{2.47}$$

(4) if $f(t,x)=-f(-t,x)$ *for* $t\in\mathbb{R}$, $x\in D$, *then* $\Delta_m(\tau,x_0)=0$ *for all* $m=0,1,2,\ldots,\tau\in\mathbb{R}$, $x_0\in D_f$.

This theorem can be proved in the way similar to the proof of the corresponding theorems in finite-dimensional case. It allows periodic solutions of a countable T-system to be constructed approximately under the assumption that they exist and pass through a point of $t=\tau$ for D_f.

The theorem below answers the question about the existence of a T-periodic solution of the countable system of equations (2.33).

Theorem 2.6 [103].*Let the following conditions be fulfilled for the* T*-system (2.33) defined in the region (2.12):*

(1) for certain $\tau\in\mathbb{R}$ *and integer* m, *the mapping*

$$\Delta_m:\Delta_m=\Delta_m(x_0)=\overline{f(t,x_m(t,\tau,x_0))} \tag{2.48}$$

of the region D_f *into the region* $\Delta_m D_f$ $D_f\to\Delta_m D_f$ *has a singular point* $x_0=x^0$

$$\Delta_m(x^0)=0; \tag{2.49}$$

(2) there exists a closed bounded region D_1 *contained in* D_f *and containing a point* x^0 *such that the operator* Δ_m *maps topologically* D_1 *onto* $\Delta_m D_1$*;*

(3) the inequality

$$\inf_{x\in\Gamma_{D_1}}\|\Delta_m(x)\|\ge\frac{\pi}{3}\|M^+\|q\rho_m. \tag{2.50}$$

holds at the boundary Γ_{D_1} of the region D_1.

Then system (2.33) has a T-periodic solution $x = x(t)$ such that $x(\tau) \in D_1$.

If $m = 0$, then $\Delta_0(x) = \overline{f(t,x)}$ and Theorem 2.6 enables us to prove the existence of a periodic solution of the countable system ofeuqations (2.33) starting from the corresponding averaged system

$$\frac{d\xi}{dt} = \overline{f(t,\xi)} = \Delta_0(\xi). \tag{2.51}$$

2.2. The existence of solutions periodic in time for wave equations

1. T-systems of the first class. Consider the second order nonlinear differential equation

$$x'' + \omega^2 x = f(t, x, \dot{x}). \tag{2.52}$$

Assume that the right-hand side in (2.52), which is defined in the region

$$t \in \mathbb{R},\ x \in [-a, a],\ \dot{x} \in [-b, b], \tag{2.53}$$

is T-periodic in t if it depends explicitly on time t, is continuous in variables t, x, $\dot{x}$ simultaneously, and satisfies the condition

$$|f(t, xy)| \leq M,$$

$$|f(t, x'', y'') - f(t, x', y')| \leq K_1|x'' - x'| = K^2|y'' - y'|, \tag{2.54}$$

where M, K_1, K_2 are positive constants.

Definition 2.3. A second order equation (2.52) is called a T-periodic system of the first class in the region (2.53) if the constants a, b, M, K_1, K_2, T, ω satisfy the relations

$$ae\frac{1}{2\omega}MTq;\ be\frac{1}{2}MTq;\ \omega Tq = (2p-1)\pi,\ (2p-1,\ \omega q) = 1; \tag{2.55}$$

$$\frac{1}{2}Tq\left(\frac{1}{\omega}K_1 + K_2\right) < 1, \tag{2.56}$$

p and q being fixed positive integers.

Theorem 2.7 [152].*Let a finction $f(t, x, \dot{x})$ be defined in the region (2.53), be continuous in $t, x, \dot{x}$ and T-periodic in t, and satisfy relations (2.54)–(2.55). Then for each continuous T-periodic function $v = v(t)$ such that*

$$|v(t)| \leq \frac{1}{2\omega}MTq,\ |\dot{v}(t)| \leq \frac{1}{2}MTq,\ \omega Tq = (2p-1)\pi, \tag{2.57}$$

the sequence of T-periodic functions

$$x_0(t) = v(t),$$

$$x_{m+1}(t) = \frac{1}{2\omega}\int\limits_0^{Tq} Q(\tau) f(\tau, x_m(\tau), \dot{x}_m(\tau)) \sin\omega(t-\tau)d\tau \tag{2.58}$$

$$m = 0, 1, 2, \dots,$$

where

$$Q(\tau) = \begin{cases} 1, & 0 \leq \tau \leq t, \\ -1, & t < \tau \leq Tq, \end{cases}$$

converges, when $m \to \infty$, uniformly with respect to t to a continuous T-periodic function $x^0(t)$ satisfying the equation

$$x(t) = \frac{1}{2\omega} \int_0^{Tq} Q(\tau) f(\tau, x(\tau), \dot{x}(\tau)) \sin \omega(t-\tau)\, d\tau, \tag{2.59}$$

and, hence, the differential equation (2.1).

Proof. Using (2.57) and (2.58) and proceeding by induction, we can show that $x_m(t)$ and $\dot{x}_m(t)$ assume values in the region (2.53) for all $m = 0, 1, 2\ldots$ and $t \in \mathbb{R}$.

To prove convergence of the sequence (2.58) we estimate the difference $|x_{m+1}(t) - x_m(t)|$. Taking (2.54) into account, from (2.58) we obtain

$$|x_{m+1}(t) - x_m(t)| \le \frac{1}{2\omega} Tq(K_1 |x_m(t) - x_{m-1}(t)|_0 +$$
$$+ K_2 |\dot{x}_m(t) - \dot{x}_{m-1}(t)|_0), \tag{2.60}$$

Here we used notation $|y(t)|_0 = \max_{0 \le t \le T}$.

Rewriting indentity (2.58) as

$$x_{m+1}(t) = \frac{1}{2\omega} \int_0^{Tq} Q(\tau) f(\tau, x_m(\tau), \dot{x}_m(\tau)) \sin \omega(t-\tau) d\tau \equiv$$

$$\equiv \frac{1}{\omega} \int_0^{t} f(\tau, x_m(\tau), \dot{x}_m(\tau)) \sin \omega(t-\tau)\, d\tau -$$

$$- \frac{1}{2\omega} \int_0^{Tq} f(\tau, x_m(\tau), \dot{x}_m(\tau)) \sin \omega(t-\tau)\, d\tau,$$

and differentiating it in t, we get

$$|\dot{x}_{m+1}(t) - \dot{x}_m(t)| \le$$

$$\le \frac{Tq}{2} \Big(K_1 |x_m(t) - x_{m-1}(t)|_0 + K_2 |\dot{x}_m(t) - \dot{x}_{m-1}(t)|_0 \Big). \tag{2.61}$$

Inequalities (2.60) and (2.61) can now be rewritten in the vector form

$$z_{m+1}(t) \le A z_m^0, \tag{2.62}$$

where

$$z_{m+1}(t) = \begin{pmatrix} |x_{m+1}(t) - x_m(t)| \\ |\dot{x}_{m+1}(t) - \dot{x}_m(t)| \end{pmatrix}; \quad A = \begin{pmatrix} \frac{TqK_1}{2\omega} & \frac{TqK_2}{2\omega} \\ \frac{TqK_1}{2} & \frac{TqK_2}{2} \end{pmatrix};$$

$$z_m^0(t) = \begin{pmatrix} |x_m(t) - x_{m-1}(t)|_0 \\ |\dot{x}_m(t) - \dot{x}_{m-1}(t)|_0 \end{pmatrix}; \quad z_1^0 = \begin{pmatrix} \frac{MTq}{\omega} \\ MTq \end{pmatrix}.,$$

and inequalities between vectors with nonnegative components are interpreted componentwise.

Iterating inequality (2.62) yields

$$z_{m+1}^0 \le A^m z_1^0, \tag{2.63}$$

which results in the estimate

$$\sum_{i=1}^{m} z_i^0 \leq \sum_{i=1}^{m} A_{i-1} z_1^0. \tag{2.64}$$

Since the eigenvalues of the matrix A are $\lambda_1 = 0$, $\lambda_2 = \frac{Tq}{2} \times \left(\frac{K_1}{\omega} + K_2\right) < 1$, the sequence of partial sums (2.64) converges uniformly and

$$\lim_{m\to\infty} \sum_{f=1}^{m} A_{f-1} z_1^0 = \sum_{f=1}^{\infty} A_{f-1} z_1^0 \equiv (E-A)_1 z_1^0. \tag{2.65}$$

The limit representation (2.65) means uniform convergence of the sequence $(x_m(t), \dot{x}_m(t))$. Set

$$\lim_{m\to\infty} x_m(t) = x^0(t), \qquad \lim_{m\to\infty} \dot{x}_m(t) = \dot{x}^0(t).$$

By using (2.63) we then obtain, for the deviation of $(x_m(t), \dot{x}_m(t))$ from $(x^0(t), \dot{x}^0(t))$, the following estimate

$$\begin{pmatrix} |x^0(t) - x_m(t)| \\ |\dot{x}^0(t) - \dot{x}_m(t)| \end{pmatrix}; \leq A^m (E-A)^{-1} z_1^0. \tag{2.66}$$

Passing to the limit in the recurrent relation (2.58) when $m \to \infty$, we see that the limit function $x^0(t)$ satisfies (2.59), which completes the proof of Theorem 2.7.

2. $2T$-systems of the first class.

Definition 2.4. A second order equation (2.52) is called a $2T$-periodic system of the first class in the region (2.53) if the constants a, b, M, K_1, K_2, T, ω satisfy the relations

$$a \mathrm{e} \frac{MTq}{2}, \; b \mathrm{e} \frac{MTq}{2}, \; \omega T q = r\pi, \; r = 2k,$$

$$q = 2s - 1, \; (\omega q, r) = 1; \tag{2.67}$$

$$\frac{Tq}{2}\left(\frac{K_1}{\omega} + K_2\right) < 1 \tag{2.68}$$

and for each function $v = v(t)$ satisfying $v(t+T) = -v(t)$ the following condition

$$f(t+T, v(t+T), \dot{v}(t+T)) = f(t, v(t), \dot{v}(t)), \tag{2.69}$$

is fulfilled, r and q being fixed positive integers.

Theorem 2.8. *Let equation (2.52) be a $2T$-periodic system of the first class. Then for any continuous $2T$-periodic function $v = v(t)$ such that*

$$|v(t)| \leq \frac{MTq}{2\omega}, \; |\dot{v}(t)| \leq \frac{MTq}{2}, \omega T q = r\pi,$$

$$r = 2k, \; q = 2s-1, \; (\omega q, r) = 1, \; v(t+T) = -v(t), \tag{2.70}$$

the sequence of $2T$-periodic functions

$$x_0(t) = v(t),$$

$$x_{m+1}(t) = \frac{1}{2\omega} \int_0^{Tq} Q(\tau) f(\tau, x_m(\tau), \dot{x}_m(\tau)) \sin \omega(t-\tau)\, d\tau, \tag{2.71}$$

$$m = 0, \; 1, \; 2, \ldots,$$

convereges, when $m \to \infty$, uniformly with respect to t to a continuous $2T$-periodic function $x^0(t)$ satisfying the integral equation

$$x(t) = \frac{1}{2\omega} \int_0^{Tq} Q(\tau) f(\tau, x(\tau), \dot{x}(\tau)) \sin \omega(t-\tau)\, d\tau.$$

Theorem 2.8 can be proved similarly to Theorem 2.7.

3. T-systems of the second class.

Definition 2.5. A second order equation (2.52) is called as a T periodic system of the second class in the region (2.53) if the constants a, b, M, K_1, K_2, T, ω satisfy the relations

$$a e \frac{1}{\omega} MT, \ b e MT, \ T = 2\pi n_0/\omega; \tag{2.72}$$

$$T\left(\frac{1}{\omega} K_1 + K_2\right) < 1, \tag{2.73}$$

where n_0 is a fixed positive integer.

Theorem 2.9. *Let a function $f(t, x, \dot{x})$ be defined in the region (2.53), be continuous in variables $t, x, \dot{x}$, $2\pi n_0/\omega$-periodic in t and satisfy relations (2.54), (2.72) and (2.73). Then for any function $y(t) = c\cos(\omega t + \varphi)$ such that*

$$|y(t)| \le a - \frac{1}{\omega} MT, \ |\dot{y}(t)| \le b - MT, \tag{2.74}$$

the sequence of T-periodic functions

$$x_0(t) = y(t), \ x_{m+1}(t) = x_0(t) +$$

$$+ \frac{1}{\omega} \int_0^T Q_1(\tau) f(\tau, x_m(\tau, c, \varphi), \dot{x}_m(\tau, c, \varphi)) \sin \omega(t-\tau)\, d\tau, \tag{2.75}$$

$$m = 0, \ 1, \ 2, \ldots,$$

$$Q_1(\tau) = \begin{cases} 1 - \frac{t}{T}, & 0 \le \tau \le t, \\ -\frac{t}{T}, & t < \tau \le T, \end{cases}$$

converges, when $m \to \infty$, uniformly with respect to t, c, φ to a continuous $2\pi n_0/\omega$-periodic function $\tilde{x}(t, c, \varphi)$ satisfying the integral equation

$$x(t) = x_0(t) + \frac{1}{\omega} \int_0^t (f(\tau, x(\tau), \dot{x}(\tau)) \sin \omega(t-\tau) -$$

$$- \frac{1}{T} \int_0^T f(\tau, x(\tau), \dot{x}(\tau)) \sin \omega(t-\tau) d\tau)\, d\tau \equiv$$

$$\equiv x_0(t) + \frac{1}{\omega} \int_0^T Q_1(\tau) f(\tau, x(\tau), \dot{x}(\tau)) \sin \omega(t-\tau)\, d\tau. \tag{2.76}$$

As the sequence (2.75) is of the form (2.58) and $|Q_1(\tau)| \le 1$, $0 \le \tau \le T$, the proof of Theorem 2.9 is similar to that of Theorem 2.7.

From equation (2.76) it is seen that any of its solutions which satisfies

$$\Delta(c,\varphi) \equiv \frac{1}{T}\int\limits_0^T f\Big(\tau, \tilde{x}(\tau,c,\varphi), \dot{\tilde{x}}(\tau,c,\varphi)\Big)\sin\omega(t-\tau)d\tau = 0, \tag{2.77}$$

is a $2\pi n_0/\omega$-periodic solution of equation (2.52provided this equation is a T-system of the second class. Therefore, the existence of a $2\pi n_0/\omega$-periodic solution of equation (2.51) relates to the existence of zeros of the function $\Delta(c,\varphi)$. In §2.1 of the present chapter it was shown that this problem was solved in [103].

Remark 2.2. Let the conditions in Theorem 2.9 and condition (2.77) be fulfilled. Writing $x_0(t) = c\cos(\omega t+\varphi)$ as $x_0(t) = C_1\cos\omega t + C_2\sin\omega t$ and using (2.76), we obtain the following representation for a periodic solution $\tilde{x}(t)$ of the wave equation (2.52)

$$\tilde{x}(t) = \left(C_1 - \frac{1}{\omega}\int\limits_0^t f(\tau,\tilde{x}(\tau),\dot{\tilde{x}}(\tau))\sin\omega\tau d\tau\right)\cos\omega t +$$

$$+\left(C_2 + \frac{1}{\omega}\int\limits_0^t f(\tau,\tilde{x}(\tau),\dot{\tilde{x}}(\tau))\cos\omega\tau d\tau\right)\sin\omega t \equiv$$

$$\equiv z_1(t)\cos\omega t + z_2(t)\sin\omega t, \ \ , z_1(0) = \tilde{x}(0), \ \ , \omega z_2(0) = \dot{\tilde{x}}(0),$$

where the functions $z_1(t)$ and $z_2(t)$ are $2\pi n_0/\omega$-periodic.

Considering, for example, a second order wave equation with small parameter, i. e. the equation of the form

$$\ddot{x} + \omega^2 x = \varepsilon f(t,x,\dot{x}), \tag{2.78}$$

and changing the variables

$$x = z_1\cos\omega t + z_2\sin\omega t, \qquad \dot{x} = -\omega z_1\sin\omega t + \omega z_2\cos\omega t \tag{2.79}$$

in equation (2.78), we get, therefore, the standard system

$$\begin{aligned} \dot{z}_1 &= -\frac{\varepsilon}{\omega}f_1(t,z_1,z_2)\sin\omega t, \\ \dot{z}_2 &= \tfrac{\varepsilon}{\omega}f_1(t,z_1,z_2)\cos\omega t \end{aligned} \quad \Leftrightarrow \dot{z} = \varepsilon F(t,z), \tag{2.80}$$

Theorem 2.4 now shows that the following assertion holds for this system.

Theorem 2.10. *Let equation (2.78) be a T-periodic system of the second class. For equation (2.78) to have a $2\pi n_0/\omega$-periodic solution for all sufficiently small values of ε it is necessary that the averaged system of equations*

$$\dot{\xi} = \varepsilon F_0(\xi) \equiv \varepsilon\frac{1}{T}\int\limits_0^T F(\tau,\xi)d\tau,$$

where $F(t,z)$ is given by (2.80) have an equilibrium state $F_0(\xi_0)) = \overline{F(t,\xi_0)} = 0$ and it is sufficient that this state be isolated and of zero index.

4. Periodic solutions of nonlinear second order wave differential equations with retardation.

Consider an equation of the form

$$\frac{d^2x}{dt^2}+\omega^2 x = f\left(t, x(t), x(t-\Delta), \frac{dx}{dt}, \frac{dx(t-\Delta)}{dt}\right). \tag{2.81}$$

To investigate periodic solutions of this equation, we can always exploit the methods presented in items 1–3 provided all the conditions of solution existence are fulfilled. The results of [70] can be used to prove the existence of a $2\pi n_0/\omega$-periodic solution of euqation (2.81).

5. Periodic solutions of nonlinear wave integro-differential equations of the second order.

To investigate periodic solutions of the wave integro-differential equation

$$\ddot{x}+\omega^2 x = f\left(t, x, \dot{x}, \int\limits_0^{h(t)} E(t, s, x(s), \dot{x}(s))\ , ds\right) \tag{2.82}$$

we can use methods set forth in items 1–3. The use of these methods does not involve major difficulties when the integral term in equation (2.82) is of Fredhom type ($h(t) = a,\ , a = const$) or, more generally, when the function $h(t)$ is periodic in t [103]. Using these methods for investigating periodic solutions of the integro-differential equation (2.82), the integral term of which is of Volterra type ($h(t) = t$), involves certain difficulties [103] connected mainly to the equality

$$\int\limits_0^T E\left(t, s, x(s), \dot{x}(s)\right) ds = 0$$

to be held.

Remark 2.3. In subsequent chapters we shall use the methods for investigating periodic solutions of wave ordinary differential equations of the second order ($T(2T)$)-periodic systems of the first class), which were presented in items 1–3, to study periodic solutions of wave partial differential and integro-differential equations of the second order of hyperbolic type.

2.3. Periodic solutions of autonomous wave differential equations

The present section is concerned with the existence of periodic solutions of the autonomous wave differential equation of the second order

$$\frac{d^2y}{dt^2}+\omega^2 y = g\left(y, \frac{dy}{dt}\right), \qquad y(t+T)=y(t). \tag{2.83}$$

Changing the variable

$$t=\frac{\tau}{\omega}(1-a), \quad a \neq 1, \quad a = const. \tag{2.84}$$

in equation (2.83) allows us to rewrite (2.83) as

$$\frac{d^2y}{d\tau^2}+(1-a)^2 y = \left(\frac{1-a}{\omega}\right)^2 g\left(y, \frac{\omega}{1-a}\frac{dy}{d\tau}\right) \equiv \tilde{g}\left(y, \frac{dy}{d\tau}\right). \tag{2.85}$$

2π-periodic solutions of (2.85) can now be found by the formulas

$$y(\tau) = z(\tau)\cos\tau + v(\tau)\sin\tau, \quad \frac{dy}{d\tau} = -z(\tau)\sin\tau + v(\tau)\cos\tau. \tag{2.86}$$

Combining the identity $\frac{d}{d\tau}(y(\tau)) \equiv \frac{dy}{d\tau}$, (2.85) and (2,86) yields

$$\frac{dz}{d\tau}\cos\tau + \frac{dv}{d\tau}\sin\tau = 0,$$

$$-\frac{dz}{d\tau}\sin\tau + \frac{dv}{d\tau}\cos\tau = (2a - a^2)(z\cos\tau + v\sin\tau) + G(\tau, a, z, v),$$

$$G(\tau, a, z, v) = \tilde{g}(z\cos\tau + v\sin\tau, -z\sin\tau + v\cos\tau). \tag{2.87}$$

Solving system (2.87) with respect to $\frac{dz}{d\tau}$ and $\frac{dv}{d\tau}$ we obtain

$$\frac{dz}{d\tau} = (a^2 - 2a)\left(\frac{z}{2}\sin 2\tau + v\sin^2\tau\right) - G(\tau, a, z, v)\sin\tau,$$

$$\frac{dv}{d\tau} = (2a - a^2)\left(z\cos^2\tau + \frac{v}{2}\sin 2\tau\right) + G(\tau, a, z, v)\cos\tau \tag{2.88}$$

or, shortly,

$$\frac{dx}{d\tau} = f(\tau, a, x), \tag{2.89}$$

where x and f are two-dimensional vectors and $f(\tau + 2\pi,\ a, x) = f(\tau, a, x)$.

Thus, the investigation of 2π-periodic solutions of the wave equation (2.85) reduces to the investigation of 2π-periodic solutions of the first order nonautonoumous system of equations (2.89). This problem was studied in [103], and §1 contains the brief account of its results. On the other hand, in the case of the averaged system

$$\frac{d\xi}{d\tau} = f_0(a, \xi) \equiv \overline{f(\tau, a, \xi)}, \quad f(\tau, a, \xi) = \frac{1}{2\pi}\int_0^{2\pi} f(\tau, a, \xi)d\tau, \tag{2.90}$$

we obtain a system of two equations in three variables $\xi = \xi_0, f(a, \xi_0) = 0$ to determine the "quasistatic" solution ξ_1, ξ_2, a. This system

$$f_0(a, \xi) = 0 \tag{2.91}$$

might have a "quasistatic" solution $a, a = a_0 \neq 1$ for a certain value of the parameter $\xi = \xi_0, f(a_0, \xi_0) = 0$. Hence, the value of $a_0 \neq 1$ determines the period T of the desired $2\pi(1 - a_0)/\omega$-periodic solution of the autonomous equation (2.83.).

There is another feature of system (2.91). Let $\xi = (\xi_1, \xi_2)$. Using (2.88) and (2.90), we can transform (2.91) into the system

$$\frac{1}{2}(a^2 - 2a)\xi_2 - \overline{G(\tau, a, \xi_1, \xi_2)\sin\tau} = 0,$$

$$\frac{1}{2}(2a - a^2)\xi_1 + \overline{G(\tau, a, \xi_1, \xi_2)\cos\tau} = 0. \tag{2.92}$$

Setting $\xi_2 \equiv 0$, we arrive at the system of two equations

$$\overline{G(\tau, a, \xi_1, 0)\sin\tau} = 0,$$

$$\frac{1}{2}(2a-a^2)\xi_1+\overline{G(\tau,a,\xi_1,0)\cos\tau}=0 \tag{2.93}$$

in two variables ξ_1 and a. If (2.93) has a solution (ξ_1^0, a_0), it enables an isolated point of the mapping $f_0(a,\xi)$ to be determined. In particular, when the first equation in (2.93) has a solution $\xi_1=\xi_1^0\neq 0$ for any value of a, the second equation in (2.93) determines a value of the parameter a needed to find the desired period $T=2\pi(1-a)/\omega$.

As an example, consider the autonomous second order wave equation with a small parameter

$$\frac{d^2y}{dt^2}+\omega^2y=\varepsilon g\left(y,\frac{dy}{dt}\right),\quad \varepsilon\ll 1. \tag{2.94}$$

The existence of periodic solutions of equation (2.94) in the class of analytic functions has been studied well with the help of the Poincaré method, and in the class of smooth functions by the Krylov-Bogolyubov-Mitropol'skii asymptotic method and, recently, by the Samoilenko numerical-analytical method [14, 74, 103]. By using the above method for investigating periodic solutions of the autonomous wave equation of the second order (2.83), we shall establish the condition ensuring the existence of periodic solutions of equation (2.94). To do this, we exploit the theory of 2π-periodic systems of the second class, the second basic theorem of Bogolyubov [14] and the Samoilenko numerical-analytical method. We shall study those periodic solutions of (2.94) which turn, when $\varepsilon=0$, into periodic solutions of the generating equation $d^2y/dt^2+\omega^2y=0$. In general, a period of such a periodic solution depends on the parameter ε:

$$T=T(\varepsilon)=\frac{2\pi}{\lambda(\varepsilon)},\quad \lim_{\varepsilon\to 0}\lambda(\varepsilon)=\omega, \tag{2.95}$$

We refer to $\lambda(\varepsilon)$ as frequency. We set

$$\lambda(\varepsilon)=\omega/(1-\varepsilon\nu),\quad \varepsilon\nu\neq 1, \tag{2.96}$$

and change the variable in equation (2.94)

$$t=\tau(1-\varepsilon\nu)/\omega. \tag{2.97}$$

Then (2.94) can be rewritten as

$$\frac{d^2y}{d\tau^2}+(1-\varepsilon\nu)^2y=\varepsilon(1-\varepsilon nu)^2\omega^{-2}g\left(y,\frac{\omega}{1-\varepsilon\nu}\frac{dy}{d\tau}\right). \tag{2.98}$$

Obviously, the value $t=T$ corresponds to the value $\tau=2\pi$, i.e. the period of the desired periodic solution of equation (2.98) with respect to the new variable can now be fixed and it equals 2π.

Remark 2.4. The investigation of T-periodic systems of the second class carried out in §2.2 of this chapter (see Theorem 2.4) is applicable both to nonautonomous and autonomous cases, namely, to the cases where the right-hand side of the second order wave equation explicitly depends or does not depend on time t. On the other hand, according to the definition of systems of the second class, a solution of such a system has the same period as a solution of the corresponding homogeneous equation. Therefore, rewriting equation (2.98) as

$$\frac{d^2y}{d\tau^2}+y=\varepsilon(2\nu-\varepsilon\nu^2)y+$$
$$+\varepsilon(1-\varepsilon\nu)^2\omega^{-2}g\left(y,\frac{\omega}{1-\varepsilon\nu}\frac{dy}{d\tau}\right)\equiv Y\left(y,\frac{dy}{d\tau}\right), \tag{2.99}$$

we see that a solution of the homogeneous equation $\frac{d^2y}{d\tau^2}+y=0$ has period τ in variable 2π. Thus, the existence of a periodic solution of (2.98) can be studied by using the operator P_1 constructed in §2.1 of the present chapter, i. e. with the help of the integral equation

$$y(\tau)=y_0(\tau)+\int_0^{2\pi} Q_1(\tau)Y\left(y(\tau),\frac{dy(\tau)}{d\tau}\sin(t-\tau)d\tau,\right)$$

where $y_0(\tau)=C_1\cos\tau+C_2\sin\tau$ and the function $Q_1(\tau)$ is given by (2.6).

We shall first obtain a standard system of the first order corresponding to equation (2.98). Changing the variable in equation (2.98) according to (2.86) and using the identity $\frac{d}{d\tau}(y(\tau))=\frac{dy}{d\tau}$, we get

$$\frac{dz}{d\tau}\cos\tau+\frac{dv}{d\tau}\sin\tau=0,$$

$$-\frac{dz}{d\tau}\sin\tau+\frac{dv}{d\tau}\cos\tau+\varepsilon(\varepsilon\nu^2-2\nu)(z\cos\tau+v\sin\tau)=$$

$$=\varepsilon(1-\varepsilon\nu)^2\omega^{-2}g\left(z\cos\tau+v\sin\tau,\frac{\omega}{1-\varepsilon\nu}(-z\sin\tau+v\cos\tau)\right)\equiv$$

$$\equiv\varepsilon\tilde{g}(\tau,\nu,z,v). \tag{2.100}$$

Equality (2.100) allows us to obtain the following first order standard system

$$\frac{dz}{d\tau}=\varepsilon\alpha(\varepsilon)\left(\frac{z}{2}\sin 2\tau+v\sin^2\tau\right)-\varepsilon\tilde{g}(\tau,\nu,z,v)\sin\tau\equiv$$

$$\equiv\varepsilon f_1(\tau,\nu,z,v),$$

$$\frac{dv}{d\tau}=-\varepsilon\alpha(\varepsilon)\left(z\cos^2\tau+\frac{v}{2}\sin 2\tau\right)+\varepsilon\tilde{g}(\tau,\nu,z,v)\cos\tau= \tag{2.101}$$

$$=\varepsilon f_2(\tau,\nu,z,v),$$

$$\alpha(\varepsilon)=\varepsilon\nu^2-2\nu$$

or, in a short form,

$$\frac{dx}{d\tau}=\varepsilon f(\tau,\nu,x), \tag{2.102}$$

Here x and f are two-dimensional vectors, $x=(z,v), f=(f_1,f_2)$.

By using the second basic theorem of Bogolyubov [14, 74] and Theorem 2.4 we arrive at the following assertions.

Theorem 2.11. *Let, for a certain value $\nu=\nu_0$, equation (2.99) be a 2π-system of the second class and the right-hand side of system (2.102) satisfy the condition of the second basic theorem of Bogolyubov. Then the autonomous equation (2.94) has a $2\pi(1-\varepsilon\nu_0)/\omega$-periodic solution*

$$x(t)=z\left(\frac{\omega t}{1-\varepsilon\nu_0}\right)\cos\frac{\omega t}{1-\varepsilon\nu_0}+v\left(\frac{\omega t}{1-\varepsilon\nu_0}\right)\sin\frac{\omega t}{1-\varepsilon\nu_0},$$

which satisfies all the stability conditions that follow from the statement of the second basic theorem of Bogolyubov.

Theorem 2.12. *Let, for a certain value $\nu=\nu_0$, equation (2.99) be a 2π-system of the second class in the region $-a\le y\le a, -b\le\dot{y}\le b$. For the autonomous equation (2.94)* *to have, for sufficiently small ε, a $2\pi(1-\varepsilon\nu_0)/\omega$-periodic solution continuously depending on $\varepsilon = 0$ in a neighbourhood of $\nu = \nu_0$ it is necessary that, for , the averaged system*

$$\frac{d\xi}{d\tau} = \overline{\varepsilon f(t, \nu_0, \xi)} \equiv \varepsilon \frac{1}{2\pi} \int_0^{2\pi} f(t, \nu_0, \xi) dt,$$

where the function $f(\tau, \nu, x)$ is defined by the right-hand side of (2.102), have an equilibrium state $\xi = \xi_0$ such that $-a < \xi_1^0 \cos\tau + \xi_2^0 \sin\tau < a$, $-b < \xi_1^0 \sin\tau + \xi_2^0 \cos\tau < b$ and it is sufficient that this state be isolated and of nonzero index.

As an example, consider the Van der Pol equation

$$\ddot{y} + y = \varepsilon(1 - y^2)\dot{y}. \tag{2.103}$$

System (2.93) ($f(\overline{\tau, \nu_0 \xi}) = 0$) constructed for equation (2.103) can be written as

$$\overline{-\frac{1}{2}\xi_1 + \frac{1}{8}\xi_1^3 + \frac{1}{2}\xi_1 \cos 2t - \frac{1}{8}\xi_1^3 \cos 4t} = 0,$$

$$\overline{\frac{1}{2}(2\nu_0 - \varepsilon\nu_0^2)\xi_1 + \frac{1}{2}\left(\xi_1 - \frac{1}{2}\xi_1^3\right) \sin 2t + \frac{1}{8}\xi_1^3 \sin 4t} = 0. \tag{2.104}$$

Averaging this system, we obtain

$$\xi_1\left(1 - \frac{1}{4}\xi_1^2\right) = 0, \quad \nu_0(2 - \varepsilon\nu_0) = 0. \tag{2.105}$$

Since for $\xi_1^0 = 0$ we get a trivial solution of the Van der Pol equation (2.103), the only periodic solution of this equation is a 2π-periodic one corresponding to $\xi_1^0 = 2$, $\xi_2 = 0$, $\nu_0 = 0$. This result is in a good agreement with the theory of asymptotic methods [14, 74].

CHAPTER 3

PERIODIC SOLUTIONS OF THE FIRST CLASS SYSTEMS

This chapter singles out some new classes of periodic boundary value problems for the second order wave differential equations of hyperbolic type (T-systems of the first class, i. e. periodic boundary value problems such that their solutions do not include solutions of the corresponding homogeneous periodic boundary value problem).

3.1. Linear systems

This section is concerned with the problem of the existence of classical periodic solutions of a linear boundary value problem given as

$$u_{tt} - u_{xx} = g(x,t), \quad (x,t) \in I \times \mathbb{R}; \tag{3.1}$$

$$u(0,t) = u(\pi,t) = 0, \quad t \in \mathbb{R}; \tag{3.2}$$

$$u(x,t+T) = u(x,t), \quad (x,t) \in I \times \mathbb{R}, \tag{3.3}$$

in specific functional spaces tightly connected to the set I.

?? continuous and bounded on $I \times \mathbb{R} = \mathbb{R}^2$ ($I \times \mathbb{R} = [0,\pi] \times \mathbb{R}$) is denoted by C (C_π, respectively). The symbol $C^j(C^j_\pi)$ stands for the space of functions $u \in C(u \in C_\pi)$ such that $D^k_t D^l_x u \in C(D^k_t D^l_x u \in C_\pi)$ for all $k + j \le l$. Let G_x ($G_{\pi t}$) denote the space of functions continuous and bounded on $\mathbb{R}^2$ ($[0,\pi] \times \mathbb{R}$, respectively) with the first derivative in x (t, respectively) continuous and bounded. The expression $(r,q) = 1$ means that the numbers r and q are coprime.

We next define the periods $T_k, k = 1,2,3$ and the functional spaces corresponding to them $u \in C$:

$$T_1 = 2\pi(2p-1)/q, \ q - \text{ even}, (2p-1,q) = 1,$$

$$A^0_1 = \{u : u(x,t) = u(x,t+T_1) = -u(x,t) = u(\pi - x,t)\};$$

$$T_2 = 2\pi(2p-1)/q, \ q - \text{ odd}, (2p-1,q) = 1,$$

$$A^0_2 = \{u : u(x,t) = u(\pi - x, t + T_2/2) = u(x,t+T_2) =$$

$$= -u(-x,t) = u(x+2\pi,t)\};$$

$$T_3 = 2\pi r/q, \ r - \text{ even}, \ q - \text{ odd}, \ (r,q) = 1,$$

$$A^0_3 = \{u : u(x,t) = -u(x,t+T_3/2) = -u(-x,t) = u(x+2\pi,t)\},$$

The symbols denote classes of functions $u \in A^0_k$, $k = 1,2,3$, satisfying the conditions

$$\int\limits_{-T_1}^{0} d\tau \int\limits_{\tau}^{\tau + T_1 q/2} \{u(x+\theta-\tau,\tau) + u(x-\theta+\tau,\tau)\} d\theta = 0; \tag{3.4}$$

$$\int\limits_{\tau}^{\tau + T_k q/2} \{u(x+\theta-\tau,\tau) + u(x-\theta+\tau,\tau)\} d\theta = 0, \ k = 2,3. \tag{3.5}$$

Lemma 3.1. Every function $u(x,t) \in A_k^0$, that can be expanded in uniformly converging Fourier series $u(x,t) = \sum_{n=1}^{\infty} u_n(t) \sin nx$, satisfies the conditions (3.4) and (3.5), i. e. $u(x,t) \in \tilde{A}_k^0$, $k = 1,2,3..$

Proof. In view of the condition in Lemma 3.1 and the notation T_k, $k = 1,2,3$, we have

$$\int_{\tau}^{\tau+T_k q/2} \{u(x+\theta-\tau,\tau) + u(x-\theta+\tau,\tau)\}d\theta =$$

$$\int_{\tau}^{\tau+T_k q/2} \sum_{n=1}^{\infty} u_n(\tau)\{\sin n(x+\theta-\tau) + \sin n(x-\theta+\tau)\}d\theta =$$

$$\sum_{n=1}^{\infty} u_n(\tau)\left\{\frac{\cos n(x-\theta+\tau)}{n} - \frac{\cos n(x+\theta-\tau)}{n}\right\}_{\tau}^{\tau+T_k q/2} =$$

$$\sum_{n=1}^{\infty} u_n(\tau)\left\{\frac{\cos n(x-T_k q/2)}{n} - \frac{\cos n(x+T_k q/2)}{n}\right\} = 0.$$

which proves the lemma.

Many mathematicians excepting the authors of [21] have used the Fourier method to prove the existence of a solution of the above linear problem (1.1) [158, 159, 161–179]. We now make use of the theory of integral equations to investigate the existence of continuous and classical periodic solutions in the functional spaces A_k^0 and in the spaces A_k from [21]. However, it should be pointed out that there is a class of functions that do not expand in the Fourier series but satisfy (3.4).

Lemma 3.2. Every continuous function $u \in A_1^0$ that is even with respect to the variable t satisfies the condition (3.4), i. e. $u(x,t) \in A_1^0$.

Proof. We shall show that the function

$$\Phi(x,\tau) = \int_{\tau}^{\tau+c_1} \{u(x+\theta-\tau,\tau) + u(x-\theta+\tau,\tau)\}d\theta, c_1 = T_1 q/2,$$

is odd and T_1-periodic in τ. These properties imply the fulfilment of the condition (3.4).

Indeed, from the definition of $\Phi(x,\tau)$ we obtain

$$\Phi(x,\tau+T_1) = \int_{\tau+T_1}^{\tau+T_1+c_1} \{u(x+\theta-\tau-T_1,\tau) + u(x-\theta+\tau+$$

$$+T_1,\tau)\}d\theta = \int_{\tau}^{\tau+c_1} \{u(x+z-\tau,\tau) + u(x-z+\tau,\tau)\}dz = \Phi(x,\tau).$$

Hence, the function $\Phi(x,\tau)$ is T_1-periodic in τ. On the other hand, we have

$$\Phi(x,-\tau) = \int_{-\tau}^{-\tau+c_1} \{u(x+\theta+\tau,-\tau) + u(x-\theta-\tau,-\tau)\}d\theta =$$

$$= -\int_{\tau+c_1}^{\tau} \{u(x-v+c_1+\tau,-\tau) + u(x+v-c_1-\tau,-\tau)\}dv =$$

$$= \int_{\tau}^{\tau+c_1} \{u(x - v + c_1 + \tau, -\tau) + u(x + v - c_1 - \tau, -\tau)\}dv.$$

By using the condition of Lemma 3.2 with $c_1 = T_1 q/2$ we obtain $\Phi(x, -\tau) = -\Phi(x, \tau)$, which establishes the lemma.

For $g \in c \cap A_k^0$ we have

$$(P_0 g)(x,t) = \frac{1}{2}\int_0^t dt \int_{\tau}^{t} \{g(x+\theta-\tau,-\tau) + g(x-\theta+\tau,\tau)\}d\theta; \tag{3.6}$$

$$(P_k g)(x,t) = \frac{1}{2}(P_0 g)(x,t) - \frac{1}{4}\int_t^{c_k} d\tau \int_{\tau}^{t} \{g(x+\theta-\tau,\tau) + g(x-$$

$$-\theta+\tau,\tau)\}d\theta \equiv \frac{1}{4}\int_0^{c_k} Q(\tau)d\tau \int_{\tau}^{t} \sum_{i=0}^{1} g(x + (-1)^i(\theta-\tau),\tau)d\theta; \tag{3.7}$$

$$Q(\tau) = \begin{cases} 1, 0 \le \tau \le t, \\ -1, t < \tau \le c_k, \end{cases} \quad c_k = T_k q/2,\ k = 1,2,3. \tag{3.8}$$

In the sequel, $L(X,Y)$ stands for the space of linear bounded mappings from X into Y. The symbols $Q_T^1, Q_T^2, Q_\pi, Q_{2\pi}$ denote the spaces of functions satisfying, on $I \times \mathbb{R}$, the following conditions $u(x,t+T) = u(x,t)$; $u(x,i+T/2) = -u(x,t)$; $u(\pi - x,t) = u(x,t)$; $u(x,t) = -u(-x,t)$, $u(x+2\pi,t) = u(x,t)$.

Lemma 3.3. If $u \in Q_\pi$ and $u(x,t) = -u(-x,t)$, then $u \in Q_{2\pi}$.

Proof. Indeed, we have $u(x+2\pi,t) = u(\pi - (-\pi - x),t) = -u(\pi + x,t) = -u(\pi - (-x),t) = u(x,t)$, which is the claim.

Lemma 3.4. Let $g \in G_x \cap \tilde{A}_k^0$, $k = 1,2,3$. Then $u = P_k g$ satisfies (3.1) and (3.2). Furthermore,

$$P_k \in L(C \cap Q_{T_k}^1 \cap Q_\pi, C^1 \cap Q_{T_k}^1 \cap Q_\pi),$$

$$P_k \in L(G_x \cap Q_{T_k}^1 \cap Q_\pi, C^2 \cap Q_{T_k}^1 \cap Q_\pi);$$

$$P_k \in L(C \cap \tilde{A}_k^0, C^1 \cap A_k^0), P_k \in L(G_x \cap \tilde{A}_k^0, C^2 \cap A_k^0),\ k = 2,3.$$

Proof. It is easily seen that the function $u = P_k g$ satisfies equation (3.1). Since $g \in \tilde{A}_k^0$, which means that g is odd and 2π-periodic in x, the function $P_k g$ satisfies the boundary condition $u(0,t) = u(\pi,t) = 0$. By virtue of (3.6) and (3.7) we now have $P_k \in L(C \cap Q_\pi, C^1 \cap Q_\pi)$, $k = 1,2,3$, $P_k \in L(C \cap Q_{2\pi}, C^1 \cap Q_{2\pi})$, $k = 2,3$, and

$$(P_k g)(x,t+T) = \frac{1}{4}\int_0^{t+T} d\tau \int_{\tau}^{t+T} \{g(x+\theta-\tau,\tau) + g(x-\theta+$$

$$+\tau,\tau)\}d\theta - \frac{1}{4}\int_{t+T}^{c_k} d\tau \int_{\tau}^{t+T} \{g(x+\theta-\tau,\tau) + g(x-\theta+\tau,\tau)\}d\theta =$$

$$= \frac{1}{4}\int_{-T}^{t} dz \int_{z+T}^{t+T} \{g(x+\theta-z-T, z+T) + g(x-\theta+z+T, z+$$

$$+T)\}d\theta - \frac{1}{4}\int_{\tau}^{c_k-T} dz \int_{z+T}^{t+T} \{g(x+\theta-z-T, z+T) + g(x-\theta+$$

$$+z+T, z+T)\}d\theta = \int_{-T}^{t} dz \int_{z}^{t} \{g(x+v-z, z+T) + g(x-$$

$$-v+z, z+T)\}dv - \frac{1}{4}\int_{t}^{c_k-T} dz \int_{z}^{t} \{g(x+v-z, z+T) + g(x-$$

$$-v+z, z+T)\}dv = \frac{1}{4}\int_{-T}^{0} dz \int_{z}^{t} \sum_{i=0}^{1} g(x+(-1)^i(v-z), z+$$

$$+T)dv + \frac{1}{4}\int_{0}^{t} dz \int_{z}^{t} \sum_{i=0}^{1} g(x+(-1)^i(v-z), z+T)dv-$$

$$-\frac{1}{4}\int_{t}^{c_k} dz \int_{z}^{t} \sum_{i=0}^{1} g(x+(-1)^i(v-z), z+T)dv - \frac{1}{4}\int_{c_k}^{c_k-T} \times$$

$$\times dz \int_{z}^{t} \sum_{i=0}^{1} g(x+(-1)^i(v-z), z+T)dv = I^1 + I^2 + I_k^3 -$$

$$-\frac{1}{4}\int_{0}^{-T} d\tau \int_{\tau+c_k}^{t} \sum_{i=0}^{1} g(x+(-1)^i(v-\tau-c_k), \tau+c_k+T)dv =$$

$$I^1 + I^2 + I_k^3 + \frac{1}{4}\int_{-T}^{0} d\tau \int_{\tau+c_k}^{t} \sum_{i=0}^{1} g(x+(-1)^i(v-\tau-c_k), \tau+$$

$$+c_k+T)dv + \frac{1}{4}\int_{-T}^{0} d\tau \int_{\tau}^{t} \sum_{i=0}^{1} g(x+(-1)^i(v-\tau-c_k), \tau+$$

$$+c_k+T)dv = I^1 + I^2 + I_k^3 + I_k^4 + I_k^5. \tag{3.9}$$

Setting $T = T_k$, $k = 1, 2, 3$, in (3.9) and taking account of $c_k = T_k q/2$, we see that $I^1 = -I_k^5, I^2 + I_k^3 = (P_k g)(x,t), I_k^4 = 0$, i. e. $P_k \in L(C \cap Q^1_{T_k}, C^1 \cap Q^1_{T_k}), k = 1, 2, 3$. For $T = T_k/2$ and $q \in \tilde{A}_k^0, k = 2, 3$, we obtain $P_k \in L(C \cap \tilde{A}_k^0, C^1 \cap A_k^0), k = 2, 3$ from (3.9).

The second statement of the lemma can be proved in a similar fashion. This completes the proof.

Lemma 3.4 enables us to prove the following assertions.

Theorem 3.1. *For $g \in G_x \cap \tilde{A}_k^0$, the function $u = P_k g$ is the only function in the space $C^2 \cap A_k^0$ satisfying (3.1)–(3.3) for $T = T_k$, $k = 1, 2, 3$.*

Theorem 3.2. *For $g \in G_x \cap Q_\pi (g \in G_x \cap Q_{2\pi})$, the function $u = P_0 g$ is a function belonging to the space $C^2 \cap Q_\pi (C^2 \cap Q_{2\pi})$, satisfying equation (3.1) and assuming equal values at $x = 0$ and $x = \pi$, $u(0,t) = u(\pi,t)$.*

Theorem 3.2 follows from (3.6).

As mentioned above, the work [21] uses other operators different from P_k to investigate classical periodic solutions of the wave equation (3.1). These operators are of the form

$$(S_1 g)x,t) = -\frac{1}{2}\int_0^x \int_{t-x+\xi}^{t+x-\xi} g(\xi,\tau)d\xi d\tau; \tag{3.10}$$

$$(S_2 g)(x,t) = -\frac{1}{2}\int_x^\pi \int_{t+x-\xi}^{t-x+\xi} g(\xi,\tau)d\xi d\tau. \tag{3.11}$$

and are obtained by solving the Cauchy problems (see Chapter 1)

$$u^1_{tt} - u^1_{xx} = g(x,t), u^1(0,t) = 0, u^1_x(0,t) = \mu_1(t); \tag{3.12}$$

$$u^2_{tt} - u^2_{xx} = g(x,t), u^2(0,t) = 0, u^2_x(0,t) = \mu_2(t). \tag{3.13}$$

In other words, the following statement is true: in the class of smooth functions, every solution of the Cauchy problem (3.12) can be represented as

$$u^1(x,t) = \frac{1}{2}\int_{t-x}^{t+x} \mu_1(\alpha)d\alpha + (S_1 g)(x,t) \equiv$$

$$\equiv \int_0^x \frac{\mu_1(t+\xi)+\mu_1(t-\xi)}{2}d\xi + (S_1 g)(x,t), \tag{3.14}$$

and every solution of (3.13) as

$$u^2(x,t) = \frac{1}{2}\int_{t+\pi-x}^{t-\pi+x} \mu_2(\alpha)d\alpha + (S_2 g)(x,t) \equiv$$

$$\equiv -\int_x^\pi \frac{\mu_2(t+\pi-\xi)+\mu_2(t-\pi+\xi)}{2}d\xi + (S_2 g)(x,t). \tag{3.15}$$

In [21] the operators S_1 and S_2 in (3.10) and (3.11) are defined, for the above-mentioned periods T_k, $k = 1,2,3$, on the following spaces of functions $u \in C_\pi$:

$$A_1 = \{u : u(x,t) = u(x,t+T_1) = u(\pi - x, t)\};$$

$$A_2 = \{u : u(x,t) = u(\pi - x, t + T_2/2) = u(x, t+T_2)\};$$

$$A_3 = \{u : u(x,t) = -u(x, t+T_3/2).$$

Although $A^0_k \subset A_k$, $k = 1,2,3$ is true, the results presented above and obtained in [21] differ. This follows from the fact that our results hold in the spaces $G_x \cap A^0_k$, $k = 1,2,3$, whereas [21] establishes the existence of classical periodic solutions in the spaces $G_t \cap A_k$, $k = 1,2,3$. Moreover, the results of [21] need to be made more precise. We do this below.

Theorem 3.3. *For $g \in G_{\pi t} \cap A_k$, $k = 1,2,3$, the function $u = Sg \equiv \frac{1}{2}(S_1 g + S_2 g)$ satisfies equation (1.1) and assumes equal values $x = 0$ at $x = \pi$ and $u(0,t) = u(\pi,t)$. Furthermore,*

$$S \in L(C_\pi \cap Q^1_{T_k} \cap Q_\pi, C^1 \cap Q^1_{T_k} \cap Q_\pi), S \in L(G_{\pi t} \cap Q^1_{T_k} \cap$$

$$\cap Q_\pi, C_\pi^2 \cap Q_{T_k}^1 \cap Q_\pi),$$
$$(Sg)_t(0,\pi) = (Sg)_t(\pi,t),\ (Sg)_x(0,t) = -(Sg)_x(\pi,t).$$

This theorem is a consequence of the expressions (3.10) abd (3.11).

Since $Sg \in Q_\pi$ satisfies (1.1) and assumes equal values at $x = 0$ and $x = \pi$, the action of the operator S at the points $x = 0$ and $x = \pi$ can be annulled only by a function φ in Q_π that satisfies the homogeneous equation $\varphi_{pp} - \varphi_{xx} = 0$ and the following equalities

$$\varphi(0,t) = \varphi(\pi,t),\ \varphi_t(0,t) = \varphi_t(\pi,t),\ \varphi_x(0,t) = -\varphi_x(\pi,t).$$

One can easily check that

$$\varphi(x,t) = b(t+x) + b(t-x+\pi), \tag{3.16}$$

for $b \in C_\pi^2 \cap Q_\pi$ is such a function.

Thus, we shall seek periodic solutions of the problem (3.1)–(3.3) in the form $u = \frac{1}{2}(S_1 g + S_2 g) + \varphi \equiv Sg + \varphi$. The function $\varphi \in Q_\pi$ is chosen to satisfy $u(0,t) = 0$. From the definition of S_1 and S_2 and the condition $U(0,t) = 0$ we then obtain the equality

$$b(t+\pi) + b(t) = \frac{1}{4}\int_0^\pi d\xi \int_{t-\xi}^{t+\xi} g(\xi,\tau)d\tau. \tag{3.17}$$

Differentiating it we get

$$h(t+\pi) + h(t) = \frac{1}{4}\int_0^\pi \{g(\xi,t+\xi) - g(\xi,t-\xi)\}d\xi, \tag{3.18}$$

where $h = b'$. Since the condition A_k, $k = 1$ is fulfilled in the space $T_k q/2 = (2p-1)\pi$, $k = 1$, the condition $g \in A_k$, $k = 1$ is evidently fulfilled for $h(t) = h(t + (2p-1)\pi)$. Therefore, replacing t by $t + j\pi$ in (3.18), multiplying each expression by $(-1)^j$ and summing these expressions with respect to $j = 0, 1, 2, ..., 2p-2$, we obtain

$$h(t) = \frac{1}{8}\int_0^\pi \sum_{j=0}^{2p-2} (-1)^j \{g(\xi, t+g\pi+\xi) - g(\xi, t+g\pi-\xi)\}d\xi, \tag{3.19}$$

in view of the above relation $h(t + (2p-1)\pi) = h(t)$.

By virtue of (3.18), the function

$$w \to \frac{1}{4}\int_0^t d\xi \int_{w-\xi}^{w+\xi} g(\xi,\tau)d\tau - \int_0^{w+\pi} h(\tau)d\tau - \int_0^w h(\tau)d\tau$$

is constant. Since $h = b'$, it is seen that

$$b(t) = \int_0^t h(\tau)d\tau + \frac{1}{2}\left\{\frac{1}{4}\int_0^\pi d\xi \int_{\omega-\xi}^{\omega+\xi} g(\xi,\tau)d\tau - \int_0^{\omega+\pi} h(\tau)d\tau - \right.$$
$$\left. \int_0^\omega h(\tau)d\tau\right\}$$

does not depend on ω and satisfies (3.17). Let

$$(R_1 g)(x,t) = \frac{1}{2}(S_1 g + S_2 g)(x,t) + \frac{1}{4}\int_0^{\pi} d\xi \int_{\omega+\xi}^{\omega-\xi} g(\xi,\tau) +$$

$$+ \int_{\omega}^{t+x} h(\tau)d\tau + \int_{\omega+\pi}^{t-x+\pi} h(\tau)d\tau \equiv \frac{1}{2}(Sg)(x,t) + \frac{1}{4}\int_0^{\pi} \times$$

$$\times d\xi \int_{\omega-\xi}^{\omega+\xi} g(\xi,\tau)d\tau + \frac{1}{8}\sum_{g=0}^{2p-2}(-1)^j \left\{ \int_{\omega}^{t+x} \mu(\tau+j\pi)d\tau + \right.$$

$$\left. + \int_{\omega+\pi}^{t-x+\pi} \mu(\tau+j\pi)d\tau \right\}, \tag{3.20}$$

Here

$$\mu(t) = \int_0^{\pi} \{g(\xi, t+\xi) - g(\xi, t-\xi)\}d\xi; \tag{3.21}$$

the function $h(\tau)$ is given by (3.19), and ω is an arbitrary number.

Theorem 3.4. *For any $g \in G_{\pi t} \cap A_1$, the function $u = R_1 g$ is the only function in the space $C_\pi^2 \cap A_1$ that satisfies (3.1)–(3.3). Furthermore, $R_1 \in L(C_\pi \cap A_1, C_\pi^1 \cap A^1)$, $R_1 \in L(G_{\pi t} \cap A_1, C_\pi^2 \cap A_1)$.*

In [21] an attempt was made to prove that the operator

$$R_2 g = \frac{1}{2}(S_1 g + S_2 g) + Z_2 g,$$

where

$$(Z_2 g)(x,t)\frac{1}{4}\int_0^{\pi}\int_{\omega-\xi}^{\omega+\xi} g(\xi,\tau)d\xi d\tau + \frac{1}{8}\sum_{j=1}^{p-1}(-1)^j \left\{ \int_{\omega}^{t+\omega} \mu(\tau+ \right.$$

$$\left. +j(\pi+T_2/2))d\tau + \int_{\omega}^{t-x} \mu(\tau+(j+1)(\pi+T_2/2))d\tau \right\};$$

and ω is an arbitrary number, is suitable to study T_2-periodic solutions of the problem (3.1)–(3.3) in the space A_2.

We should note that in the class A_2, the function $\mu(t)$ given by (3.21) is identically equal to zero. We shall prove this for $p = q = 1$, $T_2 = 2\pi$. Let $g \in A_2$, $T_2 = 2\pi$. The definition of the space A_2 implies then the identity $g(x,t) = g(\pi - x, t+\pi)$. Taking account of (3.21), we obtain

$$\mu(t) = \int_0^{\pi} g(\xi, t+\xi)d\xi - \int_0^{\pi} g(\xi, t-\xi)d\xi.$$

Changing the variable $\xi = \pi - \eta$ in the first integral and using the identity $g(\pi - x, t+\pi) = g(x,t)$, we have

$$\mu(t) = -\int_0^{\pi} g(\pi-\eta, t+\pi-\eta)d\eta - \int_0^{\pi} g(\xi, t-\xi)d\xi =$$

$$= \int_0^\pi g(\eta, t-\eta)d\eta - \int_0^\pi g(\xi, t-\xi)d\xi \equiv 0.$$

Thus, the function $\mu(t)$ given by (3.21) is identical zero for $T_2 = 2\pi$, $g \in A_2$. However, we are able to show that for $g \in G_{\pi t} \cap A_2$, $p = 1$, the function

$$(R_2 g)(x,t) = \frac{1}{2}(S_1 g + S_2 g)(x,t) + \frac{\pi - x}{4\pi}\int_0^\pi d\xi \int_{t-\xi}^{t+\xi} g(\xi,\tau)d\tau +$$

$$+\frac{x}{4\pi}\int_0^\pi \int_{t-\pi+\xi}^{t+\pi-\xi} g(\xi,\tau)d\tau = (Sg)(x,t) + (\widetilde{S}g)(x,t) \qquad (3.22)$$

is the only function in the space $C^2_\pi \cap A_2$ that satisfies (3.1)–(3.3).

We first prove that the function $u = R_2 g$ satisfies the equation $u_{tt} - u_{xx} = g(x,t)$. To do this, it suffices to show that the function $u_0 = Sg$ satisfies the homogeneous equation $u_{0tt} - u_{0xx} = 0$. It is easy to check that $u_{0xx} = 0$. We next prove that $u_{0tt} = 0$. Computing the first partial derivative u_{0t} yields

$$u_{0t} = \frac{\pi - x}{4\pi}\int_0^\pi \{g(\xi, t+\xi) - g(\xi, t-\xi)\}d\xi + \frac{x}{4\pi}\int_0^\pi \{g(\xi, t+$$

$$+\pi - \xi) - g(\xi, t-\pi+\xi)\}d\xi \equiv \frac{\pi - x}{4\pi}\mu(t) + \frac{x}{4\pi}\nu(t).$$

As shown above, if $g \in A_2$ then $\mu(t) = 0$. We now prove that $\nu(t) = 0$. By using the equality $g(x,t) = g(\pi - x, t+\pi)$, we obtain

$$\nu(t) = \int_0^\pi \{g(\xi, t+\pi-\xi) - g(\xi, t-\pi+\xi)\}d\xi =$$

$$= \int_0^\pi g(\pi-\eta, \pi+t-\pi+\eta)d\eta - \int_0^\pi g(\xi, t-\pi+\xi)d\xi \equiv 0,$$

and therefore $u_{0t} \equiv\!= 0$. Thus, the function $u_0 \widetilde{S}g$ satisfies the homogeneous equation $u_{0tt} - u_{0xx} = 0$ and hence the function $u = R_2 g$ in (3.22) satisfies equation (3.1) ($u_{tt} - u_{xx} = g(x,t)$).

By the immediate check we see that the function $u = R_2 g$ satisfies the boundary condition $u(0,t) = u(\pi,t) = 0$.

Finally, we shall prove that $R_2 g \in A_2$ for $T_2 = 2\pi$, $g \in A_2$. Indeed, taking account of the definition of A_2, we obtain

$$(R_2 g)(\pi - x, t+\pi) = \frac{1}{4}\int_0^{\pi - x} d\xi \int_{t+x+\xi}^{t+2\pi-x-\xi} g(\xi,\tau)d\tau - \frac{1}{4}\times$$

$$\times \int_{\pi-x}^{\pi} d\xi \int_{t+2\pi-x-\xi}^{t+\xi+x} g(\xi,\tau)d\tau + \frac{x}{4\pi}\int_0^\pi d\xi \int_{t+\pi-\xi}^{t+\pi+\xi} g(\xi,\tau)d\tau + \frac{\pi - x}{4\pi}\int_0^\pi \times$$

$$\times d\xi \int_{t+\xi}^{t+2\pi-\xi} g(\xi,\tau)d\tau = -\frac{1}{4}\int_{x}^{\pi} d\eta \int_{t+x-\eta}^{t-x+\eta} g(\pi-\eta,\pi+\theta)d\theta -$$

$$-\frac{1}{4}\int_{0}^{x} d\eta \int_{t-x+\eta}^{t+x-\eta} g(\pi-\eta,\theta+\pi)d\theta + \frac{x}{4\pi}\int_{0}^{\pi} d\eta \int_{t-\pi+\eta}^{t+\pi-\eta}$$

$$g(\pi-\eta,\pi+\theta)d\theta + \frac{\pi-x}{4\pi}\int_{0}^{\pi} d\xi \int_{t-\eta}^{t+\eta} g(\pi-\eta,\pi+\theta)d\theta \equiv (R_2 g)(x,t).$$

and hence $(R_2 g) \in A_2$.

The following assertion summarizes the above results.

Theorem 3.5. *For $g \in G_{\pi t} \cap A_2$, $p = q = 1$, the function $u = R_2 g$ is the only function in the space $C_\pi^2 \cap A_2$ that satisfies (3.1)–(3.3). Furthermore, $R_2 \in L(C_\pi \cap A_2, C_\pi^1 \cap A_2)$, $R_2 \in L(G_{\pi t} \cap A_2, C_\pi^2 \cap A_2)$.*

We now turn to the space A_3 and show that it appears when investigating the existence of periodic solutions of the Cauchy problem (3.12) or (3.13) in the class of functions satisfying $u(0,t) = u(\pi,t) = 0$. Indeed, if the functions $g(x,t)$ and $\mu_j(t)$ $i = 1,2$ are T-periodic with respect to t, then the solutions $u^1(x,t)$ and $u^2(x,t)$ of the Cauchy problems (3.12) and (3.13) are always T-periodic with respect to t. Moreover, the solution $u^1(x,t)$ of (3.12) always satysfies the first boundary condition $u^1(0,t) = 0$ and the solution $u^2(x,t)$ of (3.13) satisfies the second boundary condition $u^2(\pi,t) = 0$ (see (3.10) and (3.11)).

Thus, for these solutions to be solutions of the linear problem (3.1)–(3.3) it suffices to choose the unknown functions $\mu_j(t)$ so that the condition $u^1(\pi,t) = 0$, $u^2(0,t) = 0$ be fulfilled. Both problems reduce to the same difference equation, which is always solvable in the space A_3. Indeed, substituting $x = \pi$ into equality (3.14) and keeping (3.10) in mind, we get

$$u^1(\pi,t) = \frac{1}{2}\int_{t-\pi}^{t+\pi} \mu_1(\alpha)d\alpha - \frac{1}{2}\int_{0}^{\pi} d\xi \int_{t-\pi+\xi}^{t+\pi-\xi} g(\xi,\tau)d\tau.$$

Therefore, $u^1(\pi,t) = 0$ holds if and only if the condition

$$\int_{t-\pi}^{t+\pi} \mu_1(\alpha)d\alpha - \int_{0}^{\pi} d\xi \int_{t-\pi+\xi}^{t+\pi-\xi} g(\xi,\tau)d\tau = 0,$$

is fulfilled. After differentiating with respect to t, this condition turns into the defference equation

$$\mu_1(t+\pi) - \mu_1(t-\pi) - \int_{0}^{\pi} \{g(\xi,t+\pi-\xi) - g(\xi,t-\pi+$$
$$+\xi)\}d\xi = 0. \tag{3.23}$$

We shall show that the difference equation (3.23) is solvable only in the space A_3. Replacing t by $t + 2\pi j + \pi$ in (3.23) and summing these expressions with respect to $j = 0,1,2,\dots,k-1$, we have

$$\mu_1(t+2\pi k) - \mu_1(t) = \sum_{j=0}^{k-1}\int_{0}^{\pi} \{g(\xi,2\pi(j+1)+t-\xi) - g(\xi,2\pi j+$$

$$+t+\xi)\}d\xi.$$

The last equation is solvable, provided the condition

$$\mu_1(t+2\pi k) = -\mu_1(t), \tag{3.24}$$

is satisfied. In this case we have

$$\mu_1(t) = -\sum_{j=0}^{k-1}\int_0^{\pi}\{g(\xi, t+2\pi(j+1)-\xi) - g(\xi, t+2\pi j+\xi)\}d\xi. \tag{3.25}$$

Thus, for the linear problem (3.1)–(3.3) to have a $2\pi r/q$-periodic solution $u(x, t+T/2) = -u(x,t)$ it is necessary and sufficient that r be even, $r = 2k$, and q odd, which just defines the space A_3. We thus arrive at the following assertion.

Theorem 3.6. *For $g \in G_{\pi t} \cap A_3$, the function*

$$(R_3 g)(x,t) = \frac{1}{2}\int_{t-x}^{t+x}\mu_1(\alpha)d\alpha + (S_1 g)(x,t), \tag{3.26}$$

where $\mu_1(t)$ is given by (3.25), is the only function in the space $C_\pi^2 \cap A_3$ satisfying (3.1)–(3.3). Furthermore, $R_3 \in L(C_\pi \cap A_3, C_\pi^1 \cap A_3)$, $R_3 \in L(C_{\pi t} \cap A_3, C_\pi^2 \cap A_3)$.

3.2. Nonlinear systems

This section deals with the existence of a solution of the nonlinear boundary value problem given as

$$u_{tt} - u_{xx} = F[u, u_t, u_x]; \tag{3.27}$$

$$u(0,t) = u(\pi, t) = 0, \ t \in \mathbb{R}; \tag{3.28}$$

$$u(x, t+T) = u(x,t), \ t \in \mathbb{R}, \ 0 \le x \le \pi. \tag{3.29}$$

1. T-periodic nonlinear systems of the first class in the functional space G_x. Suppose the operator $F[u, u_t, u_x]$ given by

$$F[u, u_t, u_x](x,t) = f(x, t, u(x,t), u_t(x,t), u_x(x,t)), \tag{3.30}$$

admits an odd 2π-periodic continuation with respect to variable $x \in [0, \pi]$ to the whole real line $\mathbb{R} = (-\infty, +\infty)$.

Consider the system of integral equations

$$u(x,t) = \frac{1}{4}\int_0^{Tq} Q(\tau)d\tau \int_\tau^t \sum_{i=0}^{1} F[u, u_t, u_x](x + (-1)^i(\theta - \tau), \tau)d\theta,$$

$$u_t(x,t) = \frac{1}{4}\int_0^{Tq} Q(\tau)\sum_{i=0}^{1} F[u, u_t, u_x](x + (-1)^i(t-\tau), \tau)d\tau,$$

$$u_x(x,t) = \frac{1}{4}\int_0^{Tq} Q(\tau)\sum_{i=0}^{1}(-1)^1 F[u, u_t, u_x](x + (-1)^i(t-\tau), \tau)d\tau, \tag{3.31}$$

where $Q(\tau)$ is given by (3.8).

Suppose the function (3.4), the right-hand side of equation (3.1), is defined in the region

$$\mathbb{R}_x \times I \times D \equiv I^1 \times I^2 \times I^3 : x \in \mathbb{R}_x,\ t \in [0, Tq] = I,$$

$$u \in [-b, b] = I^1,\ u_t \in [-c, c] = I^2,\ u_x \in [-d, d] = I^3, \tag{3.32}$$

is jointly continuous in variables and satisfies the condition

$$\|f(x,t,u,u_t,u_x)\| \le M,\ \|f(x,t,u'',u_t'',u_x'') - f(x,t,u',u_t',u_x')\| \le$$

$$\le K\{\|u''-u'\| + \|u_t''-u_t'\| + \|u_x''-u_x'\|\}. \tag{3.33}$$

Definition 3.1. A T-system of the first class in the region (3.22) is the second order equation (3.27) with the constants b, c, d, M, K, T, q related by the inequalities

$$b\epsilon\frac{M(Tq)^2}{4},\ c\epsilon\frac{MTq}{2},\ d\epsilon\frac{MTq}{2},\ KTq\left(\frac{Tq}{2}+2\right) < 1 \tag{3.34}$$

T and q being certain fixed reals.

It is not difficult to prove the following theorem.

Theorem 3.7. *Let the function $f(x,t,u,u_t,u_x)$ be defined in the region (3.32) be continuous in variables x, t, u, u_t, u_x and satisfy inequalities (3.33) and (3.34). Then, for any continuous function $v = v(x,t)$ such that*

$$\|v(x,t)\| \le \frac{M(Tq)^2}{4},\ \|v_t(x,t)\| \le \frac{MTq}{2},\ \|v_x(x,t)\| \le \frac{MTq}{2}, \tag{3.35}$$

the sequence of functions

$$u^0(x,t) = v(x,t),\ u_t^0(x,t) = v_t(x,t),\ u_x^0(x,t) = v(x,t),$$

$$u^{m+1}(x,t) = \frac{1}{4}\int_0^{Tq} Q(\tau)d\tau \int_\tau^t \sum_{i=0}^{1} F[u^m, u_t^m, u_x^m](x+$$

$$+(-1)^j(\theta-\tau),\tau)d\theta, \tag{3.36}$$

$$u_t^{m+1}(x,t) = \frac{1}{4}\int_0^{Tq} Q(\tau)\sum_{i=0}^{1} F[u^m, u_t^m, u_x^m](x+(-1)^j(t-\tau),\tau)d\tau,$$

$$u_x^{m+1}(x,t) = \frac{1}{4}\int_0^{Tq} Q(\tau)\sum_{i=0}^{1}(-1)^i F[u^m, u_t^m, u_x^m](x+(-1)^j(t-\tau),\tau)d\tau$$

converges, when $m \to \infty$, uniformly with respect to $(x,t) \in \mathbb{R} \times I$ to a continuous function $(u_0(x,t), u_{0t}(x,t), u_{0x}(x,t)) \in D$ satisfying the system of integral equations (3.31).

Theorem 3.8. *Let the condition in Theorem 3.7 be fulfilled. If $F[u,u_t,u_x](x,t) \in C^1$ with respect to variable x, then every continuous differentiable solution of (3.31) is a classical solution of equation (3.27).*

Proof. Differentiating the first equality in (3.31) with respect to x and t gives $(u)_x' = u_x$, $(u)_t' = u_t$. By using the second and third equalities in (3.31) we then obtain

$$u_{tt} = (u_t)_t' = f((x,t), u_t(x,t), u_x(x,t))+$$

$$+\frac{1}{4}\int_0^{Tq} Q(\tau)\sum_{i=0}^{1}(-1)^i\frac{\partial F[u,u_t,u_x](x+(-1)^i(t-\tau),\tau)}{\partial x}d\tau,$$

$$u_{xx}\equiv(u_x)'_x=\frac{1}{4}\int_0^{Tq} Q(\tau)\sum_{i=0}^{1}(-1)^i\frac{\partial F[u,u_t,u_x](x+(-1)^i(t-\tau),\tau)}{\partial x}d\tau.$$

Hence $u_{tt}-u_{xx}=f(x,t,u(x,t),u_t(x,t),u_x(x,t))$, which completes the proof.

Theorem 3.9. *Let the condition of Theorem 2.1 be fulfilled and let a certain solution $(u(x,t),u_t(x,t),u_x(x,t))$ of system (3.31) satisfy the conditions*

$$1)\ F[u,u_t,u_x](-x,t)=-F[u,u_t,u_x](x,t);$$

$$2)\ F[u,u_t,u_x](x+2\pi,t)=F[u,u_t,u_x](x,t);$$

i. e. the operator $F[u,u_t,u_x]$ admits an odd 2π-periodic continuation with respect to variable x.

Then the function $u(x,t)$ satisfies the boundary condition (3.28), $u(0,t)=u(\pi,t)=0$.

Proof. By the conditions in Theorem 3.9 and the first equation in (3.31), we have

$$u(0,t)=\frac{1}{4}d\tau\int_\tau^t\{F[u,u_t,u_x](\theta-\tau,\tau)+F[u,u_t,u_x](-(\theta-$$

$$-\tau),\tau)\}d\theta-\frac{1}{4}\int_t^{Tq}d\tau\int_\tau^t\{F[u,u_t,u_x](\theta-\tau,\tau)+$$

$$+F[u,u_t,u_x](-(\theta-\tau),\tau)\}d\theta\equiv 0,$$

$$u(\pi,t)=\frac{1}{4}\int_0^t d\tau\int_\tau^t\{F[u,u_t,u_x](\pi+\theta-\tau,\tau)+$$

$$+F[u,u_t,u_x](\pi-\theta+\tau,\tau)\}d\theta-\frac{1}{4}\int_t^{Tq}d\tau\int_\tau^t\{F[u,u_t,u_x](\pi+\theta-$$

$$-\tau,\tau)+F[u,u_t,u_x](\pi-\theta+\tau,\tau)\}d\theta=\frac{1}{4}\int_\tau^t d\tau\int_\tau^t\{F[u,u_t,$$

$$u_x](\pi+\theta-\tau,\tau)+F[u,u_t,u_x](-\pi-\theta+\tau,\tau)\}d\theta-\frac{1}{4}\times$$

$$\int_t^{Tq}d\tau\int_\tau^t\{F[u,u_t,u_x](\pi+\theta-\tau,\tau)+F[u,u_t,u_x](-\pi-$$

$$-\theta+\tau,\tau)\}d\theta\equiv 0.$$

This completes the proof.

Remark 3.1. According to the condition (2) in Theorem 3.8, for the boundary condition $u(0,t)=u(\pi,t)=0$ to be fulfilled it is necessary that both the right-hand side $f(x,t,u,u_t,u_x)$ of equation (3,27) and its solution $u(x,t)$ be 2π-periodic in variable x.

Theorem 3.10. *Let the condition of Theorem 2.1 be fulfilled. If for a solution* $(u(x,t), u_t(x,t), u_x(x,t))$ *of (3.31), the function* $f(x,t,u(x,t),u_t(x,t),u_x(x,t))$ *can be expanded into the Fourier series of the form*

$$f(x,t,u(x,t),u_t(x,t),u_x(x,t)) = \sum_{k=1}^{\infty} f_k(t) \sin kx, \tag{3.37}$$

where the functions $f_k(t)$ *are periodic on the interval* $[0, Tq]$ *and satisfy the condition*

$$f_k(t) \leq \frac{C}{k^{\nu}}, \quad C = \text{const}, \tag{3.38}$$

then the function $u(x,t)$ *is a classical solution of the following boundary value problem*

$$u_{tt} - u_{xx} = f(x,t,u,u_t,u_x), \quad u(0,t) = u(\pi,t) = 0. \tag{3.39}$$

Proof. From (3.37) we conclude that the function $F[u, u_t, u_x](x,t) = f(x,t,u(x,t),u_t(x,t),u_x(x,t))$ satisfies both the conditions (1) and (2) in Theorem 3.8. The function $u(x,t)$ satisfies, therefore, the boundary condition $u(0,t) = u(\pi,t) = 0$. By uniform convergence of the series in (3.37) and the first equation in (3.31), the function $u(x,t)$ admits the representation

$$u(x,t) = \frac{1}{4}\int_0^t d\tau \int_\tau^t \sum_{k=1}^{\infty} f_k(\tau)[\sin k(x+\theta-\tau) + \sin k(x-$$

$$-\theta+\tau)]d\theta - \frac{1}{4}\int_0^{Tq} d\tau \int_\tau^t \sum_{k=1}^{\infty} f_k(\tau)[\sin k(x+\theta-\tau) + \sin k(x-$$

$$-\theta+\tau)]d\theta = \frac{1}{2}\int_0^t \sum_{k=1}^{\infty} \frac{1}{w_k} f_k(\tau) \sin kx \sin w_k(t-\tau) d\tau -$$

$$-\frac{1}{2}\int_0^{Tq} \sum_{k=1}^{\infty} \frac{f_k(\tau)}{w_k} \sin kx \sin w_k(t-\tau)d\tau = \sum_{k=1}^{\infty} \sin kx \frac{1}{2w_k} \int_0^{Tq} Q(\tau) f_k \times$$

$$\times(\tau) \sin w_k(t-\tau) d\tau, \tag{3.40}$$

where

$$w_k = k, \quad k = 1, 2, \ldots \tag{3.41}$$

In view of (3.38) and (3.40), both the function $u(x,t)$ and its derivative u_{xx} are continuous for any $(x,t) \in \mathbb{R} \times I$ and the derivative u_{xx} can be obtained by differentiating the Fourier series (3.40). Equation (3.27) then implies that the derivative u_{tt} is continuous for any $(x,t) \in \mathbb{R} \times I$.

The function $u(x,t)$ representable by the series (3.40) can thus be differentiated twice with respect to variable t. After computing the derivatives u_{tt} and u_{xx} we verify that $u(x,t)$ is a classical solution of equation (3.27) and the boundary value problem (3.39) on the whole. This proves the theorem.

Definition 3.2. A function $(u(x,t), u_t(x,t), u_x(x,t))$ satisfying the boundary condition $u(x,t) \in C^1$ is called a smooth generalized or, simply, smooth solution of the boundary value problem (3.39) if a continuous solution $u(0,t) = u(\pi,t) = 0$ of system (3.31) exists.

Remark 3.2. The condition $kTq = (2p-1)\pi$ cannot hold for any positive integers k, p, q. We can therefore assert that for any initial function

$$u^0(x,t) \equiv v(x,t) = \sum_{k=1}^{\infty} v_k(t) \sin kx, \quad v_k(t) \neq 0, \quad k = 1, 2, \dots,$$

T-periodic in variable t and satisfying (3.35) and for a $(2p-1)\pi/g$-periodic right-hand side $f(x,t,u,u_t,u_x)$ of equation (3.27), the solution $u(x,t)$ of the boundary value problem (3.39) cannot be periodic in variable t.

Definition 3.3. The boundary value problem (3.39) is called a T-periodic system of the first class if the right-hand side $f(x,t,u,u_t,u_x)$ of (3.39) is a function T-periodic in variable t, equation (3.39) is a T-system of the first class (see Definition 3.1) and the condition $Tq = (2p-1)\pi$, $(q, 2p-1) = 1$ holds for the period T, p and q being fixed positive integers.

Theorem 3.11. *Let the condition of Theorem 3.7 be fulfilled and let the boundary value problem (3.39) be the T-periodic system of the first class.*

If for any smooth initial $(2p-1)\pi/q$-periodic function

$$|u^0(x,t)| \equiv |v(x,t)| = |\sum_{k=1}^{\infty} v_{2k-1}(t) \sin(2k-1)x| \leq \frac{1}{4} M (Tq)^2, \tag{3.42}$$

$$|v_t(x,t)| \leq \frac{1}{2} MTq, \quad |v_x(x,t)| \leq \frac{1}{2} MTq$$

the right-hand side $f(x,t,u,u_t,u_x)$ of (3.39) can be expanded into the Fourier series

$$f(x,t,v(x,t)v_t(x,t),v_x(x,t)) = \sum_{k=1}^{\infty} f_{2k-1}^0(t) \sin(2k-1)x, \tag{3.43}$$

the Fourier coefficients $f_{2k-1}^0(t)$ of which are continuous T-periodic functions satisfying the condition

$$|f_{2k-1}^0(t)| \leq \frac{c}{(2k-1)^\nu}, \quad \nu > 1, \quad c = \text{const}, \tag{3.44}$$

then she boundary value problem (3.39) has a smooth $(2p-1)\pi/q$-periodic solution $u(x,t) \in C^1$.

Proof. Consider the recurrent relations in (3.36) constructed for the initial function (3.42). As the condition of Theorem 3.7 holds, the system of integral equations (3.31) has a unique continuous solution defined as the limit of the sequence $(u^m(x,t), u_t^m(x,t), u_x^m(x,t))$ constructed by (3.36). On the other hand, taking into account (3.42) and uniform convergence of the series in (3.43) and using the transformations similar to (3.40), we can prove by induction that the elements of the sequence $u^m(x,t)$ can be represented as

$$u^{m+1}(x,t) = \sum_{k=1}^{\infty} \frac{\sin(2k-1)x}{2\omega_{2k-1}} \int_0^{Tq} Q(\tau) f_{2k-1}^m(\tau) \sin \omega_{2k-1}(t-\tau) d\tau, \tag{3.45}$$

where $\omega_{2k-1} = 2k-1$, f_{2k-1}^m are the coefficients in the expansion of $f(x,t,u^m(x,t),u_t^m(x,t),u_x^m(x,t))$ into the Fourier series (3.43). Lemma 2.1 in Chapter 2 combined with the condtion of Theorem 3.11 now shows that all the elements of the sequence (3.45) are functions of period $T = (2p-1)\pi/q$ with respect to t.

Since the functions $u^m(x,t)$ are $T = (2p-1)\pi/q$-periodic in t, the limit function $u(x,t)$ is also periodic in t with the same period T. From (3.45) we conclude that $u(x,t)$ satisfies the boundary condition $u(0,t) = u(\pi, t)$. Whence it follows that the limit function $u(x,t)$ is a smooth T-periodic solution of (3.39), which proves the theorem.

Definition 3.4. The boundary value problem (3.39) is called a $2T$-periodic system of the first class if the right-hand side $f(x,t,u,u_t,u_x)$ of (3.39) is a function $2T$-periodic in variable t, equation (3.39) is a T-system of the first class (see Definition 3.1), the condition $Tq = r\pi$, $r = 2k$, $q = 2s-1, (r,q) = 1$ holds for the period T, and for any function $v(x,t) \in C^1$ satisfying $v(x,t+T) = -v(x,t)$, the equality

$$f(x,t+T,v(\tau,t+T),v_t(x,t+T),v_x(x,t+T)) =$$
$$= -f(x,t,v(x,t),v_t(x,t),v_x(x,t)),$$

holds, r and q being fixed positive integers.

Theorem 3.12. *Let the boundary problem (3.39) be the $2T$-periodic system of the first class and let the condition of Theorem 3.7 be fulfilled. If for a smooth $2T$-periodic initial function*

$$|u^0(x,t)| \equiv |v(x,t)| = |\sum_{k=1}^{\infty} v_k(t)\sin kx| \le \frac{1}{4}M(Tq)^2,$$

$$|v_t(x,t)| \le \frac{MTq}{2},\ |v_x(x,t)| \le \frac{MTq}{2},\ v(x,t+T) = -v(x,t),$$

the right-hand side of (3.39) can be expanded into the Fourier series

$$f(x,t,v,v_t,v_x) = \sum_{k=1}^{\infty} f_k^0(t)\sin kx,\ f_k^0(t+T) = -f_k^0(t),$$

the coefficients of which are continuous $2T$-periodic functions satisfying $|f_k^0(t)| \le ck^{-\gamma}$, $\gamma > 1$, then the boundary value problem (3.39) has a smooth periodic solution $u(x,t) \in C^1$ of period $2T = 2r\pi/q$.

Theorem 3.12 can be proved in the same manner as Theorem 3.11.

Setting $T = T_k/2$ and using the above results, the following theorem can be proved by the method of successive approximations, the theory of Fourier series not being used.

Theorem 3.13. *Let $k = 1,2,3$. Suppose the function $F[u,u_t,u_x](x,t) = f(x,t,u(x,t),u_t(x,t),u_x(x,t))$ belongs, for any function $u \in C^2 \cap A_k^0$, $|u(x,t)| \le \frac{1}{16}M(T_kq)^2$, $|u_t(x,t)| \le \frac{1}{4}MT_kq$, $|u_x(x,t)| \le \frac{1}{4}MT_kq$, to the space $G_x \cap \tilde{A}_k^0$ and the condition of Theorem 3.7 are fulfilled for $T = T_k/2$. Then for any function $v \in G_x \cap A_k^0$ such that $|v(x,t)| \le \frac{1}{16}M(T_kq)^2$, $|v_t(x,t)| \le \frac{1}{4}MT_kq$, $|v_x(x,t)| \le \frac{1}{4}MTq$, the sequence of functions (3.36) with $T = T_k/2$ converges, when $m \to \infty$, uniformly with respect to $(x,t) \in \Pi$ to a function $(\tilde{u}_k, \tilde{u}_{kt}, \tilde{u}_k) \in C \cap A_k^0$ that satisfies the system of integral equations (3.31) with $T = T_k/2$, and hence, the function $\tilde{u}_k$ is a smooth T_k-periodic solution of the boundary value problem (3.27)–(3.29).*

Remark 3.3. The results of investigations obtained earlier [21, 180–183] concerning the existence of periodic solutions of the second order differential equations with a small parameter

$$u_{tt} - a^2u_{xx} = g(x,t) + \varepsilon f_1(x,t,u,u_t,u_x,\varepsilon),\ u(0,t) = u(\pi,t) = 0, \tag{3.46}$$

are the consequences of the above statements. Indeed, the condition defining T-systems of the first class are, in general, of the form

$$be\frac{M(Tq)^2}{4},\ c > \frac{MTq}{2},\ de\frac{MTq}{2a},\ KTq\left(\frac{Tq}{2} + 1 + \frac{1}{a}\right) < 1. \tag{3.47}$$

For certain functions $g(x,t)$ and $f_1(x,t,u,u_t,u_x)$ bounded and T_k-periodic in the region $x \in \mathbb{R},\ t \in \mathbb{R},\ |u| \le b,\ |u_t| \le c,\ |u_x| \le d$, the condition (3.47) is always satisfied, provided ε is sufficiently small, that is, if $|g(x,t)| \le M_1$ and $|f(x,t,u,u_t,u_x)| \le M_2$, then for certain values of M_2 and $\varepsilon, \varepsilon \ll 1$, the conditions

$$be\frac{1}{4}M_1(Tq)^2 + \frac{\varepsilon}{4}M_2(Tq)^2, \quad ce\frac{1}{2}M_1Tq + \frac{\varepsilon}{2}M_2Tq,$$

$$de\frac{1}{2a}M_1Tq + \frac{\varepsilon}{2a}M_2(Tq)^2, \quad \varepsilon TKq\left(\frac{Tq}{2} + 1 + \frac{1}{a}\right) < 1.$$

are always fulfilled, the values of b, c, d and $T = T_k/2$ being fixed.

2. T-periodic nonlinear systems of the first class in the functional space $G_{\pi t}$. The notion of a nonlinear T-system of the first class in the functional space $G_{\pi t}$ (the space of continuous functions $u(x,t)$ bounded on $[0,\pi] \times \mathbb{R}$ together with the first derivative with respect to t) can be introduced in a fashion similar to that of item 1 by using the operators $R_k,\ k = 1,2,3$ defined in Section 1.

We shall consider a particular case of the boundary value problem (2.20) with $a = 1$.

Theorem 3.14. *Let k be one of the numbers $1,2,3$. Assume that for the function*

$$F[u,u_t,u_x](x,t) = g(x,t) + \varepsilon f_1(x,t,u(x,t),u_t(x,t),u_x(x,t),\varepsilon),$$

the function g belongs to $G_{\pi t} \cap A_k$, and that the function f_1 belongs to $G_{\pi t} \cap A_k \cap C_{[0,\varepsilon_0]}$ for $u \in C^1_\pi \cap A_k$ and the derivatives $\partial f_1/\partial u,\ \partial f_1/\partial u_t,\ \partial f_1/\partial u_x$ satisfy a Lipschitz condition locally in $\Omega = [0,\pi] \times \mathbb{R} \times D \times [0,\varepsilon_0]$. Then there exist $\rho > 0$ and $\varepsilon_0 > 0$ such that for any $\varepsilon(|\varepsilon| < \varepsilon_0)$, in the ball $C^2_\pi \cap A_k,\ k = 1,2,3$, with the centre in $R_k g$ and radius ρ there exists a unique function $u(x,t)$ satisfying (3.46) and T_k-periodic in variable t.

This theorem can be proved by the contraction mapping principle applied to the equation $u = R_k(g + \varepsilon f_1),\ k = 1,2,3$.

Remark 3.4. By using the Schauder principle one can prove the existence of a solution of the integral equations [17, 18, 37, 38]

$$u(x,t) = \frac{1}{4}\int\limits_0^{T_kq/2} Q(\tau)d\tau \int\limits_{x-t+\tau}^{x+t-\tau} F[u,u_t,u_x](\eta,\tau)d\eta,$$

$$u(x,t) = (R_3F[u,u_t,u_x])(x,t)$$

and, hence, the existence of a solution of the wave differential equation (3.27) under weaker assumptions about the function $F[u,u_t,u_x](x,t) = f(x,t,u(x,t),u_t(x,t),u_x(x,t))$. To do this, we state the Schauder principle as follows.

Theorem 3.15[45]. *A continuous mapping $\widetilde{P}$ sending a closed convex set Ω of a Banach space X into a compact set $\Delta \cap \Omega$ has a fixed point.*

For the problem in question we can formulate the following theorem.

Theorem 3.16. *Let k be one of the numbers $1,2,3$ and let the following conditions be fulfilled*

(1) the function $f(x,t,u)$ (the right-hand side of the equation $u_{tt} - u_{xx} = f(x,t,u)$) is defined in the region $Q = \{(x,t) \in \mathbb{R}^2, u \in [-b,b]\}$, is continuous jointly in variables x,t,u, T_k-periodic in t, 2π-periodic in x, satisfies $|f(x,t,u)| \le M$ ($M = \text{const}$) and, for any $(x,t) \in \mathbb{R}^2$, is continuous in u uniformly with respect to $(x,t) \in \mathbb{R}^2$;

(2) the constants b, M, T_k, q relate by the inequality $b > \frac{1}{16}M(T_kq)^2$;

(3) the function $F[u](x,t) = f(x,t,u(x,t))$ belong to the space $C \cap \tilde{A}^0_k$ for any function $u \in C \cap A^0_k$.

Then the boundary value problem

$$u_{tt} - u_{xx} = f(x,t,u), \ (x,t) \in \mathbb{R}^2,$$

$$u(0,t) = u(\pi,t) = 0, \ u(x,t+T_k) = u(x,t), \ (x,t) \in \mathbb{R}^2, \tag{3.48}$$

has at least one continuous T_k-periodic solution.

Proof. In Theorem 3.16, set $X = C^{T_k,2\pi}(\mathbb{R}^2)$ where $C^{T_k,2\pi}(\mathbb{R}^2)$ denotes the set of continuous functions $u(x,t)$ which are 2π-periodic in variable x and T_k-periodic in variable t. Take also the set of those $u \in C^{T_k,2\pi}(\mathbb{R}^2)$ which satisfy

$$||u|| \leq b, \ ||u|| \max_{[0,2\pi]\times[0,T_k]} |u(x,t)|.$$

as the set Ω. The operator $\widetilde{P}_k$ is defined as

$$z = \widetilde{P}_k(u), \ z(x,t) = \frac{1}{4}\int_0^{C_k} Q(\tau)d\tau \int_{x-t+\tau}^{x+t-\tau} f(\eta,\tau,u(\eta,\tau))d\eta. \tag{3.49}$$

From the conditions 1–3 of Theorem 3.16 it follows that $\widetilde{P}_k(u) \in \Omega$. Indeed, by Lemma 3.4 and condition 3 in Theorem 3.16 the operator $\widetilde{P}_k$ sends every T_k-periodic function $u(x,t) \in \Omega \cap C^{T_k,2\pi}$ into the T_k-periodic function $z(x,t)$ satisfying

$$||z|| = ||\widetilde{P}_k(u)|| \leq \frac{M}{16}(T_k q)^2 < b, \ u \in \Omega. \tag{3.50}$$

We should now check that the operator $\widetilde{P}_k$ is continuous. Indeed, let $u_n \to u_0$, $u_n \in \Omega$, $z_n = \widetilde{P}_k(u_n)$, $n = 1, 2, \ldots$. As the function $f(x,t,u)$ is continuous in u uniformly with respect to $(x,t) \in \mathbb{R}^2$, for any $\varepsilon > 0$ there exists $\eta > 0$ such that

$$|f(x,t,u) - f(x,t,u')| < \varepsilon, \ (x,t) \in [0,2\pi] \times [0,T_k]. \tag{3.51}$$

provided $|u - u'| < \eta$. Since $|u_n - u_0| \to 0$, the inequality $|u_n - u_0| < \eta$ holds for sufficiently large n, $n e n_0$ and, therefore, the inequality

$$|u_n(x,t) - u_0(x,t)| < \eta, \ (x,t) \in [0,2\pi] \times [0,T_k].$$

holds too.

In view of (3.51), we can write

$$|f(x,t,u_n(x,t)) - f(x,t,u_0(x,t))|,\varepsilon \ \forall (x,t) \in [0,2\pi] \times [0,T_k].$$

for these values of n. We therefore obtain

$$||z_n - z_0|| = \max_{[0,2\pi]\times[0,T_k]} |z_n(x,t) - z_0(x,t)| \leq$$

$$\leq \frac{1}{4} \max_{[0,2\pi]\times[0,T_k]} \int_0^{T_k q/2} d\tau \Big| \int_{x-t+\tau}^{x+t-\tau} |f(\eta,\tau,u_n(\eta,\tau)) - f(\eta,\tau,u_0(\eta,\tau))| d\eta \Big| \leq$$

$$\leq \frac{1}{16}(T_k q)^2 \varepsilon.$$

for $n \geq n_0$. Hence $z_n = \widetilde{P}_k(u_n) \to z_0 = \widetilde{P}_k(u_0)$, $n \to \infty$.

Since the set Ω is closed and convex, it is necessary to establish only relative compactness of the set $\tilde{P}_k(\Omega)$ in order to apply Theorem 3.15. To do this, we prove that the functions in $\tilde{P}_k(\Omega)$ are equicontinuous (the fact that they are uniformly bounded follows from $\tilde{P}_k(\Omega) \cap \Omega$,, Ω being the bounded set).

By virtue of (3.49) and (3.8) we have

$$|z(x_2,t_2) - z(x_1,t_1)| \leq \frac{1}{4}| \int_0^{t_2} d\tau \int_{x_2-t_2+\tau}^{x_2+t_2-\tau} f(\eta,\tau,u(\eta,\tau))d\eta -$$

$$- \int_0^{t_1} d\tau \int_{x_1-t_1+\tau}^{x_1+t_1-\tau} f(\eta,\tau,u(\eta,\tau))d\eta| + \frac{1}{4}| \int_{c_k}^{t_2} d\tau \int_{x_2-t_2+\tau}^{x_2+t_2-\tau} f(\eta,\tau,u(\eta,\tau))d\eta -$$

$$- \int_{c_k}^{t_1} d\tau \int_{x_1-t_1+\tau}^{x_1+t_1-\tau} f(\eta,\tau,u(\eta,\tau))d\eta| = I_1 + I_2.$$

For I_1, we obtain the following estimate

$$I_1 \leq A|t_2 - t_1| + \frac{1}{4} \int_0^{t_1} | \int_{x_1-t_1+\tau}^{x_2-t_2+\tau} f(\eta,\tau,u(\eta,\tau))d\eta| d\tau +$$

$$+ \frac{1}{4} \int_0^{t_1} | \int_{x_1+t_1-\tau}^{x_2+t_2-\tau} f(\eta,\tau,u(\eta,\tau))d\eta| d\tau \leq \alpha|t_2 - t_1| - \beta|x_2 - x_1|, \tag{3.52}$$

where α_1 and β are positive constants.

For I_2, we obtain, in a similar manner, the estimate

$$I_2 \leq \alpha_1|t_2 - t_1| + \beta_1|x_2 - x_1|,$$

α_1 and β_1 being certain positive constants.

As $|z(x_2,t_2) - z(x_1,t1)| \leq I_1 + I_2$, from (3.52) and (3.53) we conclude that the set $\tilde{P}_k(\Omega)$ is equicontinuous. Therefore, all the conditions in Theorem 3.15 are fulfilled and the theorem implies that the integral equation

$$u(x,t) = \frac{1}{4} \int_0^{c_k} Q(\tau)\, d\tau \int_{x-t+\tau}^{x+t-\tau} f(\eta,\tau,u(\eta,\tau))\, d\eta, \quad c_k = T_k q/2,$$

has at least one continuous T_k-periodic solution, which means there exists a continuous T_k-periodic solution of the boundary value problem (3.48). This proves the theorem.

Remark 3.5. By using the operators T_k, a similar result can be obtained for (3.48) in the space $C_\pi \cap A_k$.

CHAPTER 4

PERIODIC SOLUTIONS OF THE SECOND CLASS SYSTEMS

In this chapter, we continue studying periodic boundary value problems for the second order wave differential equations of hyperbolic type. In particular, we single out and study T-systems of the second class, i. e. the periodic boundary value problems solutions of which contain solutions of the corresponding homogeneous periodic problem.

4.1. Some preliminaries

If the right-hand side of the equation $u_{tt} - u_{xx} = f(x, t, u, u_t, u_x)$ is periodic of period $T = 2\pi n_0$, n_0 a fixed positive integer, or if this right-hand side does not depend explicitly upon time, the operators P_k and $R_k, k = 1, 2, 3$ constructed in Chapter 3 cannot be used for investigating the existence of periodic solutions (of period $T = 2\pi n_0$) of the second class systems. This follows from that these operators were constructed for the particular functional classes A_k^0 and A_k, in which the homogeneous periodic boundary value problem $u_{tt} - u_{xx} = 0$, $u(0, t) = u(\pi, t) = 0$, $u(x, t + 2\pi n_0) = u(x, t)$ has only trivial (zero) solution. On the other hand, the above-mentioned homogeneous periodic boundary value problem has infinitely many $2\pi n_0$-periodic solutions. Note however that, for a class of operators $F[u, u_t, u_x]$ given as $F[u, u_t, u_x](x, t) = f(x, t, u(x, t)u_t(x, t), u_x(x, t))$, the results in Chapters 1–3 allows us to construct formally an operator sending a function periodic in t with period $T = 2\pi n_0$ into a periodic function of the same period. For instance, if we suppose that $F[u, u_t, u_x](x, t) = \sum_{k=1}^{\infty} f_k(t) \sin kx$, then

$$(\tilde{P}f) = p(t + x) - p(t - x) +$$

$$+\frac{1}{2}\int_0^t \left\{ \int_\tau^t [F[u, u_t, u_x](x + \theta - \tau, \tau) + F[u, u_t, u_x](x - \theta + \tau, \tau)]\, d\theta - \right.$$

$$\left. -\frac{1}{T}\int_0^T \int_\tau^t [F[u, u_t, u_x](x + \theta - \tau, \tau) + F[u, u_t, u_x](x - \theta + \tau, \tau)]\, d\theta \right\} \times$$

$$\times d\tau \equiv \sum_{k=1}^{\infty} (C_k^1 \cos kt + C_k^2 \sin kt) \sin kx +$$

$$+\frac{1}{2}\int_0^T Q_1(\tau) \left\{ \int_\tau^t \sum_{i=0}^{1} F[u, u_t, u_x](x + (-1)^i(\theta - \tau), \tau)\, d\theta \right\} d\tau. \tag{4.1}$$

is such operator.

In connection with the investigation of $2\pi n_0$-periodic solutions with the help of integral equation (4.1) several problems appear. The first problem is to find out a conditon for the equality

$$\frac{1}{T}\int_0^T \int_\tau^t \sum_{i=0}^{1} F[u, u_t, u_x](x + (-1)^i(\theta - \tau), \tau)\, d\theta = 0, \tag{4.2}$$

to be held. The next one is to find out constants C_k^1 and C_k^2 which make equality (4.2) hold. Another one is to investigate convergence of the series

$$\sum_{k=1}^{\infty}(X_k^1 \cos kt + C_k^2 \sin kt) \sin kx \tag{4.3}$$

for particular values of C_k^1 and $C_k^2, k = 1, 2, \ldots$.

Most of the Soviet and foreign mathematicians have proved the existence of 2π-periodic solutions of the boundary value problem $u_{tt} - u_{xx} = f(x, t, u, u_t, u_x), u(0, t) = u(\pi, t) = 0$ just in the form of series (4.3) and, what is essential, these solutions are non-classical. In the subsequent sections this boundary value problem is investigated by other methods and a class of discontinuous functions is given rather than a Fourier series representing the generalized solution of the boundary value problem.

4.2. The structure of generalized periodic solutions of the second order wave equation of the first kind

Consider the following boundary value problem

$$u_{tt} - u_{xx} = f(x, t, u, u_t, u_x), u(0, t) = u(\pi, t) = 0 \tag{4.4}$$

Suppose that the function $f(x, t, u, u_t, u_x)$ is defined in the region

$$I \times \mathbb{R} \times D = I \times \mathbb{R} \times I^1 \times I^2 \times I^2 : x \in [0, \pi] = I, \times \in \mathbb{R};$$

$$u \in [a, b] = I^1, \quad u_t \in [-c, c] = I^2, \quad u_x \in [-c, c], \tag{4.5}$$

is continuous with respect to variables x, t, is 2π-periodic and satisfies the condition

$$|f(x, t, u, u_t, u_x)| \leq M,$$

$$|f(x, t, u'', u_t'', u_x'') - f(x, t, u', u_t', u_x')| \leq K\{|u'' - u'| +$$

$$+ |u_t'' - u_t'| + |u_x'' - u_x'|\}, \tag{4.6}$$

M and K being positive constants.

Consider a class of continuous 2π-periodic functions $\mu(t)$ that can be expanded into the Fourier series

$$\mu(t) = \frac{\alpha_0}{2} + \sum_{k=1} \alpha_k' \cos kt + \alpha_k'' \sin kt.$$

Let $\alpha = \{\alpha_0, \alpha_1', \alpha_1'', \alpha_2', \alpha_2'', \ldots\}$ denote the collection of the Fourier coefficients determining a function $\mu(t)$ such that

$$|\mu(t)| = c - M\pi, \ c > M\pi, \tag{4.7}$$

and let $\mathfrak{D}_\mu = \{\alpha\}$ denote the set of the Fourier coefficients satisfying (4.7).

Definition 4.1. The boundary value problem (4.4) is called 2π-hyperbolic differential system of the firts kind in the region (4.5) if the constants a, b, c, M relate by the inequalities

$$c > M\pi, \ a < -\left(c\pi - \frac{M\pi^2}{2}\right) < c\pi - \frac{M\pi^2}{2} < B \tag{4.8}$$

and $f(x, t, o, o, o) = 0$ holds.

Lemma 4.1. *Let the function $f(x,t,u,u_t,u_x)$ be defined in the region (4.5), be continuous with respect to x and t, 2π-periodic in t and satisfy inequalities (4.6) and (4.8). Then, for any 2π-periodic continuous function $|\mu(t)| = c - M\pi_,$, the following sequences of functions 2π-periodic in t*

$$u_1^0(x,t,\alpha) = \mu(t+x),\ u_2^0(x,t,\alpha) = -\mu(t+x),$$

$$u^0(x,t,\alpha) = \int_0^x \frac{\mu(t+\xi)+\mu(t-\xi)}{2} d\xi,$$

$$u_1^{m+1}(x,t,\alpha) = \mu(t+x) - \int_0^x \tilde{F}[u^m, u_1^m, u_2^m](\eta, t+x-\eta)d\eta,$$

$$u_2^{m+1}(x,t,\alpha) = -\mu(t-x) + \int_0^x \tilde{F}[u^m, u_1^m, u_2^m](\eta, t-x+\eta)d\eta, \tag{4.9}$$

$$u^{m+1}(x,t,\alpha) = \int_0^x \frac{u_1^{m+1}(\eta,t,\alpha)+u_2^{m+1}(\eta,t,\alpha)}{2} d\xi \equiv$$

$$\equiv \int_0^x \frac{\mu(t+\xi)+\mu(t-\xi)}{2} d\xi - \frac{1}{2}\int_0^x d\xi \int_0^x \Big\{\tilde{F}[u^m, u_1^m, u_2^m](\eta, t+x-\eta)+$$

$$+\tilde{F}[u^m, u_1^m, u_2^m](\eta, t-x+\eta)\Big\} d\eta,\ m = 0,1,2,\ldots;$$

$$\tilde{F}[u, u_1, u_2](x,t) = f(x,t,u(x,t),u_t(x,t),u_x(x,t)),$$

$$u_1 = u_t + u_x,\ u_2 = u_t - u_x,\ \Leftrightarrow\ u_t = \frac{u_1+u_2}{2},\ u_x = \frac{u_1-u_2}{2} \tag{4.10}$$

converge, as $m \to \infty$, uniformly with respect to

$$x, t, \alpha \in I \times \mathbb{R} \times \mathfrak{D}_\mu \tag{4.11}$$

to continuous functions $u_{10}(x,t,\alpha)$, $u_{20}(x,t,\alpha)$ and $u_0(x,t,\alpha)$ defined in the region (4.11), 2π-periodic in t and satisfying the system of integral equations

$$u_1(x,t,\alpha) = \mu(t+x) - \int_0^x \tilde{F}[u, u_1, u_2](\eta, t+x-\eta)d\eta,$$

$$u_2(x,t,\alpha) = -\mu(t+x) + \int_0^x \tilde{F}[u, u_1, u_2](\eta, t-x+\eta)d\eta, \tag{4.12}$$

$$u(x,t,\alpha) = \int_0^x \frac{u_1(\eta,t,\alpha)+u_2(\eta,t,\alpha)}{2} d\xi \equiv \int_0^x \frac{\mu(t+\xi)-\mu(t-\xi)}{2} d\xi -$$

$$-\frac{1}{2}\int_0^x d\xi \int_0^x \Big\{\tilde{F}[u, u_1, u_2](\eta, t+x-\eta) + \tilde{F}[u, u_1, u_2](\eta, t-x+\eta)\Big\} d\eta.$$

Furthermore, the following inequalities hold

$$|u^1(x,t,\alpha) - u_0(x,t,\alpha)| \le MK\left(2+\frac{\pi}{2}\right)\frac{\pi^3}{3!},$$

$$|u^m(x,t,\alpha) - u_0(x,t,\alpha)| \le (3K)^{m-2}\gamma MK^2\frac{\pi^{m+2}}{(m+2)!}, \tag{4.13}$$

$$\gamma = \left(2+\frac{\pi}{2}\right)\left(2+\frac{\pi}{3}\right), \quad m = 2,3,\ldots$$

P r o o f. Proceeding by induction we show that $u^m(x,t,\alpha), u_t^m(x,t,\alpha) = \frac{1}{2}(u_1^m(x,t,\alpha)+u_2^m(x,t,\alpha))$, $u_x^m = \frac{1}{2}(u_1^m(x,t,\alpha) - u_2^m(x,t,\alpha))$ assume values in the region (4.5) for all $m = 0,1,2,\ldots$ and $x,t \in I\times\mathbb{R}$. Indeed, for $m=0$, from inequalities (4.6) and (4.7) and representation (4.9) it follows that

$$\left|\frac{u_1^1(x,t,\alpha) \pm u_2^1(x,t,\alpha)}{2} - \frac{\mu(t+x) \mp \mu(t-x)}{2}\right| \le Mx \le M\pi. \tag{4.14}$$

Whence $|u_t^1(x,t,\alpha)| \le c$, $|u_x^1(x,t,\alpha)| \le c$ for all $(x,t,\alpha) \in I\times\mathbb{R}\times\mathfrak{D}_\mu$, provided $|\mu(t)| \le c - m\pi$.

From the inequality

$$|u^1(x,t,\alpha) - u^0(x,t,\alpha)| \le \frac{1}{2}\int_0^x d\xi \int_0^\xi |\tilde F[u^0,u_1^0,u_2^0](\eta,t+\xi-\eta)+$$

$$\tilde F[u^0,u_1^0,u_2^0](\eta,t-\xi+\eta)|d\eta \le M\frac{x^2}{2!} \le M\frac{\pi^2}{2!}, \tag{4.15}$$

inequalities (4.7) and (4.8) and the representation

$$u^0(x,t,\alpha) = \frac{1}{2}\int_0^x (\mu(t+\xi)+\mu(t-\xi))d\xi$$

we conclude that $a \le u^1(x,t,\alpha) \le b$ for all $(x,t,\alpha) \in I\times\mathbb{R}\times\mathfrak{D}_\mu$.

Suppose that $a \le u^k(x,t,\alpha) \le b, |u_t(x,t,\alpha)| \le c$, $|u_x^k(x,t,\alpha)| \le c$, c - a positive integer. By using defining recurrence relations for $u_1^{k+1}(x,t,\alpha), u_2^{k+1}(x,t,\alpha), u^{k+1}(x,t,\alpha)$ we obtain

$$\left|\frac{u_1^{k+1}(x,t,\alpha) \pm u_2^{k+1}(x,t,\alpha)}{2} - \frac{\mu(t+x) \mp \mu(t-x)}{2}\right| \le Mx \le M\pi,$$

$$\left|\frac{u^{k+1}(x,t,\alpha) - u^0(x,t,\alpha)}{2} \le M\frac{x^2}{2!}\right| \le M\frac{\pi^2}{2!}$$

Thus $a \le u^{k+1}(x,t,\alpha) \le b$, $|u_t^{k+1}(x,t,\alpha)| \le c$, $|u_x^{k+1}(x,t,\alpha)| \le c$ for all $x,t,\alpha \in I\times\mathbb{R}\times\mathfrak{D}_\mu$. Whence it follows that the functions $u^m(x,t,\alpha), u_t^m(x,t,\alpha)$ and $u_x^m(x,t,\alpha)$ satisfy the inequalities $a \le u^m(x,t,\alpha) \le b$, $|u_t^m(x,t,\alpha)| \le c$, $|u_x^m(x,t,\alpha)| \le c$ for all $m\epsilon 0$ and $x,t,\alpha \in I\times\mathbb{R}\times\mathfrak{D}_\mu$.

We shall prove now that the seqiences (4.9) converge uniformly for all $x,t,\alpha \in I\times\mathbb{R}\times\mathfrak{D}_\mu$ and hence the limit functions are continuous in this set. To do this, we make use the fact that convergence of the sequences (4.9) is equivalent to convergence of the series

$$u_i^0 + (u_i^1 - u_i^0) + (u_i^2 - u_i^1) + \cdots + (u_i^m - u_i^{m-1}) + \cdots, \quad i = 1,2,$$

$$u^0 + (u^1 - u^0) + (u^2 - u^1) + \cdots + (u^m - u^{m-1}) + \cdots \tag{4.16}$$

since $u_i^m(x,t,\alpha)$, $i = 1,2$, $u^m(x,t,\alpha)$ are just the partial sums of these series.

By the relations $u_t = (u_1 + u_2)/2$, $u_x = (u_1 - u_2)/2$ and the Lipschitz condition (4.6) we have

$$|f\left(x,t,u'', \frac{u_1'' + u_2''}{2}, \frac{u_1' - u_2'}{2}\right) - f\left(x,t,u', \frac{u_1' + u_2'}{2}, \frac{u_1' - u_2'}{2}\right)| \leq$$
$$\leq K\{|u'' - u'| + |u_1'' - u_1'| + |u_2'' - u_2'|\}. \tag{4.17}$$

Using inequality (4.6) and representation (4.9) (for $m = 0$) yields the following estimate for the terms of the series (4.16)

$$|u_i^1 - u_i^1| \leq Mx, \; i = 1,2, |u^1 - u^0| \leq M\frac{x^2}{2} \leq M\frac{\pi x}{2!}. \tag{4.18}$$

Combining (4.9), (4.10), (4.17) and (4.18), we obtain

$$|u_i^2 - u_i^1| \int_0^x |\tilde{F}[u^1, u_1^1, u_2^1](\eta, t + (-1)^i(\eta - x)) -$$

$$\tilde{F}[u^0, u_1^0, u_2^0](\eta, t + (-1)^i(\eta - x))|d\eta \leq K \int_0^x \{|u^1 - u^0| +$$

$$+|u_1^1 - u_1^0| + |u_2^1 - u_2^0|\}d\eta \leq KM\left(2 + \frac{\pi}{2}\right)\frac{x^2}{2!}, \; i = 1,2,$$

$$|u^2 - u^1| \leq \int_0^x |\frac{u_1^2 - u_2^2}{2} - \frac{u_1^1 - u_2^1}{2}|d\xi \leq \frac{1}{2}\int_0^x \{|u_1^2 - u_1^1| +$$

$$+|u_2^2 - u_2^1|\}d\xi \leq KM\left(2 + \frac{\pi}{2}\right)\frac{x^3}{3!} \leq KM\left(2 + \frac{\pi}{2}\right)\frac{\pi}{3}\frac{x^2}{2!}. \tag{4.19}$$

For the differences $|u_i^{m+i} - u_i^m|$, $i = 1,2$, $|u^{m+1} - u^m|$ $\bar{m} = 2,3,\ldots$, we can get estimates in the same manner,

$$|u_i^3 - u_i^2| \leq MK\gamma\frac{x^3}{3!}, \; \gamma = \left(2 + \frac{\pi}{2}\right)\left(2 + \frac{\pi}{3}\right), \; i = 1,2,$$

$$|u^3 - u^2| \leq MK^2\gamma\frac{x^4}{4!} \leq MK^2\gamma\frac{x^3}{3!},$$

$$|u_i^4 - u_i^3| \leq 3K\gamma MK^2\frac{x^4}{4!}, \; i = 1,2,$$

$$|u^4 - u^3| \leq 3K\gamma MK^2\frac{x^5}{5!} \leq 3K\gamma MK^2\frac{x^4}{4!},$$

. . . .

$$|u_i^{m+1} - u_i^m| \leq (3K)^{m-2}\gamma MK\frac{x^{m+1}}{(m+1)!}, \; i = 1,2; \tag{4.20}$$

$$|u^{m+1} - u^m| \leq (3K)^{m-2}\gamma MK\frac{x^{m+2}}{(m+2)!}, \; i = 2,3,\ldots, \tag{4.21}$$

....

From the estimates (4.7), (4.18)–(4.21) it follows that, for all $x, t, \alpha \in I \times \mathbb{R} \times \mathfrak{D}_\mu$, the terms of the first series in (4.16) do not exceed, in module, the corresponding terms of the convergent series with positive terms

$$(c - M\pi) + M\pi + KM\left(2 + \frac{\pi}{3}\right)\frac{\pi^2}{2!} + \gamma MK^2\frac{\pi^3}{3!} + 3K\gamma MK^2\frac{\pi^4}{4!} +$$

$$+ \cdots + (3K)^{m-2}\gamma MK^2\frac{\pi^{m+1}}{(m+1)!} + \cdots, \quad \gamma = \left(2 + \frac{\pi}{2}\right)\left(2 + \frac{\pi}{3}\right).$$

Therefore, by the Weierstrass criterion, the first of the series in (4.16) converges uniformly for all $x, t, \alpha \in I \times \mathbb{R} \times \mathfrak{D}_\mu$. In the same manner we can prove that the second series in (4.16) converges uniformly, too.

Let $u_{0i}(x, t, \alpha)$, $i = 1, 2$, $u_0(x, t, \alpha)$ denote the sums of the series (4.16) or, which is the same, the limits $m \to \infty$ of the sequences (4.9). Passing to the limit in the recurrent relations (4.9) shows that the limit functions u_0 satisfy the system of integral equations

$$u_{01}(x, t, \alpha) = \mu(t + x) - \int_0^x \tilde{F}[u_0, u_{01}, u_{02}](\eta, t + x - \eta)d\eta,$$

$$u_{02}(x, t, \alpha) = -\mu(t - x) + \int_0^x \tilde{F}[u_0, u_{01}, u_{02}](\eta, t - x + \eta)d\eta,$$

$$u_0(x, t, \alpha) = \frac{1}{2}\int_0^x \{u_{01}(\xi, t, \alpha) - u_{02}(\xi, t, \alpha)d\eta. \tag{4.22}$$

Let us estimate the differences $(u_i^m - u_{0i})$, $i = 1, 2$, and $(u^m - u_0)$. By using (4.9) and (4.22) and the Lipschitz condition (4.17), we obtain

$$|u_i^m - u_{0i}| \le K\int_0^x \{|u^{m-1} - u_0| + |u_1^{m-1} - u_{01}| + |u_2^{m-1} - u_{02}|\}d\eta,$$

$$|u^m - u_0| \le \frac{1}{2}\int (|u_1^m - u_{01}| + |u_2^m - u_{02}|)d\eta. \tag{4.23}$$

In order to derive the desired estimates from (4.23), we first estimate the differences $|u_i^0 - u_{0i}|$, $i = 1, 2$, and $|u^0 - u_0|$. By using (4.22) we obtain

$$|u_i^0 - u_{0i}| \le \int_0^x |F[u_0, u_{01}, u_{02}](\eta, t + (-1)^i(\eta - x))|d\eta \le Mx,$$

$$|u^0 - u_0| \le \frac{1}{2}\int_0^x d\xi \int_0^\xi |\tilde{F}[u_0, u_{01}, u_{02}](\eta, t + \xi - \eta) +$$

$$+ \tilde{F}[u_0, u_{01}, u_{02}](\eta, t - \xi + \eta)|d\eta \le M\frac{x^2}{2!} \le M\frac{\pi}{2}x. \tag{4.24}$$

Setting $m = 1, 2, 3, \ldots$ in (4.23) and taking account of (4.24) gives

$$|u_i^1 - u_{0i}| \le MK\left(2 + \frac{\pi}{2}\right)\frac{x^2}{2!}, \quad i = 1, 2,$$

$$|u^1 - u_0| \le MK\left(2+\frac{\pi}{2}\right)\frac{x^3}{3!} \le MK\left(2+\frac{\pi}{3}\right)\frac{\pi}{3!}\frac{x^2}{2!},$$

$$|u_i^2 - u_{0i}| \le \gamma MK^2\frac{x^3}{3!}, \ \gamma = \left(2+\frac{\pi}{2}\right)\left(2+\frac{\pi}{3}\right), \ i = 1,2,$$

$$|u^2 - u_0| \le \gamma MK^2\frac{x^4}{4!} \le \gamma MK^2\frac{x^3}{3!},$$

$$|u_i^3 - u_{0i}| \le 3K\gamma MK^2\frac{x^4}{4!}, \ i = 1,2,$$

$$|u^3 - u_0| \le 3K\gamma MK^2\frac{\pi^4}{4!}.$$

We continue in this fashion to obtain

$$|u_i^m - u_{0i}| \le (3K)^{m-2}\gamma MK^2\frac{x^{m+1}}{(m+1)!}, \ i = 1,2,$$

$$|u^m - u_0| \le (3K)^{m-2}\gamma MK^2\frac{x^{m+2}}{(m+2)!}, \ m = 2,3\ldots \tag{4.25}$$

Thus, replacing the exact solution $u_0(x,t,\alpha)$ of the system of integral equations (4.22) ((4.12)) by the approximate solution $u^m(x,t,\alpha)$ brings an error, which can be estimated by the formula

$$|u^1(x,t,\alpha) - u_0(x,t,\alpha)| \le MK\left(2+\frac{\pi}{2}\right)\frac{\pi^3}{3!}; \tag{4.26}$$

$$|u^m(x,t,\alpha) - u_0(x,t,\alpha)| \le (3K)^{m-2}\gamma MK^2\frac{\pi^{m+2}}{(m+2)!}, \tag{4.27}$$

$$\gamma = \left(2+\frac{\pi}{2}\right)\left(2+\frac{\pi}{3}\right), \ m = 2,3,\ldots$$

This completes the proof of the lemma.

Since the Cauchy problem

$$u_{tt} - u_{xx} = f(x,t,u,u_t,u_x), \ u(0,t) = 0, \ u_x(0,t) = \mu(t) \tag{4.28}$$

in the triangle $\bar{\Delta} = \{0 \le x \le \pi, \ x \le t \le 2\pi - x\}$ is equivalent, in the class of smooth functions, to the system of integral equations (4.12), Lemma 4.1 allows us to obtain the following assertion.

Theorem 4.1. *Let the condition of Lemma 4.1 be satisfied. Then, for any continuous 2π-periodic function $|\mu(t)| \le c - M\pi$, the Cauchy problem (4.28) has a unique continuous 2π-periodic solution $u_{\bar{\Delta}}(x,t)$ in the region $\Delta_\infty = \{0 \le x \le \pi, \ 2\pi n + x \le t \le 2\pi(n+1) - x, \ n = 0, \pm 1, \pm 2, ldots\}$.*

Remark 4.1. The solution $u_{\bar{\Delta}}(x,t,\alpha)$ of the Cauchy problem (4.28) satisfies the first boundary condition $u_{\bar{\Delta}}(0,t,\alpha)$, and the second boundary condition $u_{\bar{\Delta}}(\pi,\pi,\alpha)$ is evidently satisfied not by arbitrary initial function $\mu(t)$.

Theorem 4.2. *Let the condition of Lemma 4.1 and the equality $f(x,t,0,0,0) = 0$ hold. For the boundary problem (4.1) to have, for every continuous 2π-periodic function $|\mu(t)| \le c - M\pi$, a solution $\tilde{u}(x,t,\alpha) \in L^\infty$ of the form*

$$\tilde{u}(x,t,\alpha) = \begin{cases} u_{\Delta}(\pi,\pi,\alpha), & (x,t) \in \bar{\Delta}, \\ 0, & (x,t) \in \Pi_{2\pi}/\bar{\Delta} \end{cases}, \tag{4.29}$$

in the rectangle a it is necessary and sufficient that the condition

$$\tilde{\Delta}(\alpha) \equiv \int_0^{2\pi} \mu(\xi)d\xi + \int_0^{\pi} d\xi \int_0^{\xi} \{F[u, u_t, u_x](\eta, \pi + \xi - \eta)+$$

$$+F[u, u_t, u_x](\eta, \pi - \xi + \eta)\} \, d\eta = 0, \tag{4.30}$$

where

$$F[u, u_t, u_x](x,t) = f(x, t, u(x,t), u_t(x,t), u_x(x,t)). \tag{4.31}$$

be fulfilled.

P r o o f. Let the boundary value problem (4.1) have a solution of the form (4.29) in the rectangle $\Pi_{2\pi}$. Then the boundary conditions $\tilde{u}(0,t,\alpha) = 0$, $\tilde{u}(\pi,t,\alpha)$ are fulfilled, which is equivalent to the condition $u_{\tilde{\Delta}}(0,t,\alpha) = 0$ and $u_{\tilde{\Delta}}(\pi,\pi,\alpha) = 0$. Since the function $u_{\tilde{\Delta}}(x,t,\alpha)$ satisfies the system of integral equations (4.12), from the condition (4.12) and equality $u_{\tilde{\Delta}}(\pi,\pi,\alpha) = 0$ we conclude that (4.30) holds.

Now let the condition (4.30) be fulfilled for the initial function $|u_x(0,t)| \equiv |\mu(t)| \leq c - M\pi$ and the solution $u_{\tilde{\Delta}}(x,t,\alpha)$ of the Cauchy problem (4.28). Then (4.29) is a solution of the boundary value problem (4.4) in the generalized sense, and the proof is complete.

Corollary 4.1. Let the condition of Theorem 4.2 be satisfied and let the constant term α_0 in the collection of the Fourier coefficients $\alpha = (\alpha_0, \alpha_1', \alpha_1'', \alpha_2', \alpha_2'', \ldots)$ determining a 2π-periodic function $|\mu(t)| \leq c - M\pi$ is zero. For the boundary value problem (4.4) to have a solution $\tilde{u}(x,t,\alpha) \in L^\infty$ of the form (4.29) it is necessary and sufficient that the following condition be satisfied

$$\Delta(\alpha) \equiv \int_0^{\pi} d\xi \int_0^{\xi} \{F[u, u_t, u_x](\eta, \pi + \xi - \eta)+$$

$$+F[u, u_t, u_x](\eta, \pi - \xi + \eta)\} \, d\eta = 0. \tag{4.32}$$

The condition (4.32) ((4.30)) is thus a computability condition for a collection of the Fourier coefficients $\alpha = (\alpha_0, \alpha_1', \alpha_1'', \alpha_2', \alpha_2'', \ldots)$ determining a 2π-periodic function $|\mu(t)| \leq c - M\pi$, by which a 2π-periodic solution a can be constructed for a $\tilde{u}(x,t,\alpha) \in L^\infty 2\pi$-system of the first kind. Therefore, the problem on existence and construction of a generalized 2π-periodic solution ($\tilde{u} \in L^\infty$) of the 2π-hyperbolic differential system of the first kind is equivalent to the existence of a point α (collection of the Fourier coefficients) determining a function $|\mu(t)| \leq c - M\pi$ satisfying $\tilde{\Delta}(\alpha) = 0$ ($\Delta(\alpha) = 0$).

The point α satisfying $\tilde{\Delta}(\alpha) = 0$ is a singular point of the mapping [103]

$$\tilde{\Delta} : \mathfrak{D}_\mu \to \mathbb{R}, \ \tilde{\Delta}(\alpha) = \int_0^{2\pi} \mu(z)dz + \int_0^{\pi} d\xi \int_0^{\xi} \{F[u, u_t, u_x](\eta, \pi + \xi - \eta)+$$

$$+F[u, u_t, u_x](\eta, \pi - \xi + \eta)\} \, d\eta. \tag{4.33}$$

This mapping can be computed only approximately using, for example, the functions

$$\tilde{\Delta}(\alpha) = \int_0^{2\pi} \mu(z)dz + \int_0^{\pi} d\xi \int_0^{\xi} \{F[u^m, u_t^m, u_x^m](\eta, \pi + \xi - \eta)+$$

$$+F[u^m, u_t^m, u_x^m](\eta, \pi - \xi + \eta)\} \, d\eta. \tag{4.34}$$

In this connection, a problem appears: how to conclude from (4.34) about zeros of the mapping $\tilde{\Delta}(\alpha)$ (and hence about the existence of generalized periodic solutions of the 2π-systems of the first kind). As indicated in §2.1, Chapter 2, this problem has been sucessfully solved for the so-called T-systems of ordinary differential equations in [103].

Since (4.32) is a scalar equation, the function $\mu(t)$ can be sought in the form $\mu(t) = \alpha_k' \cos kt$ or $\mu(t) = \alpha_k'' \sin kt$. Then equation (4.32) is an equation for determining the desired Fourier coefficients α_k' or α_k''.

The functions $\Delta(\alpha_k')$ and $\Delta(\alpha_k'')$ can be computed only approximately by using the sequence

$$\Delta(\alpha_k^\nu) = \int\limits_0^{2\pi} d\xi \int\limits_0^{\xi} \{F[u^m, u_t^m, u_x^m](\eta, \pi + \xi - \eta) + $$

$$+F[u^m, u_t^m, u_x^m](\eta, \pi - \xi + \eta)\}\, d\eta. \tag{4.35}$$

where ν stands for one or two primes, so a problem arises: how one can conclude about zeros of the function $\Delta_m(\alpha_k^\nu)$ from zeros of the function $\Delta(\alpha_k^\nu)$. To solve it, one should take account of the fact that each function $\Delta_m(\alpha_k^\nu)$, as the limit function $\Delta(\alpha_k^\nu)$ itself, is defined and continuous on the interval $\mathfrak{D}_\mu = [-c + M\pi,\ c - M\pi]$ and satisfies the inequality

$$|\Delta_m(\alpha_k^\nu) - \Delta(\alpha_k^\nu)| \le (3K)^{m-1}\gamma M K^2 \frac{\pi^{m+3}}{(m+1)!} = d_m, \quad m e 2. \tag{4.36}$$

The last estimate results in the following assertion.

Theorem 4.3. *Let the boundary value problem (4.4) be a 2π-system of the first kind and let the function $\Delta_m(\alpha_k^\nu)$ given by (4.35) satisfy, for certain m, the inequalities*

$$\min_{\alpha_k^\nu \in \mathfrak{D}_\mu} \Delta_m(\alpha_k^\nu) \le -d_m, \quad \max_{\alpha_k^\nu \in \mathfrak{D}_\mu} \Delta_m e d_m, \tag{4.37}$$

where $\mathfrak{D}_\mu = [-c + M\pi, c - M\pi]$ and d_m is the constant from (4.36). Then the boundary value problem (4.4) has a solution $\tilde{u}(x,t,\alpha) \in L^\infty$ of the form (4.29).

P r o o f. Let $y_1 \in \mathbb{R}$ and $y_2 \in \mathbb{R}$ be points from the interval $[-c + M\pi, c - M\pi]$ such that

$$\Delta_m(y_1) = \min_{y \in \mathfrak{D}_\mu} \Delta_m(y), \ \Delta_m(y_2) = \max_{y \in \mathfrak{D}_\mu} \Delta_m(y).$$

Combining (4.36) and (4.37) we obtain

$$\Delta(y_1) = \Delta_m(y_1) + (\Delta(y_1) - \Delta_m(y_1)) \le 0,$$

$$\Delta(y_2) = \Delta_m(y_2) + (\Delta(y_2) - \Delta_m(y_2)) e 0.$$

Whence it follows that there is a point $y_0 = \alpha_k^\nu \in [y_1, y_2]$ such that $\Delta(y_0) = 0$, since the function $\Delta(y)$ is continuous. The last equality proves the theorem.

The fact that the function $\Delta(\alpha_k^\nu)$ is a scalar function of one variable allows us to use the classical theorem on zeros of an analytic function [103] and to obtain the following assertion.

Theorem 4.4. *Let the right-hand side of equation (4.4) be a polynomial in u, u_t, u_x and let inequalities (4.6)–(4.8) be satisfied. Then if the boundary value problem (2.1) has a 2π-periodic solution $\tilde{u} \in L^\infty$ of the form (4.29), then it has a periodic solution $\tilde{u} \in L^\infty$ either for any value of α_k^ν) from the interval $[-c + M\pi, c - M\pi]$ or for a finite number of these values only.*

Theorem 4.5. *Let the boundary value problem (4.4) be a 2π-hyperbolic differential system of the first kind inthe region (4.5) and let (4.6) and the following condition be fulfilled*

1) for any function $u(x,t) \in C^1$ that is odd in variable t, the function $F[u, u_t, u_x](x,t) = f(x,t,u(x,t),u_t(x,t),u_x(x,t))$ is odd in variable t.

Then, for any continuous odd 2π-periodic function $|u_x(0,t)| \equiv |\mu(t)| \leq c - M\pi$, the boundary value problem (4.4) has a 2π-periodic solution $\tilde{u}(x,t,\alpha) \in L^\infty$ of the form (4.29).

P r o o f. Proceeding by induction we shall show that if the function

$$u^0(x,t,\alpha) = \frac{1}{2}\int_0^x \{\mu(t+\xi) + \mu(t-\xi)\}\, d\xi, \quad |\mu(t) \leq c - M\pi|, \tag{4.38}$$

is picked as the zero approximation $u^0(x,t,\alpha)$ in the respresentation (4.9), then all the terms of the sequence $u^m(x,t,\alpha)$ constructed according to (4.9) are odd functions of variable t. Indeed, in view of (4.9), (4.10) and (4.31) we have

$$u^1(x,t,\alpha) = u^0(x,t,\alpha) - \frac{1}{2}\int_0^x d\xi \int_0^\xi \Big\{F[u^0, u_t^0, u_x^0](\eta, t+\xi-\eta)+$$

$$+F[u^0, u_t^0, u_x^0](\eta, t-\xi+\eta)\Big\}\, d\eta.$$

for $m = 0$. The function $\mu(t)$ is odd and so is the function $u^0(x,t,\alpha)$ given by (4.38). By the condition (1) in Theorem 4.5, the first approximation $u^1(x,t,\alpha)$ is an odd function of variable t.

Suppose that $u^k(x,t,\alpha)$, k a positive integer, is an odd function of variable t. By using the recurrence relation defining $u^{k+1}(x,t,\alpha)$, i.e.

$$u^1(x,t,\alpha) = u^0(x,t,\alpha) - \frac{1}{2}\int_0^x d\xi \int_0^\xi \Big\{F[u^k, u_t^k, u_x^k](\eta, t+\xi-\eta)+$$

$$+F[u^k, u_t^k, u_x^k](\eta, t-\xi+\eta)\Big\}\, d\eta.$$

and the condition (1) in Theorem 4.5, we conclude that the function $u^{k+1}(x,t,\alpha)$ is odd in variable t.

Thus, if the conditions of Theorem 4.5 are satiafied, then all the terms of the sequence $(u^m(x,t,\alpha))$ are odd fucntions of variable t.

We shall now prove that the condition (4.32) holds in this case. Indeed, according to the third equation in (4.9), $\Delta_m(\alpha) \equiv u^{m+1}(\pi,\pi,\alpha)$ assumes the values of the form

$$\Delta_m(\alpha) = \int_0^\pi d\xi \int_0^\xi \{F[u^m, u_t^m, u_x^m](\eta, \pi+\xi-\eta)+$$

$$+F[u^m, u_t^m, u_x^m](\eta, \pi-\xi+\eta)\}\, d\eta \equiv \int_0^{2\pi} d\xi \int_0^\pi \{F[u^m, u_t^m, u_x^m](\eta, \pi+$$

$$+\xi-\eta) + F[u^m, u_t^m, u_x^m](\eta, \pi-\xi+\eta)\}\, d\eta.$$

Whence, by using the condition (1) in Theorem 4.5, we deduce that $\Delta_m(\alpha) \equiv u^{m+1}(\pi, \pi, \alpha)$ for all $m = 0, 1, 2, \ldots$. Hence $\Delta(\alpha) = 0$ and the boundary value problem (4.4) has therefore a solution $\tilde{u} \in L^\infty$ of the form (4.29). The proof is complete.

Note that applying the Schauder principle [45] to the operators (4.12) enables us to prove the existence of a solution of the boundary value problem (4.4), which is a 2π-hyperbolic differential system of the first kind, under a weaker assumption about the function $f(x, t, u, u_t, u_x)$.

4.3. The structure of generalized periodic solutions of the second order wave equation of the second kind

Definition 4.2. The boundary value problem (4.4) is called a 2π-hyperbolic differential system of the second kind in the region (4.5) if the constants a, b, c, M, K relate by the inequalities

$$c > 2M\pi, \quad a \leq -c\pi < |c\pi| \leq b; \tag{4.39}$$

$$2K\pi(2+\pi) < 1. \tag{4.40}$$

Let the rectangle $\Pi_{2\pi}$ be partitioned into the three parts by the lines and . As shown in §1.9, Chapter 1, equation (4.4) is equivalent, in the class of smooth functions, to the system of integral equations (4.12) for all and to the following systems

$$u_1(x,t,\alpha) = \mu(t+x) - \int_0^x \tilde{F}[u, u_1, u_2](\eta, t+x-\eta)d\eta,$$

$$u_2(x,t,\alpha) = -\mu(t-x) + \int_0^\pi \tilde{F}[u, u_1, u_2](\eta, 2\pi+t-\eta-x)d\eta -$$

$$- \int_x^\pi \tilde{F}[u, u_1, u_2](\eta, t-x+\eta)d\eta,$$

$$u(x,t,\alpha) = \int_\pi^x \frac{u_1(\xi,t,\alpha) - u_2(\xi,t,\alpha)}{2} d\xi \quad \forall (x,t) \in \bar{\Delta}_1; \tag{4.41}$$

$$u_1(x,t,\alpha) = \mu(t+x) - \int_0^\pi \tilde{F}[u, u_1, u_2](\eta, 2\pi+t+x+\eta)d\eta +$$

$$+ \int_0^\pi \tilde{F}[u, u_1, u_2](\eta, t+x-\eta)d\eta,$$

$$u_2(x,t,\alpha) = -\mu(t-x) - \int_0^x \tilde{F}[u, u_1, u_2](\eta, t-x+\eta)d\eta,$$

$$u(x,t,\alpha) = \int_\pi^x \frac{u_1(\xi,t,\alpha) - u_2(\xi,t,\alpha)}{2} d\xi \quad \forall (x,t) \in \bar{\Delta}_2; \tag{4.42}$$

where $\mu(t) \in C^1$ is an arbitrary function and $F[u, u_1, u_2](x,t)$ is given by (4.10).

Lemma 4.2. *Let the function $f(x, t, u, u_t, u_x)$ be defined in the region (4.5) be continuous with respect to x, t, periodic in t with period 2π and satisfy inequalities (4.6), (4.39) and (4.40). Then, for any continuous 2π-periodic function*

$$|\mu(t)| \equiv |\frac{\alpha_0}{2} + \sum_{k=1} \alpha_k{}' \cos kt + \alpha_k \sin kt| \leq c - 2M\pi \tag{4.43}$$

the sequences of functions (4.9) 2π-periodic in t and the following sequences of functions

$$\tilde{u}_1^0(x, t, \alpha) = \mu(t + x), \quad \tilde{u}_2^0(x, t, \alpha) = -\mu(t - x),$$

$$\tilde{u}^0(x, t, \alpha) = \int\limits_{\pi}^{x} \frac{\mu(t + \xi) + \mu(t - \xi)}{2} d\xi,$$

$$\tilde{u}_1^{m+1}(x, t, \alpha) = \mu(t + x) - \int\limits_0^x \tilde{F}[\tilde{u}^m, \tilde{u}_1^m, \tilde{u}_2^m](\eta, t + x - \eta) d\eta,$$

$$\tilde{u}_2^{m+1}(x, t, \alpha) = -\mu(t - x) + \int\limits_0^{\pi} \tilde{F}[\tilde{u}^m, \tilde{u}_1^m, \tilde{u}_2^m](\eta, 2\pi + t - \eta - x) d\eta -$$

$$- \int\limits_x^{\pi} \tilde{F}[\tilde{u}^m, \tilde{u}_1^m, \tilde{u}_2^m](\eta, t - x + \eta) d\eta,$$

$$\tilde{u}^{m+1}(x, t, \alpha) = \frac{1}{2} \int\limits_{\pi}^{x} \{\tilde{u}_1^{m+1}(\xi, t, \alpha) - \tilde{u}_2^{m+1}(\xi, t, \alpha)\} d\xi, \; m = 0, 1, 2, \ldots; \tag{4.44}$$

$$\tilde{\tilde{u}}_1^0(x, t, \alpha) = \mu(t + x), \quad \tilde{\tilde{u}}_2^0(x, t, \alpha) = -\mu(t - x),$$

$$\tilde{\tilde{u}}^0 = \int\limits_{\pi}^{x} \frac{\mu(t + \xi) + \mu(t - \xi)}{2} d\xi,$$

$$\tilde{\tilde{u}}_1^{m+1} = \mu(t + x) - \int\limits_0^{\pi} \tilde{F}[\tilde{\tilde{u}}^m, \tilde{\tilde{u}}_1^m, \tilde{\tilde{u}}_2^m](\eta, 2\pi + t + x + \eta) d\eta +$$

$$+ \int\limits_x^{\pi} \tilde{F}[\tilde{\tilde{u}}^m, \tilde{\tilde{u}}_1^m, \tilde{\tilde{u}}_2^m](\eta, t + x - \eta) d\eta,$$

$$\tilde{u}_2^{m+1}(x, t, \alpha) = -\mu(t - x) + \int\limits_0^x \tilde{F}[\tilde{\tilde{u}}^m, \tilde{\tilde{u}}_1^m, \tilde{\tilde{u}}_2^m](\eta, 2\pi + t - \eta - x) d\eta,$$

$$\tilde{\tilde{u}}^{m+1}(x, t, \alpha) = \frac{1}{2} \int\limits_{\pi}^{x} [\tilde{\tilde{u}}_1^{m+1}(\xi, t\alpha) - \tilde{\tilde{u}}_2^{m+1}(\xi, t, \alpha)\} d\xi, \; m = 0, 1, 2, \ldots; \tag{4.45}$$

converge, as $m \to \infty$, uniformly with respect to

$$x, t, \alpha \in I \times \mathbb{R} \times \mathfrak{D}_\mu \tag{4.46}$$

to the functions $(u_0(x,t,\alpha), u_{10}(x,t,\alpha), u_{20}(x,t,\alpha))$, $(\tilde{u}_0(x,t,\alpha), \tilde{u}_{10}(x,t,\alpha), \tilde{u}_{20}(x,t,\alpha))$, $(\tilde{\tilde{u}}_0(x,t,\alpha), \tilde{\tilde{u}}_{10}(x,t,\alpha), \tilde{\tilde{u}}_{20}(x,t,\alpha))$ *defined in the region (4.46), 2π-periodic in t and satisfying the systems of integral equations (4.12), (4.41) and (4.42), respectively. Furthermore, the inequalities hold*

$$|u^1(x,t,\alpha) - u_0(x,t,\alpha)| \le MK\left(2+\frac{\pi}{2}\right)\frac{\pi^3}{3!},$$

$$|u^m(x,t,\alpha) - u_0(x,t,\alpha)| \le (3K)^{m-2}\gamma MK^2\frac{\pi^{m+2}}{(m+2)!}, \tag{4.47}$$

$$\gamma = \left(2+\frac{\pi}{2}\right)\left(2+\frac{\pi}{3}\right); \quad m = 2,3,\ldots;$$

$$|\tilde{u}^m(x,t,\alpha) - u_0(x,t,\alpha)| \le 2M\pi^2\gamma_1^m, \tag{4.48}$$

$$|\tilde{\tilde{u}}^m(x,t,\alpha) - u_0(x,t,\alpha)| \le 2M\pi^2\gamma_1^m, \tag{4.49}$$

$$\gamma_1 = 2\pi K(2+\pi) < 1, \quad m = 1,2,\ldots$$

P r o o f. Let the boundary value problem (4.4) be a 2π-hyperbolic differential system of the second kind. The condition (4.39) then holds and hence the condition (4.8) holds, too. By Lemma 4.1, it follows that the sequences of functions (4.9), constructed for $|\mu(t)| \le c - 2M\pi < c - M\pi$, converge uniformly for all $x,t,\alpha \in I \times \mathbb{R} \times \mathfrak{D}_\mu$ and the estimates in (3.4.47) hold.

We shall now prove that the sequence in (4.44) converges uniformly. First, we shall show that $\tilde{u}^1(x,t,\alpha), \tilde{u}^1_t(x,t,\alpha) = (\tilde{u}^1_1(x,t,\alpha) + \tilde{u}^1_2(x,t,\alpha))/2, \tilde{u}^1_x(x,t,\alpha) = (\tilde{u}^1_1(x,t,\alpha) - \tilde{u}^1_2(x,t,\alpha))/2$ assume values in the region (4.5) for all $x,t,\alpha \in I\times\mathbb{R}\times\mathfrak{D}_\mu$. Indeed, if $m=0$, inequlities (4.6), (4.43) and representation (4.44) imply that

$$\left|\frac{\tilde{u}^1_1(x,t,\alpha) \pm \tilde{u}^1_2(x,t,\alpha)}{2} - \frac{\mu(t+x) \mp \mu(t-x)}{2}\right| \le 2M\pi. \tag{4.50}$$

Whence $|\tilde{u}^1_t(x,t,\alpha)| \le c$, $|\tilde{u}^1_x(x,t,\alpha)| \le c$ for all $x,t,\alpha \in I \times \mathbb{R} \times \mathfrak{D}_\mu$, provided $|\mu(t)| \le c - 2M\pi$.

From the inequalities

$$|\tilde{u}^1(x,t,\alpha) - \tilde{u}_0(x,t,\alpha)| \le \frac{1}{2}\left|\int_x^\pi d\xi \int_0^\xi |\tilde{F}[\tilde{u}^0,\tilde{u}^0_1,\tilde{u}^0_2)](\eta, t+\xi-\eta)|d\eta\right| +$$

$$+\frac{1}{2}\left|\int_x^\pi d\xi \int_\xi^\pi |\tilde{F}[\tilde{u}^0,\tilde{u}^0_1,\tilde{u}^0_2)](\eta, t-\xi+\eta)|d\eta\right| +$$

$$+\frac{1}{2}\left|\int_x^\pi d\xi \int_0^\pi |\tilde{F}[\tilde{u}^0,\tilde{u}^0_1,\tilde{u}^0_2)](\eta, 2\pi+t-\xi-\eta)|d\eta\right| \le 2M\pi^2, \tag{4.51}$$

$$|\tilde{u}^0(x,t,\alpha)| = \left|\int_\pi^x \frac{\mu(t+\xi)+\mu(t-\xi)}{2}\right| \le c\pi - 2M\pi^2$$

and (4.39) it follows that $a \le \tilde{u}^1(x,t,\alpha) \le b$ for all $x,t,\alpha \in I \times \mathbb{R} \times \mathfrak{D}_\mu$.

Next, proceeding by induction, we conclude that the functions $\tilde{u}^m(x,t,\alpha), \tilde{u}^m_t(x,t,\alpha), \tilde{u}^m_x(x,t,\alpha)$

satisfy the inequalities $a \le \tilde{u}^m(x,t,\alpha) \le b$, $|\tilde{u}_t^m(x,t,\alpha)| \le c$, $|\tilde{u}_x^m(x,t,\alpha)| \le c$ for all $m \epsilon 0$ and $x,t,\alpha \in I \times \mathbb{R} \times \mathfrak{D}_\mu b$.

We are now in a position to prove that the sequences in (4.44) converge uniformly for all $x,t,\alpha \in I \times \mathbb{R} \times \mathfrak{D}_\mu b$, provided the condition (4.40) is fulfilled, and hence that the limit functions are continuous on this set. Note that convergence of sequences (4.44) is equivalent to convergence of the series

$$\tilde{u}_i^0 + (\tilde{u}_i^1 - \tilde{u}_i^0) + (\tilde{u}_i^2 - \tilde{u}_i^1) + \cdots + (\tilde{u}_i^m - \tilde{u}_i^{m-1}) + \ldots, \ i = 1,2,$$

$$\tilde{u}^0 + (\tilde{u}^1 - \tilde{u}^0) + (\tilde{u}^2 - \tilde{u}^1) + \cdots + (\tilde{u}^m - \tilde{u}^{m-1}) + \cdots \tag{4.52}$$

Let us estimate the terms of the series (4.52). By inequalities (4.6) and (4.51) and representation (4.44), we have

$$|\tilde{u}_i^1 - \tilde{u}_i^0| \le 2M\pi, \quad i = 1,2, \ |\tilde{u}_i^1 - u_i^0| \le 2M\pi^2. \tag{4.53}$$

for $m = 0$. Combining (4.10), (4.17), (4.44) and (4.53) yields

$$|\tilde{u}_i^2 - \tilde{u}_i^1| \le 2K \int_0^\pi (|\tilde{u}^1 - \tilde{u}^0| + |\tilde{u}_1^1 - \tilde{u}_1^0| + |\tilde{u}_2^1 - \tilde{u}_2^0|)d\eta \le$$

$$\le 2^2 K M\pi^2(2+\pi) = 2M\pi\gamma_1, \quad \gamma_1 = 2\pi K(2+\pi),$$

$$|\tilde{u}^2 - \tilde{u}^1| \le \frac{1}{2}\int_0^\pi (|\tilde{u}_1^2 - \tilde{u}_1^1| + |\tilde{u}_2^2 - \tilde{u}_2^1|)d\xi \le 2M\pi^2\gamma_1.$$

For the differences $|\tilde{u}_i^{m+1} - \tilde{u}_i^|, i = 1,2, |\tilde{u}^{m+1} - \tilde{u}^m|, m = 2,3,\ldots$, we obtain, in the similar fashion, the estimates

$$|\tilde{u}_i^3 - \tilde{u}_i^2| \le 2M\pi\gamma_1, \quad i = 1,2, |\tilde{u}^3 - \tilde{u}^2| \le 2M\pi^2\gamma_1^2,$$

$$|\tilde{u}_i^{m+1} - \tilde{u}_i^m| \le 2M\pi\gamma_1^m, \ i = 1,2, \tag{4.54}$$

$$|\tilde{u}^{m+1} - \tilde{u}^m| \le 2M\pi^2\gamma_1^2, \ m = 0,1,2,\ldots$$

The estimates (4.54) show that, for all values of $x,t,\alpha \in I \times \mathbb{R} \times \mathfrak{D}_\mu b$, the terms of the first series in (4.52) do not exceed, in module, the corresponding terms of the following series with positive terms

$$(c - 2M\pi) + 2M\pi + 2M\pi\gamma_1 + \cdots + 2M\pi\gamma_1^{m-1} + \cdots,$$

$$\gamma_1 = 2\pi K(2+\pi) < 1.$$

Therefore, the first series in (4.52) converges unifromly for all $x,t,\alpha \in I \times \mathbb{R} \times \mathfrak{D}_\mu b$.

In the same manner we can see that the second series in (4.52) converges uniformly on the same set. Denoting by $\tilde{u}_{i0}(x,t,\alpha)$, $i = 1,2, \tilde{u}_0(x,t,\alpha)$ the sums of the series in (4.52) or, what is the same, the limit functions of the sequences in (4.44), and passing to the limit in the recurrence relations, we see that the limit functions $\tilde{u}_{i0}(x,t,\alpha)$, $i = 1,2, \tilde{u}_0(x,t,\alpha)$ satisfy the system of integral equations (4.41).

Proceeding in the similar way, we can prove that the sequences of functions in (4.45) converge uniformly to the limit functions $\tilde{\tilde{u}}_{i0}(x,t,\alpha)$, $i = 1,2, \tilde{\tilde{u}}_0(x,t,\alpha)$ for all $x,t,\alpha \in I \times \mathbb{R} \times \mathfrak{D}_\mu b$ and that the limit functions satisfy the system of integral equations (4.42). For the limit functions $\tilde{u}_0(x,t,\alpha)$, $i = 1,2, \tilde{\tilde{u}}_0(x,t,\alpha)$, the following estimates hold

$$|\tilde{u}^m(x,t,\alpha) - \tilde{u}_0(x,t,\alpha)| \le 2M\pi^2\gamma_1^m,$$

$$|\tilde{\tilde{u}}^m(x,t,\alpha) - \tilde{\tilde{u}}_0(x,t,\alpha)| \le 2M\pi^2\gamma_1^m, \quad \gamma_1 = 2\pi K(2+\pi) < 1; \tag{4.55}$$
$$m = 1, 2, \ldots$$

This proves the lemma.

Now we turn to the question of existence of continuous solutions of the boundary value problem (4.4) in the rectangular $\Pi_{2\pi} = \{0 \le x \le \pi, \ 0 \le t \le 2\pi\}$. Let the rectangular $\Pi_{2\pi}$ be partitioned into the three pieces (see Fig. 2) $\bar{\Delta}_1 = \{(x,t) \in \Pi_{2\pi} : 0 \le x \le \pi, x \le t \le 2\pi - x\}, \bar{\Delta}_1 = \{(x,t) \in \Pi_{2\pi} : 0 \le x \le \pi, x \le t \le \pi\}, \bar{\Delta}_2 = \{(x,t) \in \Pi_{2\pi} : 0 \le x \le \pi, 2\pi - x \le t \le 2\pi\}$ by the characteristics $t = x$ and $t = 2\pi - x$.

By Theorem 4.1, for any continuous initial function $u_x(0,t) = \mu(t)$, there exists a unique continuous solution $u_{\bar{\Delta}}(x,t,\alpha)$ of the Cauchy problem (4.28), provided the condition of Lemma 4.1 is fulfilled. Moreover, if the condition of Theorem 4.1 is satisfied, the solution $u_{\bar{\Delta}}(x,t,\alpha)$ of the Cauchy problem (4.28) can immediately be obtained from the system of the integral equations

$$u(x,t) = \frac{1}{2}\int_0^x [\mu(t+\xi) + \mu(t-\xi)]\,d\xi -$$

$$-\frac{1}{2}\int_0^x d\xi \int_0^\xi \{F[u,u_t,u_x](\eta, t+\xi-\eta) + F[u,u_t,u_x](\eta, t-\xi+$$

$$+\eta)\}\,d\eta \equiv \frac{1}{2}\int_{t-x}^{t+x} u_x(0,z)dz - \frac{1}{2}\int_0^x d\eta \int_{t-x+\eta}^{t+x-\eta} F[u,u_t,u_x](\eta,\xi)d\xi, \tag{4.56}$$

$$u_t(x,t) = \frac{1}{2}[\mu(t+x) - \mu(t-x)] -$$

$$-\frac{1}{2}\int_0^x \{F[u,u_t,u_x](\eta, t+x-\eta) - F[u,u_t,u_x](\eta, t-x+\eta)\}\,d\eta,$$

$$u_t(x,t) = \frac{1}{2}[\mu(t+x) + \mu(t-x)] -$$

$$-\frac{1}{2}\int_0^x \{F[u,u_t,u_x](\eta, t+x-\eta) + F[u,u_t,u_x](\eta, t-x+\eta)\}\,d\eta,$$

where

$$u_1 = u_t + u_x, \ u_2 = u_t - u_x, \ F[u,u_t,u_x](x,t) \equiv$$
$$\equiv \tilde{F}[u,u_1,u_2](x,t) = f(x,t,u(x,t),u_t(x,t),u_x(x,t)). \tag{4.57}$$

If the initial function $\mu(t)$ and the solution $u_{\bar{\Delta}}(x,t,\alpha)$ satisfy (4.30), then the boundary condition $u_{\bar{\Delta}}(0,t,\alpha) = 0, 0 \le t \le 2\pi, u_{\bar{\Delta}}(\pi,\pi,\alpha) = 0$ is fulfilled. Since the values of the solution $u_{\bar{\Delta}}(x,t,\alpha)$ and its derivatives at the sides OC and AC of the triangle $\bar{\Delta}$ (see Fig. 2) are known, one can prove the existence of a continuous solution of the boundary value problem (4.4) on the sets $\bar{\Delta}_1$ and $\bar{\Delta}_2$ in the following way.

Suppose $(x,t) \in \bar{\Delta}_1$. Then on $\bar{\Delta}_1$ equation (4.4) is equivalent, in the class of smooth functions, to the system of integral equations

$$u_1(x,t) = u_1(0,t+x) - \int_0^x F[u,u_t,u_x](\eta, t+x-\eta)d\eta,$$

$$u_2(x,t) = u_2(\pi, t_\pi^1) - \int\limits_x^\pi F[u, u_t, u_x](\eta, t - x + \eta)d\eta, \tag{4.58}$$

$$u(x,t) = u(x,x) + \frac{1}{2}\int\limits_x^t \{u_1(x,\tau) + u_2(x,\tau)\}\, d\tau,$$

where $u_t = (u_1 + u_2)/2$, $u_x = (u_1 - u_2)/2, \pi$ and $t_\pi^1 = \pi - t + x$ is the point at which the characteristic $\xi = \tau - t + x$ meets the line $\xi - \pi$; the function F is given by (4.57).

From the boundary condition $u(\pi, t) = 0 \forall t \in [0, 2\pi]$ it follows that $u_t \equiv (u_1 + u_2)/2 = 0$ for $x = \pi$. Thus $u_2(\pi, t_\pi^1)$, and system (4.58) can be rewritten as

$$u_1(x,t) = u_1(0, t + x) - \int\limits_0^x \tilde{F}[u, u_t, u_x](\eta, t + x - \eta)d\eta,$$

$$u_2(x,t) = -u_1(0, t - x) + \int\limits_0^\pi \tilde{F}[u, u_t, u_x](\eta, 2\pi + t - x - \eta)d\eta -$$

$$- \int\limits_x^\pi \tilde{F}[u, u_t, u_x](\eta, t - x + \eta)d\eta, \tag{4.59}$$

$$u(x,t) = \frac{1}{2}\int\limits_0^{2x} u_1(0, z)dz - \frac{1}{2}\int\limits_0^x d\eta \int\limits_\eta^{2x-\eta} \tilde{F}[u, u_t, u_x](\eta, \xi)d\xi +$$

$$+ \frac{1}{2}\int\limits_x^t \{u_1(x,\tau) + u_2(x,\tau)\}\, d\tau.$$

Note that by the notation $u_t = (u_1 + u_2)/2, u_x = (u_1 - u_2)/2$, we have $u_1 = u_t + u_x$. Since $u_t(0,t) = 0$, we have $u_1(0,t) \equiv u_t(0,t) + u_x(0,t) = u_x(0,t) = \mu(t)$, the known function.

Next, suppose that $f(x, t, u, u_t, u_x) \in C^1$ with respect to all the variables, $u_1(0,t) \equiv \mu(t) \in C^1$ and system (4.59) has a smooth solution. Differentiating the third equation in (4.59) with respect to t we then obtain

$$\frac{\partial^2 u}{\partial t^2} = \frac{d}{dt}\frac{u_1(x,t) + u_2(x,t)}{2}, \tag{4.60}$$

and differentiating the same equation with respect to x and changing the order of integration, we obtain

$$\frac{\partial u}{\partial t} = \frac{u_1(x,t) - u_2(x,t)}{2}. \tag{4.61}$$

By using (4.61), the third equation in (4.52) can be written as

$$u(x,t) = \frac{1}{2}\int\limits_\pi^x \{u_1(\xi,t) + u_2(\xi,t)\}\, d\xi, \tag{4.62}$$

and system (4.58) can be written in the form of (4.41). In this case, the condition $u(\pi,t) = 0$ is evidently satisfied. Combining (4.59) and (4.61), we obtain

$$\frac{\partial^2 u}{\partial x^2} = \frac{d}{dx}\frac{u_1(x,t) - u_2(x,t)}{2} \equiv$$

$$\equiv -f(x,t,u(x,t),u_t(x,t),u_x(x,t)) + \frac{d}{dt}\frac{u_1(x,t)-u_2(x,t)}{2}. \tag{4.63}$$

Hence, in the class of smooth functions, every solution of the system of integral equations (4.59) ((4.41)) is a smooth solution of equation (4.4) on the set $\bar{\Delta}_1$.

Note that the system of integral equations (4.59) ((4.41)) is of the Fredholm type because of the term

$$\int_0^\pi F[u,u_t,u_x](\eta, 2\pi+t-x-\eta)d\eta$$

Therefore, system (4.59) has a unique continuous solution $u_{\bar{\Delta}}(x,t,\alpha)$ only for a particular choice of the constants M, K, π [45].

Now suppose that $(x,t)\in\bar{\Delta}_2$. Then, in the class of smooth functions on this set, equation (4.4) satisfies the system of integral equations (4.42), which has a unique continuous solution $u_{\bar{\Delta}}(x,t,\alpha)$, provided the condition of Lemma 4.2 holds. By the preceding, the following assertion is true.

Theorem 4.6. *Let the condition of Lemma 3.1 be fulfilled. Then, for any contunuous 2π-periodic function $|u_x(0,t)|\equiv|\mu(t)|\leq c-2M\pi$ satisfying equation (4.30), which is equivalent to the equation*

$$u_0(\pi,\pi,\alpha)\equiv\int_0^{2\pi}\mu(z)dz-\int_\eta^{2\pi-\eta}F[u_0,u_{0t},u_{0x}](\eta,\xi)d\xi=0, \tag{4.64}$$

the boundary value problem (4.4) has a unique solution $u(x,t,\alpha)$ continuous in the rectangle $\Pi_{2\pi}=\{0\leq x\leq\pi,\ 0\leq t\leq 2\pi\}$,

$$u(x,t,\alpha)=\begin{cases} u_0(x,t,\alpha), & \text{если}(x,t)\in\bar{\Delta}, \\ \tilde{u}_0(x,t,\alpha), & \text{если}(x,t)\in\Pi_{2\pi}\ \{\bar{\Delta}\cup\bar{\Delta}_2\}, \\ \tilde{\tilde{u}}_0(x,t,\alpha), & \text{если}(x,t)\in\Pi_{2\pi}\ \{\bar{\Delta}\cup\bar{\Delta}_1\}, \end{cases}$$

where $u_0(x,t,\alpha), \tilde{u}_0(x,t,\alpha)$ and $\tilde{\tilde{u}}_0(x,t,\alpha)$ are the solutions of the systems of integral equations (4.12), (4.41) and (4.42), respectively.

Theorem 4.7. *Let the condition of Lemma 4.2 and the condition (1) in Theorem 4.5 be fulfilled. Then, for every contunuous odd 2π-periodic function $|u_x(0,t)|\equiv|\mu(t)|\leq c-2M\pi$, the boundary value problem (4.4) has a unique continuous solution $u(x,t,\alpha)$ defined in $\Pi_{2\pi}$.*

P r o o f. Choosing the function

$$u^0(x,t,\alpha)=\frac{1}{2}\int_0^x(\mu(t+\xi)+\mu(t-\xi))d\xi$$

which is odd with respect to t, as the zero approximation $u^0(x,t,\alpha)$ for the system of integral equations (4.12) and taking account of the condition of Theorem 4.7, we see that the theorem is true.

Remark 4.2. The continuous solution $u(x,t,\alpha)$ of the boundary value problem (4.4) constructed in the rectangle $\Pi_{2\pi}$ according to Theorem 4.6 does not satisfy, in general, the periodicity condition, i. e. $u(x,0,\alpha)\neq u(x,2\pi,\alpha)$ for all $x,t,\alpha\in I\times\mathbb{R}\times\mathfrak{D}_\mu$. Indeed, consider the difference $u_1(x,t,\alpha)-u_2(x,t,\alpha)$, which determines, according to the third equations in (3.3) and (3.4), the functions $\tilde{u}_0(x,t,\alpha)$ and $\tilde{\tilde{u}}_0(x,t,\alpha)$ in the regions $\bar{\Delta}_1$ and $\bar{\Delta}_2$. By (4.41) and (4.42), we have

$$u_1(x,t,\alpha)-u_2(x,t,\alpha)=\mu(t+x)-\mu(t-x)-$$

$$-\int_0^x \tilde F[u,u_1,u_2](\eta,t+x-\eta)d\eta+\int_x^\pi \tilde F[u,u_1,u_2](\eta,t-x+\eta)d\eta-$$

$$-\int_0^\pi \tilde F[u,u_1,u_2](\eta,2\pi+t-x-\eta)d\eta \ \ \forall(x,t)\in\bar\Delta_1 \tag{4.65}$$

$$u_1(x,t,\alpha)-u_2(x,t,\alpha)=\mu(t+x)+\mu(t-x)-$$

$$-\int_0^x \tilde F[u,u_1,u_2](\eta,t+\eta-x)d\eta+\int_x^\pi \tilde F[u,u_1,u_2](\eta,t+x-\eta)d\eta-$$

$$-\int_0^\pi \tilde F[u,u_1,u_2](\eta,2\pi+t+x+\eta)d\eta \ \forall(x,t)\in\bar\Delta_2. \tag{4.66}$$

Assuming $\tilde F[u,u_1,u_2]$ to be a 2π-periodic function of t and using equalities (4.65) and (4.66) and the equation

$$u(x,t,\alpha)=\frac{1}{2}\int_\pi^x \{u_1(\xi,t,\alpha)-u_2(\xi,t,\alpha)\}\,d\xi,$$

we see that $u(x,0,\alpha)=u(x,2\pi,\alpha)$ if and only if the equality

$$\int_0^\pi \left[\tilde F[u,u_1,u_2](\eta,x-\eta)-\tilde F[u,u_1,u_2](\eta,\eta-x)\right]d\eta+$$

$$+\int_0^\pi \left[\tilde F[u,u_1,u_2](\eta,-x-\eta)-\tilde F[u,u_1,u_2](\eta,x+\eta)\right]d\eta=0. \tag{4.67}$$

holds. Hence, the periodicity condition $u(x,0,\alpha)=u(x,2\pi,\alpha)$ in the class of continuous 2π-periodic solutions of the boundary value problem (4.4) implies the additional condition (4.67) to be fulfilled. In particular, (3.29) is always fulfilled if the function $F[u,u_t,u_x](x,t)$ is even with respect to t, or the following equality

$$\int_0^\pi \{F[u,u_t,u_x](\eta,x-\eta)-F[u,u_t,u_x](\eta,x+\eta)\}\,d\eta=0, \tag{4.68}$$

holds if $F[u,u_t,u_x](x,t)$ is odd with respect to t.

The first case leads to the following assertions.

Theorem 4.8. *Let the condition of Lemma 4.2 and the following condition be fulfilled*

(1) for any function $u(x,t)\in C^1$ that is even with respect to t, the function $F[u,u_t,u_x](x,t)$ is even with respect to t.

Then, for any continuous even 2π-periodic function $|u_x(0,t)|=|\mu(t)|\le c-2M\pi$, the boundary value problem (4.4) has a 2π-periodic solution $\bar u\in L^\infty$ of the form

$$\bar u(x,t,\alpha)=\begin{cases} u_0(x,t,\alpha), & \text{if } (x,t)\in\bar\Delta\ \{(x,t):t=x,\ 0<x\le\pi\},\\ \tilde u_0(x,t,\alpha), & \text{if } (x,t)\in\bar\Delta_1\ \{(0,0)\},\\ \tilde{\tilde u}_0(x,t,\alpha), & \text{if } (x,t)\in\bar\Delta_2\ \{(x,t):t=2\pi-x,0\le x\le\pi\}.\end{cases} \tag{4.69}$$

P r o o f. Let the rectangle $\Pi_{2\pi} = \{0 \le x \le \pi,\ 0 \le t \le 2\pi\}$ be partitioned into the three pieces $\bar{\Delta}, \bar{\Delta}_1$ and $\bar{\Delta}_2$ by the lines $t = x$ and $t = 2\pi - x, 0 \le x \le \pi$ (see Fig. 2). Consider the systems of integral equations (4.12), (4.41) and (4.42) on the sets $\bar{\Delta}, \bar{\Delta}_1$ and $\bar{\Delta}_2$, respectively. For the indicated systems of equations, let $(u_i^m(x,t,\alpha), i = 1,2, u^m(x,t,\alpha)), (\tilde{u}_i^m(x,t,\alpha), i-1,2, \tilde{u}^m(x,t,\alpha))$ and $\tilde{\tilde{u}}_i^m(x,t,\alpha), i = 1,2, \tilde{\tilde{u}}^m(x,t,\alpha)$ be the sequences of functions constructed according to (4.9), (4.44) and (4.45).

By Theorem 4.8, for every continuous 2π-periodic function $|\mu(t)| \le c - 2M\pi$, the sequences $(u^m(x,t,\alpha)), (\tilde{u}^m(x,t,\alpha))$ and $(\tilde{\tilde{u}}^m(x,t,\alpha))$ converge uniformly to the continuous functions $u_0(x,t,\alpha), \tilde{u}(x,t,\alpha)$ and $\tilde{\tilde{u}}(x,t,\alpha)$. Then, almost everywhere on $\Pi_{2\pi}$, the sequence

$$\bar{u}^m(x,t,\alpha) = \begin{cases} u_m(x,t,\alpha), & \text{if } (x,t) \in \bar{\Delta}\ \{(x,t) : t = x,\ 0 < x \le \pi\}, \\ \tilde{u}_m(x,t,\alpha), & \text{if } (x,t) \in \bar{\Delta}_1\ \{(0,0)\}, \\ \tilde{\tilde{u}}_m(x,t,\alpha), & \text{if } (x,t) \in \bar{\Delta}_2\ \{(x,t) : t = 2\pi - x, 0 \le x \le \pi\}. \end{cases}$$

converges uniformly to the discontinuous bounded function (4.69), i. e. to the function belonging to L^∞.

Let the initial function $\mu(t)$ be even. Then one can readily see that the sequences $(u^m(x,t,\alpha)), (\tilde{u}^m(x,t,\alpha))$ and $(\tilde{\tilde{u}}^m(x,t,\alpha))$ are even functions of variable t.

Thus, by the condition (1) in Theorem 4.8 and Remark 4.2, we see that the sequence $\tilde{u}^m(x,t,\alpha)$ satisfies the periodicity condition $\tilde{u}^m(x,0,\alpha) = \tilde{u}^m(x,2\pi,\alpha))$ and hence, the limit function (3.31) is 2π-periodic with respect to t, as required to prove.

Theorem 4.9. *Suppose we are given a 2π-hyperbolic differential system of the first kind satisfying the condition of Lemma 4.1. Then for any continuous 2π-periodic function $|\mu(t)| \le c - 2M\pi$, the boundary problem (4.4) has a 2π-periodic solution $\bar{u} \in L^\infty$ of the form*

$$\bar{u}(x,t,\alpha) = \begin{cases} u_{\bar{\Delta}}(x,t,\alpha) \equiv u_0(x,t,\alpha) & (x,t) \in \bar{\Delta}\ \{(\pi,\pi), \\ 0 & (x,t) \in \Pi_{2\pi}\ \{(\bar{\Delta}\ \{(\pi,\pi)\}\}. \end{cases}$$

Remark 4.3. The problems that can be described by the equations with a small parameter

$$u_{tt} - u_{xx} = \varepsilon f(x,t,u,u_t,u_x),$$

$$u(o,t) = u(\pi,t) = 0, \quad u(x,t+2\pi) = u(x,t), \tag{4.70}$$

where the function $f(x,t,u,u_t,u_x)$ satisfies the conditions (4.6) and $f(x,t,0,0,0) \equiv 0$ in the region (4.5), are always 2π-periodic systems of the first kind, provided ε is sufficiently small. For example, the equation

$$u_{tt} - u_{xx} = \varepsilon(\alpha u + \beta u^3), \tag{4.71}$$

the right-hand side of which satisfies the condition (1) in Theorem 4.5, has infinitely many 2π-periodic solutions of the form (4.29) if ε is sufficiently small. In other words, for every odd function $\mu(t) = \alpha_k'' \sin kt, |\alpha_k''| c - \varepsilon M\pi$ and $f(x,t,u,u_t,u_x) \equiv \alpha u + \beta u^3$, the boundary problem (4.70) has a 2π-periodic solution of the form (4.29). Clearly, that solution depends on the real parameter α_k''.

Furthermore, in the class of smooth functions, the system of integral equations (4.12) is equivalent to the Cauchy problem (4.28) and $f(x,t,0,0,0) \not\equiv$ for $\mu(t) = 0$. Therefore, depending upon the choice of the initial approximation $u^0(x,t,\alpha)$, the solution $u(x,t,\alpha)$ of system (4.12) with $\mu(t) \equiv 0$ may satisfy or may not satisfy the equality $u(\pi,\pi,0) = 0$. We note that there is no additional condition for the equality $u(\pi,\pi,0) = 0$ to hold. However, for $\mu(t) \equiv 0$, there are 2π-periodic continuous solutions of the boundary value problem (4.70) and even infinitely many. These solutions are of a different kind than solutions of the form (4.29). This can easily be explained with the help of the following example.

Consider the boundary value problem, described by the equation

$$u_{tt} - u_{xx} = \varepsilon(\alpha u_t + \beta u_t^3), \; u(0,t) = u(\pi,t) = 0,$$
$$u(x,t+2\pi) = u(x,t). \tag{4.72}$$

Pick the function $u^0(x,t) = \alpha \cos t \cos x$ as the zero approximation. This function obviously belongs to the class $A_2 = \{u^0 : u^0(x,t) = u^0(x,t+2\pi) = u^0(\pi - x, t+\pi)\}$ (see §3.1, Chapter 3). Then the function $f(x,t,u^0,u_t^0,u_x^0) \equiv \alpha u_t^0 + \beta u_t^{03} = -(\alpha \sin t \cos x + \alpha^3 \sin^3 t \cos^3 x)$, which determines the first approximation $u^1(x,t)$ in the problem (4.72), belongs to the class A_2, too. As shown in the last chapter, there exist operators of other forms which are distinct from the operators (4.12) and transform A_2 into A_2. It is noteworthy that, in the space A_2, there exist also classical 2π-periodic solutions of the boundary value problem (4.72). By the results of Chapter 3, the boundary value problem (4.72) has infinitely many smooth 2π-periodic solutions, which is in perfect accordance with the result in [169], provided the parameter α in the initial approximation $u^0(x,t) = \alpha \cos t \cos x$ and the parameter $\varepsilon, \varepsilon << 1$ are properly chosen. This also holds true for the boundary value problem

$$u_{tt} - u_{xx} = \varepsilon(1 - u^2)u_t, \quad u(0,t) = u(\pi,t) = 0,$$
$$u(x,t+2\pi) = u(x,t). \tag{4.73}$$

However, if the right-hand side of equation (4.72) ((4.73)) is of the form $f(x,t,u,u_t,u_x) \equiv g(x,t) + \varepsilon(\alpha u_t + \beta u_t^3), g(x,t) \in A_2$, then, as shown in the last chapter, the solution is unique. Note that the solutions indicated above are the solutions of the second order wave equations of the first kind. In this case, the periodicity condition (4.68) also holds, i. e. in the space A_2, (4.68) holds automatically for $\mu(t) \equiv 0$. However, the question about the condition ensuring the general periodicity condition (4.67) hold is still unsettled. If the function $\mu(t)$ is chosen so that the solution of the boundary value problem (4.4) satisfies the condition $u(\pi,\pi,\alpha) = 0$, then, as shown above, we obtain a continuous solution in $\Pi_{2\pi}$ of the boundary value problem (4.4), which generally speaking does not satisfy the periodicity condition $u(x,0,\alpha) \neq u(x,2\pi,\alpha)$. Obviously, such solution is discontinuous at the lines $t = 2\pi k, k = 0, \pm 1, \pm 2, \ldots$. In this case, the following proposition is true.

Theorem 4.10. *Let the condition of Lemma 4.2 be satisfied. Then, for any continuous 2π-periodic function $|u_x(0,t)| = |\mu(t)| \leq c - 2M\pi$, the boundary value problem (4.4) has a 2π-periodic solution $\bar{\bar{u}} \in L^\infty$ of the form*

$$\bar{\bar{u}}(x,t,\alpha) =$$
$$= \begin{cases} u_0(x,t,\alpha), & \text{if} \;\; (x,t) \in \bar{\Delta} \; \{(x,t) : t = x, \; 0 < x \leq \pi\}, \\ \tilde{u}_0(x,t,\alpha), & \text{if} \;\; (x,t) \in \bar{\Delta}_1 \; \{(0,0)\}, \\ \tilde{\tilde{u}}_0(x,t,\alpha), & \text{if} \;\; (x,t) \in \bar{\Delta}_2 \; \{(x,t) : t = 2\pi - x, 0 < x < \pi, t = 2\pi\}, \\ \tilde{u}(x,t,\alpha), & \text{if} \;\; (x,t) \in \{t = 2\pi, \; 0 < x \leq 2\pi\}, \end{cases} \tag{4.74}$$

where $u_0(x,t,\alpha)$ is a solution of the system of integral equations (4.12) in the triangle $\bar{\Delta}$, $\tilde{u}_0(x,t,\alpha)$ is a solution of system (4.41) in the triangle $\bar{\Delta}_1$ and $\tilde{\tilde{u}}_0(x,t,\alpha)$ a solution of (4.42) in $\bar{\Delta}_2$ (see Fig. 2).

Remark 4.3. In the present section, we have investigated solutions of the second class 2π-systems in the space L^∞ only. The next section deals with another kind of the second order systems, which have continuous solutions.

4.4. The structure of continuous periodic solutions of systems

In the last section, we have investigated the existence of generalized 2π-periodic solutions of the second class 2π-systems. A particular feature in this investigation is the conditions determining a nontrivial 2π-periodic solution of the homogeneous 2π-periodic boundary value problem $u_{tt}-u_{xx}=0, u(0,t)=u(\pi,t)=0, u(x,t+2\pi)=u(x,t)$ corresponding to the boundary value problem (4.4). As shown above, the initial condition $u(0,t)=0, u_x(0,t)=\mu(t), \mu(t+2\pi)=\mu(t)$ plays an essential part. In the present section, we shall study another class of functions leading to 2π-periodic solutions of the second class 2π-systems.

Consider, for the period $T=2\pi$, the functional space

$$\tilde{A}_2=\{u: u(x,t)=u(x,t+2\pi)=-u(-x,t)=u(x+2\pi,t)\},$$

If the elements of this space $\tilde{A}_2$ satisfy the conditions

$$\int_0^{2\pi}[u(x+\theta-\tau,\tau)+u(x-\theta+\tau,\tau)]d\theta=0; \tag{4.75}$$

$$\Phi(x,t+2\pi,\tau)=\Phi(x,t,\tau+2\pi)=\Phi(x,t,\tau), \tag{4.76}$$

where

$$\Phi(x,t,\tau)=\int_\tau^t[u(x+\theta-\tau,\tau)+u(x-\theta+\tau,\tau)]d\theta, \tag{4.77}$$

then this functional space is denoted by $\tilde{A}_{22}$.

Lemma 4.3. *Every function $u(x,t)\in\tilde{A}_2$ can be expanded into the uniformly converging Fourier series*

$$u(x,t)=\sum_{k=1}^{\infty}u_k(t)\sin kx, \tag{4.78}$$

and satisfies the conditions (4.75) and (4.76), that is, $u(x,t)\in\tilde{A}_{22}$.

P r o o f. Indeed, by the condition of the lemma and expression (4.78), we have

$$\int_0^{2\pi}\sum_{k=1}^{\infty}u_k(\tau)\{\sin k(x+\theta-\tau)+\sin k(x-\theta+\tau)\}\,d\theta=$$

$$=\sum_{k=1}^{\infty}u_k(\tau)\left\{\frac{\cos k(x-\theta+\tau)}{k}-\frac{\cos k(x+\theta-\tau)}{k}\right\}\Big|_0^{2\pi}\equiv 0.$$

The condition (4.75) is thus fulfilled. Next, by (4.77) and (4.78), we have

$$\Phi(x,t,\tau)=\int_\tau^t\sum_{k=1}^{\infty}u_k(\tau)\{\sin k(x+\theta-\tau)-\sin k(x-\theta+\tau)\}d\theta=$$

$$=\sum_{k=1}^{\infty}u_k(\tau)\left\{\frac{\cos k(x-t+\tau)}{k}-\frac{\cos k(x+t-\tau)}{k}\right\}.$$

Whence it follows that (4.76) is fulfilled and the proof is complete.

Now consider the problem described by the equations

$$u_{tt}-u_{xx}=\varepsilon f(x,t,u,u_t,u_x),$$

$$u(0,t)=u(\pi,t)=0,\quad u(x,t,+2\pi)=u(x,t). \tag{4.79}$$

Lemma 4.4. *Let* $u(x,t) \in C^1 \cap \tilde{A}_2$ *for every function* $f[u, u_t, u_x](x,t) = f(x,t,u(x,t),u_t(x,t),u_x(x,t)) \in \tilde{A}_{22}$. *Then the operator of the form*

$$(\tilde{P}F)(x,t) = \int_0^t \Big\{ \int_\tau^t [F[u,u_t,u_x](x+\theta-\tau,\tau)+$$

$$+F[u,u_t,u_x](x-\theta+\tau,\tau)]\,d\theta-$$

$$-\frac{1}{2\pi}\int_0^{2\pi} d\tau \int_\tau^t [F[u,u_t,u_x](x+\theta-\tau,\tau)+F[u,u_t,u_x](x-\theta+$$

$$+\tau,\tau)]\,d\theta\Big\}\,d\tau \equiv \int_0^{2\pi} Q_1(\tau)d\tau \int_\tau^t \sum_{i=0}^{1} F[u,u_t,u_x](x+(-1)^i(\theta-\tau),\tau)d\theta, \qquad (4.80)$$

where

$$Q_1(\tau) = \begin{cases} 1-\frac{1}{2\pi}, & 0 \le \tau \le t, \\ -\frac{t}{2\pi}, & t < \tau \le 2\pi, \end{cases} \qquad (4.81)$$

transforms the space $\tilde{A}_2$ into itself.

P r o o f. By virtue of (4.78) and the condition of Lemma 4.4, we have

$$(\tilde{P}F)(x,t+2\pi) = \int_0^{t+2\pi} \Big\{ \int_\tau^{t+2\pi} \sum_{i=0}^{1} F[u,u_t,u_x](x+(-1)^i(\theta-\tau),\tau)d\theta-$$

$$-\frac{1}{2\pi}\int_s^{t+2\pi} ds \int_s^{t+2\pi} \sum_{i=0}^{1} F[u,u_t,u_x](x+(-1)^i(\theta-s),\tau)d\theta \Big\}\,d\tau =$$

$$= \int_0^{t+2\pi} \Big\{ \int_\tau^{1} \sum_{i=1}^{1} F[u,u_t,u_x](x+(-1)^i(\theta-\tau),\tau)d\theta-$$

$$-\frac{1}{2\pi}\int_0^{t+2\pi} ds \int_s^{t} \sum_{i=1}^{1} F[u,u_t,u_x](x+(-1)^i(\theta-s),\tau)d\theta \Big\}\,d\tau+$$

$$+\int_0^{t+2\pi} \Big\{ \int_t^{t+2\pi} \sum_{i=0}^{1} F[u,u_t,u_x](x+(-1)^i(\theta-\tau),\tau)d\theta-$$

$$-\frac{1}{2\pi}\int_0^{2\pi}\int_t^{t+2\pi} \sum_{i=0}^{1} F[u,u_t,u_x](x+(-1)^i(\theta-s),\tau)d\theta \Big\}\,d\tau =$$

$$= (\tilde{P}F)(x,t) + \int_t^{t+2\pi} d\tau \int_\tau^t \sum_{i=0}^{1} F[u,u_t,u_x](x+(-1)^i(\theta-\tau),\tau)d\theta-$$

$$-\int_0^{2\pi} ds \int_s^t \sum_{i=0}^{1} F[u,u_t,u_x](x+(-1)^i(\theta-s),s)d\theta+$$

$$+\int_0^{t+2\pi} d\tau \int_0^{2\pi} \sum_{i=0}^{1} F[u,u_t,u_x](x+(-1)^i(\theta-\tau),\tau)d\theta-$$

$$-\frac{t+2\pi}{2\pi}\int_0^{2\pi} ds \int_0^{2\pi} \sum_{i=0}^{1} F[u,u_t,u_x](x+(-1)^i(\theta-s),s)d\theta = (\tilde{P}F)(x,t).$$

Next, immediate check shows that $(PF)(0,t) = (PF)(\pi,t) = 0$, $(PF)(x+2\pi,t) = (PF)(\pi,t)$ and $(PF)(-x,t) = -(PF)(\pi,t)$, which proves the lemma.

Theorem 4.11. *Let the function* $v(x,t) \in C^2 \cup \tilde{A}_2$ *for every function* $F[v,v_t,v_x](x,t) = f(x,t,v(x,t),v_t(x,t)) \in C^1 \cup \tilde{A}_{22}$. *Then the function* $v = \tilde{P}F$, *where*

$$(\tilde{P}F)(x,t) = z(x,t) + \frac{\varepsilon}{2}\int_0^{2\pi} Q_1(\tau)d\tau \int_\tau^t \sum_{i=0}^{1} F[v,v_t,v_x](x+$$

$$+(-1^i)(\theta-\tau),\tau)d\theta, \quad z(x,t) = p(t+x) - p(t-x) \equiv$$

$$\equiv \int_{k=1}^{\infty} (a'_k \cos kt + a''_k \sin kt)\sin kx, \quad p(z) \in C^2, \quad p(z+2\pi) = p(z), \tag{4.82}$$

is a 2π*-periodic solution of the boundary value problem*

$$v_{tt} - v_{xx} = \varepsilon f(x,t,v,v_t,v_x)-$$

$$\frac{\varepsilon}{2\pi}\int_0^{2\pi} \sum_{i=0}^{1} F[v,v_t,v_x](x+(-1^i)(t-s),s)ds,$$

$$v(0,t) = v(\pi,t) = 0, \quad v(x,t+2\pi) = v(x,t). \tag{4.83}$$

P r o o f. The operator (4.82) can be rewritten as

$$v(x,t) \equiv (\tilde{P}F) =$$

$$= z(x,t) + \frac{\varepsilon}{2}\int_0^t \left\{ \int_{x-t+\tau}^{x+t-\tau} F[v,v_t,v_x](\eta,\tau)d\eta - \right.$$

$$\left. -\frac{1}{2\pi}ds \int_{x-t+s}^{x+t-s} F[v,v_t,v_x](\eta,s)d\eta \right\} d\tau.$$

Successively differentiating, we obtain

$$v_x = z_x + \frac{\varepsilon}{2}\int_0^t \{[Fv,v_t,v_x(x+t-\tau,\tau)-$$

$$-F[v,v_t,v_x](x-t+\tau,\tau)] - \frac{\varepsilon}{2\pi}\int_0^{2\pi} [F[v,v_t,v_x](x+t-s,s)-$$

$$-F[v,v_t,v_x](x-t+s,s)]\,ds\}\,d\tau; \tag{4.84}$$

$$v_{xx} = z_{xx} + \frac{\varepsilon}{2}\int\limits_0^t \left\{ \frac{\partial F[v, v_t, v_x](x+t-\tau, \tau)}{\partial(x+t-\tau)} - \right.$$

$$- \frac{\partial F[v, v_t, v_x](x-t+\tau)}{\partial(x-t+\tau)} - \frac{1}{2\pi}\int\limits_0^{2\pi} \left[\frac{\partial F[v, v_t, v_x](x+t-s, s)}{\partial(x+t-s)} - \right.$$

$$\left. \left. - \frac{\partial F[v, v_t, v_x](x-t+s, s)}{\partial(x+t-s)} \right] ds \right\} d\tau; \tag{4.85}$$

$$v_t = z_t - \frac{\varepsilon}{4\pi}\int\limits_0^{2\pi} ds \int\limits_{x-t+s}^{x+t-s} F[v, v_t, v_x](\eta, s)d\eta +$$

$$+ \frac{\varepsilon}{2}\int\limits_0^t \{[Fv, v_t, v_x(x+t-\tau, \tau) +$$

$$+ F[v, v_t, v_x](x-t+\tau, \tau)] - \frac{\varepsilon}{2\pi}\int\limits_0^{2\pi} [F[v, v_t, v_x](x+t-s, s) +$$

$$+ F[v, v_t, v_x](x-t+s, s)]\, ds\}\, d\tau; \tag{4.86}$$

$$v_{tt} = z_{tt} + \frac{\varepsilon}{4\pi}\int\limits_0^{2\pi} [Fv, v_t, v_x(x+t-s, s) +$$

$$+ F[v, v_t, v_x](x-t+s, s)]\, ds + \varepsilon F[v, v_t, v_x](x, t) -$$

$$- \frac{\varepsilon}{4\pi}\int\limits_0^{2\pi} [F[v, v_t, v_x](x+t-s, s) + F[v, v_t, v_x](x-t+s, s)]\, ds +$$

$$+ \frac{\varepsilon}{2}\int\limits_0^t \left\{ \frac{\partial F[v, v_t, v_x](x+t-\tau, \tau)}{\partial(x+t-\tau)} - \frac{\partial F[v, v_t, v_x](x-t+\tau)}{\partial(x-t+\tau)} - \right.$$

$$\left. - \frac{1}{2\pi}\int\limits_0^{2\pi} \left[\frac{\partial F[v, v_t, v_x](x+t-s, s)}{\partial(x+t-s)} - \frac{\partial F[v, v_t, v_x](x-t+s, s)}{\partial(x+t-s)} \right] ds \right\} d\tau. \tag{4.87}$$

Combining the notation $F[v, v_t, v_x](x, t) = f(x, t, v(x, t), v_t(x, t), v_x(x, t))$ and expressions (4.85) and (4.87) yields

$$v_{tt} - v_{xx} = \varepsilon f(x, t, v, v_t, v_x) -$$

$$- \frac{\varepsilon}{2\pi}\int\limits_0^{2\pi} [F[v, v_t, v_x](x+t-s, s) + F[v, v_t, v_x](x-t+s, s)]\, ds.$$

Next, by immediate check, we see that $v(0, t) = v(\pi, t) = 0, v(x, t + 2\pi) = v(x, t)$. The proof is complete.

Suppose that the function $f(x, t, u, u_t, u_x)$ is defined in the region

$$\mathbb{R} \times \mathbb{R} \times D \equiv I^1 \times I^2 \times I^3; x \in \mathbb{R},\ t \in \mathbb{R},\ u \in [-b, b] = I^1,$$

$$u_t \in [-c, c] = I^2, \quad u_x \in [-d, d] = I^3, \tag{4.88}$$

is continuous jointly with respect to variables x, t, u, u_t, u_x and satisfies the condition

$$|f(x,t,u,u_t,u_x)| \leq M, \ |f(x,t,u'',u''_t,u''_x) - f(x,t,u',u'_t,u'_x)| \leq$$
$$\leq K\{|u''-u'| + |u''_t - u'_t| + |u''_x - u'_x|\}. \tag{4.89}$$

Expressions (4.82), (4.84) and (4.86) allows us to write down the following system of integral equations

$$v(x,t) = z(x,t) + \frac{\varepsilon}{2}\int_0^{2\pi} Q_1(\tau)d\tau \int_\tau^t \sum_{i=0}^1 F[v, v_t, v_x](x + (-1)^i(\theta - \tau), \tau)d\theta,$$

$$v_t(x,t) = z_t(x,t) + \frac{\varepsilon}{2}\int_0^{2\pi} Q_1(\tau)\sum_{i=0}^1 F[v, v_t, v_x](x + (-1)^i(\theta - \tau), \tau)d\tau -$$
$$-\frac{\varepsilon}{4\pi}\int_0^{2\pi} ds \int_{x-t+s}^{x+t-s} F[v, v_t, v_x](\eta, s)d\eta, \tag{4.90}$$

$$v_x(x,t) =$$
$$z_x(x,t) + \frac{\varepsilon}{2}\int_0^{2\pi} Q_1(\tau)\sum_{i=0}^1 (-1^i)F[v, v_t, v_x](x + (-1)^i(t - \tau), \tau)d\tau,$$

where the function Q_1 is given by (4.81).

By using the method of successive approximations, we arrive at the following

Theorem 4.12. *Let the following condition be fulfilled*

(1) the function $f(x,t,u,u_t,u_x)$ is defined in the region (4.88), is continuous with respect to x, t and 2π-periodic in t and satisfies inequalities in (4.89);

(2) for every function $u \in \tilde{A}_2$, the function $F[u, u_t, u_x](x,t) = f(x,t,u(x,t),u_t(z,t),u_x(x,t))$ belongs to the space $\tilde{A}_{22}$.

Then for every function $z \in \tilde{A}_2$ such that

$$|z(x,t)| < b - 2\varepsilon M\pi^2, |z_t(x,t)| < c - \frac{5}{2}\varepsilon M\pi, \tag{4.91}$$
$$|z_x(x,t)| < d - 2\varepsilon M\pi,$$

and sufficiently small $\varepsilon < \varepsilon_0$, the sequence of 2π-periodic in t functions

$$v^0(x,t) = z(x,t), \quad v_t^0(x,t) = z(x,t), \quad v_x^0(x,t) = z_x(x,t),$$

$$v^{m+1}(x,t) = z(x,t) +$$
$$+\frac{\varepsilon}{2}\int_0^{2\pi} Q_1(\tau)d\tau \int_\tau^t \sum_{i=0}^1 F[v^m, v_t^m, v_x^m](x + (-1)^i(\theta - \tau), \tau)d\theta,$$

$$v_t^{m+1}(x,t) = z_t(x,t) +$$
$$+\frac{\varepsilon}{2}\int_0^{2\pi} Q_1(\tau)\int_\tau^t \sum_{i=0}^1 F[v^m, v_t^m, v_x^m](x + (-1)^i(t - \tau), \tau)d\tau -$$

$$-\frac{\varepsilon}{4\pi}\int\limits_{x-t+s}^{x+t-s} F[v^m, v_t^m; v_x^m](\eta, s)d\eta, \tag{4.92}$$

$$v_x^{m+1}(x,t) = z_x(x,t)+$$

$$+\frac{\varepsilon}{2}\int\limits_0^{2\pi} Q_1(\tau)\sum_{i=0}^{1}(-1)^i F[v, v_t, v_x](x+(-1)^i(t-\tau),\tau)d\tau$$

converges, as $m \to \infty$, uniformly with respect to ε, $\varepsilon < \varepsilon_0$ to functions $\tilde{v}(x,t)$ and $\tilde{v}_t(x,t), \tilde{v}_x(x,t)$, which are 2π-periodic in t and satisfy the system of integral equations (4.90), and hence, the function $\tilde{v}(x,t)$ is a smooth solution of the boundary value problem (4.83).

$$u_{tt} - u_{xx} = \varepsilon f(x,t,u,u_t,u_x) - \mu(x,t),$$

$$u(0,t) = u(\pi,t) = 0, \quad u(x,t+2\pi) = u(x,t). \tag{4.93}$$

Definition 4.3. A Δ-function with respect to an initial function $z(x,t)$ for the boundary value problem (4.79) is a function $\mu(x,t) \in \tilde{A}_{22}$ such that the corresponding solution of the boundary value problem (4.93) assuming value $u(x,0) = z(x,0)$ at $t=0$ is 2π-periodic, provided there is only one such function $\mu(x,t)$. Evidently, the initial value $z(x,t)$, for which the Δ-function vanishes, determines the initial value of 2π-periodic solutions of the boundary value problem (4.79).

Definition 4.4. The second order equation (4.79) is called a 2π-hyperbolic second class differential system of the third kind in the region (4.88) if the constants b, c, d, M and parameter ε relate by the inequalities $be2\varepsilon M\pi^2$, $ce\frac{5}{2}\varepsilon M\pi$ and $de2\varepsilon M\pi$ and the conditions in Theorem 4.12 hold.

By using Theorem 4.12 one can easily answer a question about the existence of a Δ-function for the points of the set $D_f \subset D$. Regarding second class 2π-systems of the third kind, the following propositions hold.

Theorem 4.13. *For any initial values $(z, z_t, z_x) \in D_f$, there exists a Δ-function*

$$\Delta(x,t) = \frac{\varepsilon}{4\pi}\int\limits_0^{2\pi}[F[v, v_t, v_x](x+t-\tau,\tau)+$$

$$+F[v, v_t, v_x](x-t+\tau,\tau)]d\tau. \tag{4.94}$$

provided ε is sufficiently small.

Theorem 4.14. *Suppose that the equation in (4.79) is a 2π-system of the third kind. The solution $u = u(x,t)$ of (4.79) satisfying $(u(x,0), u_t(x,0), u_x(x,0)) \in D_f$ is periodic if and only if the Δ-function determined by the initial value $z(x,t) = u(x,0)$ is equal to zero. Furthermore, the equality $u(x,t) = \tilde{v}(x,t)$ holds, where $\tilde{v}(x,t)$ is the limit of the sequence of 2π-periodic functions (4.92).*

The existence of periodic solutions of second class 2π-systems of the third kind is equivalent to the existence of a function $z(x,t)$ for which the Δ-function is equal to zero. The construction of these solutions is equivalent to the construction of approximations $v^m(x,t)$.

From the definition of the function $z(x,t)$ ((4.82)) it follows that

$$z(x,t,a) = \sum_{k=1}^{\infty}(a'_k \cos kt + a''_k \sin kt)\sin kx, \quad a \in \mathcal{D}_z, \tag{4.95}$$

where $a = (a'_1, a''_1, a'_2, a''_2, \ldots)$ stands for the collection of the Fourier coefficients determining the function $z(x,t)$ such that

$$|z(x,t)| < b - 2\varepsilon M\pi^2, \quad |z_t(x,t)| < c - \frac{5}{2}\varepsilon M\pi,$$

$$|z_x(x,t)| < d - 2\varepsilon M\pi, \quad |\varepsilon| \le \varepsilon_0, \tag{4.96}$$

and $\mathcal{D}_z = \{a\}$ stands for the set of the Fourier coefficients satisfying (4.96). Then the solution $\tilde{v}(x,t)$ of the system of integral equations (4.90) and the function (4.94) can be written as

$$\tilde{v} = \tilde{v}(x,t,a), \tag{4.97}$$

$$\Delta(x,t,a) = \frac{\varepsilon}{2\pi}\int_0^{2\pi} [F[\tilde{v},\tilde{v}_t,\tilde{v}_x](x+t-\tau,\tau,a)+$$

$$+F[\tilde{v},\tilde{v}_t,\tilde{v}_x](x-t+\tau,\tau,a)d\tau.$$

Clearly, the points a for which $\Delta(x,t,a) = 0$ are singular points of the mapping

$$\Delta : \mathbb{R}\times\mathbb{R}\times D_f \to \mathbb{R}^2,$$

$$\Delta(x,t,a) = \varepsilon\overline{\sum_{i=0}^{1} F[\tilde{v},\tilde{v}_t,\tilde{v}_x](x+(-1)^i(t-\tau),\tau,a)}. \tag{4.98}$$

The mapping (4.24) can be computed only approximately by computing, say, the following functions

$$\Delta_m(x,t,a) = \varepsilon\overline{\sum_{i=0}^{1} F[\tilde{v},\tilde{v}_t,\tilde{v}_x](x+(-1)^i(t-\tau),\tau,a)}. \tag{4.99}$$

In this connection, a problem arises how to conclude about the zeros of the mapping $\Delta(x,t,a)$ and hence about the periodic solutions of second class 2π-systems of the third kind starting from the mapping (4.99). For the first time, this problem was solved for the mapping (4.98) in [175] under the condition that the function F depends upon the variables x, t, u only.

On the other hand, starting from the function $z(x,t,a)$ (4.95) and assuming $m = 0$ in (4.99), we obtain

$$\Delta_0(x,t,a) = \varepsilon\overline{\sum_{i=0}^{1} F[z,z_t,z_x](x+(-1)^i(t-\tau),\tau,a)}. \tag{4.100}$$

Suppose now that the function $F[z,z_t,z_x](x,t) = f(x,t,z(x,t),z_t(x,t),z_x(x,t))$ can be expanded into the uniformly convergent Fourier series

$$F[z,z_t,z_x](x,t,a) = \sum_{k=1}^{\infty} f_k^0(t,a)\sin kx. \tag{4.101}$$

Combining (4.100) and (4.101), we get

$$\Delta_0(x,t,a) = \frac{\varepsilon}{2\pi}\int_0^{2\pi}\sum_{k=1}^{\infty} f_k^0(\tau,a)[\sin k(x+t-\tau)+\sin k(x-t+\tau)]d\tau =$$

$$= \varepsilon \sum_{k=1}^{\infty} \sin kx \frac{1}{2\pi} \int_0^{2\pi} f_k^0(\tau, a) \cos k(t - \tau) d\tau.$$

Therefore $\Delta_0(x, t, a) = 0$ holds if and only if the following conditions

$$\Delta_0^{1k} \equiv \frac{1}{2\pi} \int_0^{2\pi} f_k^0(\tau, a) \cos k\tau d\tau = 0.$$

$$\Delta_0^{2k} \equiv \frac{1}{2\pi} \int_0^{2\pi} f_k^0(\tau, a) \sin k\tau d\tau = 0. \tag{4.102}$$

are satisfied. By using Theorem 2.6 of Chapter 2 and Theorem 4.12, we obtain the following

Theorem 4.15. *Let the conditions of Theorem 4.12 be satisfied for $|\varepsilon| < \varepsilon_0$ and let the following conditions be also satisfied*

(1) the mapping

$$\Delta_0^{1k} : \Delta_0^{1k} = \overline{f_k^0(\tau, a) \cos k\tau},$$
$$\Delta_0^{2k} : \Delta_0^{2k} = \overline{f_k^0(\tau, a) \sin k\tau} \tag{4.103}$$

has a singular point $a = a^0$:

$$\Delta_0^{1k}(a^0) = 0, \quad \Delta_0^{2k}(a^0) = 0;$$

(2) there is a closed bounded region $\mathcal{D}_0 \subset \mathcal{D}_z$ containing the point a^0 such that the operators Δ_0^1 and Δ_0^2 map $\mathcal{D}_0$ topologically onto $\Delta_0^1\mathcal{D}_0$ and $\Delta_0^2\mathcal{D}_0$, respectively.

Then, for every value of the parameter ε, $|\varepsilon| < \varepsilon_0$ for which the condition

$$|z(x, t, a^0)| < b - 2\varepsilon M\pi^2, \quad |z_t(x, t, a^0)| <$$

$$< c - \frac{5}{2}\varepsilon M\pi, \quad |z_x(x, t, a^0)| < d - 2\varepsilon M\pi, \tag{4.104}$$

holds, the boundary value problem (4.79) has a 2π-periodic solution.

As an example of application of Theorem 4.5 we first consider the 2π-periodic boundary value problem for the Van der Pol autonomous equation

$$u_{tt} - u_{xx} = \varepsilon(1 - u^2)u_t,$$

$$u(0, t) = u(\pi, t) = 0, \quad u(x, t + 2\pi) = u(x, t). \tag{4.105}$$

Let the function

$$z(x, t, a) = a \cos t \sin x. \tag{4.106}$$

be initial.

Now, computing the value of the function (4.101) and using the right-hand side of the equation in (4.105), we have

$$F[z, z_t, z_x](x, t, a) = (-1)(1 - a^2 \cos^2 t \sin^2 x) a \sin t \sin x =$$

$$= -a \sin t \sin x + a^3 \cos^2 t \sin t \left(\frac{3}{4} \sin x - \frac{1}{4} \sin 3x\right) =$$

$$- \left(\frac{3}{4} a^3 \cos^2 t \sin t - a \sin t\right) \sin x - \frac{1}{4} a^3 \cos^2 t \sin t \sin 3x.$$

Thus $f_1^0 = \frac{3}{4}a^3\cos^2 t \sin t - a\sin t$, $f_2^0 = 0$, $f_3^0 = -\frac{1}{4}a^3\cos^2 t\sin t$.
Next, computing the values of the functions $f_k^0 \sin kt$ and $f_k^0\cos kt$, we have

$$f_1^0 \sin t = \frac{3}{4}a^3\cos^2 t\sin^2 t - a\sin^2 t \equiv \frac{3}{16}a^3\sin^2 2t - a\sin^2 t,$$

$$f_1^0\cos t = \frac{3}{4}\cos^2 t\sin t - a\sin t\cos t \equiv$$

$$\equiv \frac{3}{8}a^3\cos^2 t\sin 2t - \frac{1}{2}a\sin 2t, \tag{4.107}$$

$$f_3^0 \sin 3t = -\frac{1}{4}a^3 cos^2 t\sin t\sin 3t \equiv$$

$$\equiv -\frac{1}{32}a^3(1+\cos 2t - \cos 4t - \cos 6t),$$

$$f_3^0\cos 3t = -\frac{1}{4}a^3\cos^2 t\sin t\cos 3t \equiv$$

$$\equiv -\frac{1}{32}a^3(\sin 2t - \sin 4t - \sin 6t),$$

$$f_k^0\sin kt = 0,\quad f_k^0\cos kt = 0,\quad k = 2,4,5,\ldots$$

Whence, computing the mean values $\overline{f_k\cos kt}$ and $\overline{f_k\sin kt}$, we obtain the mapping (4.103)

$$\Delta_0^{11} : \Delta_0^{11} = \frac{3}{32}a^3 - \frac{1}{2}a,\quad \Delta_0^{13} : \Delta_0^{13} = -\frac{1}{32}a^3, \tag{4.108}$$

$$\Delta_0^{ik} : \Delta_0^{ik} = 0,\quad i = 1,2;\quad k = 2,3,\ldots$$

Therefore the singular point of the mapping (4.108) can be found from the system

$$\frac{3}{32}a^3 - \frac{1}{2}a = 0,\quad -\frac{1}{32}a^3 = 0. \tag{4.109}$$

This system has only zero solution $a = 0$. In this case, the initial function (4.106) is identical zero and the Van der Pol equation (4.105) has only zero 2π-periodic solution, $u \equiv 0$. Hence, in the functional class $\tilde{A}_{22}$, there is no nontrivial 2π-periodic solutions of the boundary value problem (4.105).

The above investigation enables us to determine the form of the right-hand side of equation in (4.79) for which the mapping (4.103) has a non-zero isolated singular point. To do this, it suffices to consider, for example, the generalized Van der Pol equation

$$u_{tt} - u_{xx} = \varepsilon g(x,t) + \varepsilon(1-u^2)u_t \equiv \varepsilon f(x,t,u,u_t,u_x),$$

$$u(0,t) = u(\pi,t) = 0,\quad u(x,t+2\pi) = u(x,t), \tag{4.110}$$

where

$$g(x,t) = \sum_{k=1}^{\infty} g_k(t)\sin kx. \tag{4.111}$$

If $\overline{g_3(t)\sin 3t} = \alpha \neq 0$, $g_k(t)\sin kt = 0$, $k = 1,2,4,\ldots$, $g_k(t)\cos kt = 0$, $k = 1,2,\ldots$, then the singular point of the mapping (4.103) can be found from the system

$$\frac{3}{32}a^3 - \frac{1}{2}a = 0,\quad \alpha - \frac{1}{32}a^3 = 0. \tag{4.112}$$

Solving (4.112), we obtain that equation (4.110) has a 2π-periodic solution only for $a = \frac{4\sqrt{3}}{3}$ and $\alpha = \frac{1}{32}a^3$.

Remark 4.4. Besides 2π-periodic solutions, the boundary value problem (4.110) may have the solutions of the period $T = 2\pi(1-\varepsilon\nu)$. These solutions can be studied in the same manner as we studied the solutions (of the same period) of wave autonomous differential equations of second order (see §2.3, Chapter 2). Changing variable in equation (4.79)

$$t = \tau(1-\varepsilon\nu). \tag{4.113}$$

allows us to rewrite equation (4.79) as

$$\frac{\partial^2 u}{\partial \tau^2} - (1-\varepsilon\nu)^2\frac{\partial^2 u}{\partial x^2} = \varepsilon(1-\varepsilon\nu)^2 f(x, \tau(1-\varepsilon\nu), u, u_t, u_x) \tag{4.114}$$

or

$$\frac{\partial^2 u}{\partial \tau^2} - \frac{\partial^2 u}{\partial x^2} = \varepsilon(2\nu - \varepsilon\nu^2)\frac{\partial^2 u}{\partial \tau^2} +$$
$$+\varepsilon(1-\varepsilon\nu)^2 f(x, \tau(1-\varepsilon\nu), u, u_t, u_x).$$

A solution 2π-periodic in variable τ of equation (4.114) is sought in the form

$$u(x,\tau) = \sum_{k=1}^{\infty} u_k(\tau)\sin kx.$$

Then the Fourier coefficients $u_k(\tau)$ satisfy the countable system of second order ordinary differential equations

$$\frac{\partial^2 u_k}{\partial \tau^2} + k^2 u_k = \varepsilon(2\nu - \varepsilon\nu^2)k^2 u_k + \varepsilon(1-\varepsilon\nu)^2 f_k(\tau, u_1, u_2, \ldots), \tag{4.115}$$

$$k = 1, 2, \ldots,$$

where $f_k(\tau, u_1, u_2, \ldots)$ are the Fourier coefficients of the expansion of the function in the eigenfunctions $\sin kx$.

Changing variable in system (4.115)

$$u_k = v_k \cos k\tau + w_k \sin k\tau, \quad \frac{du_k}{d\tau} = -kv_k \sin k\tau + kw_k Qsk\tau.$$

and making algebraic transformations, we arrive at the following first order countable system

$$\frac{dv_k}{d\tau} = -\varepsilon(2\nu - \varepsilon\nu^2)k(v_k\cos kt + w_k \sin kt)\sin k\tau -$$
$$-\frac{\varepsilon(1-\varepsilon\nu^2)}{k} f_k(\tau, v_1\cos\tau + w_1\sin\tau, \ldots)\sin k\tau,$$
$$\frac{dw_k}{d\tau} = \varepsilon(2\nu - \varepsilon\nu^2)k(v_k\cos kt + w_k \sin kt)\cos k\tau +$$
$$-\frac{\varepsilon(1-\varepsilon\nu^2)}{k} f_k(\tau, v_1\cos\tau + w_1\sin\tau, \ldots)\cos k\tau.$$

The averaged system corresponding to system (4.116) has the form

$$\frac{d\xi_k}{d\tau} = -\varepsilon(2\nu - \varepsilon\nu^2)\frac{k\eta_k}{2} - \frac{\varepsilon(1-\varepsilon\nu)^2}{k}\overline{f_k \sin k\tau},$$

$$\frac{d\eta_k}{d\tau} = \varepsilon(2\nu - \varepsilon\nu^2)\frac{k\xi_k}{2} + \frac{\varepsilon(1-\varepsilon\nu)^2}{k}\overline{f_k \cos k\tau}. \tag{4.116}$$

Now, assuming $\eta_k = 0, \quad k = 1, 2, \ldots, \quad \xi_1 \neq 0, \quad \xi_k = 0, \quad k = 2, 3, \ldots$, in (4.117), we arrive at the system

$$\begin{gathered} \overline{\varepsilon f_1 \sin \tau} = 0, \\ \varepsilon(2\nu - \varepsilon\nu^2)\xi_1 + \varepsilon(1-\varepsilon\nu)^2\overline{f_1 \cos\tau} = 0, \\ \overline{f_k \sin k\tau} = 0, \quad \overline{f_k \cos k\tau} = 0 \end{gathered} \tag{4.117}$$

to determine a "quasistatic" solution ξ_1^0 of the averaged system (4.117) and a value of ν, if such a solution of this system exists.

Thus, the first equation $\overline{f_1 \sin\tau} = 0$ in (4.118) determines a "quasistatic" solution $\xi_1 \neq 0$ of system (4.117), the second equation determines (for $\overline{f_1 \cos\tau} \not\equiv 0$,) the values of the parameter ν, and the rest of the equations in (4.118) detemine the form of the right-hand side of equation (4.79) for which $2\pi(1-\varepsilon\nu)$-periodic solutions of the boundary value problem (4.79) exist.

CHAPTER 5

PERIODIC SOLUTIONS OF THE SECOND ORDER INTEGRO-DIFFERENTIAL EQUATIONS OF HYPERBOLIC TYPE

In this chapter we continue our investigation of periodic boundary value problems and consider the second order wave integro-differential equations of hyperbolic type. Studying the existence of classical and continuous periodic solutions of these equations, we single out and investigate T-systems of the first and second class. The latter leads us to -periodic systems of the firts and second kind.

5.1. Some preliminaries

The results obtained in §2.1, Chapter 2 concerning properties of the operators P and P_1 can be exploit in the investigation of periodic solutions of both autonomous and nonautonomous nonlinear wave equations of second order of the following form $\ddot{x} + \omega^2 x = f(t, x, \dot{x}, A(x, \dot{x}))$. We shall give several examples illustrating practical use of the operator P.

Consider the second order ordinary nonlinear integro-diffrential equation

$$\ddot{x} + \omega^2 x = f\left(t, x, \dot{x}, \int\limits_0^{h(t)} \varphi(t, s, x(s), \dot{x}(s))ds\right). \tag{5.1}$$

Suppose that the functions $f(t, x, \dot{x}, v)$, where

$$v(t) = \int\limits_0^{h(t)} \varphi(t, s, x(s), \dot{x}(s))ds;$$

$\varphi(t, s, x(s), \dot{x}(s))$ and $h(t)$ are defined in the region

$$t \in \mathbb{R}_t, \;\; s \in \mathbb{R}_s, \;\; x \in [-a, a] = I^1,$$

$$x \in [-b, b] = I^2, \;\; v \in \mathbb{R}^1_v, \tag{5.2}$$

are continuous jointly with respect to variables $t, s, x, \dot{x}, v$, T-periodic in t and satisfy the inequalities

$$|f(t, x, \dot{x}, v)| \leq M, \;\; |f(t, x'', \dot{x}'', v'') - f(t, x', \dot{x}', v')| \leq$$

$$\leq K_1|x'' - x'| + K_2|\dot{x}'' - \dot{x}'| + K_3|v'' - v'|; \tag{5.3}$$

$$|\varphi(t, s, x, \dot{x})| \leq M_1, \;\; |\varphi(t, s, x'', \dot{x}'') - \varphi(t, s, x', \dot{x}')| \leq$$

$$\leq L_1|x'' - x'| + L_2|\dot{x}'' - \dot{x}'|; \tag{5.4}$$

$$h(t) \leq N, \tag{5.5}$$

Here $M, M_1, K_1, K_2, K_3, L_1, L_2$, and N are certain positive constants.

For equation (5.1), the notion of a $T(2T)$-periodic system of the first class can be introduced in the same manner as it was done above for the second order nonlinear ordinary differential equation in §2.2, Chapter 2.

Definition 5.1. The second order equation (5.1) is called a T-periodic ($2T$-periodic) system of the first class in the region (5.2) if the constants $a, b, M, K_1, K_2, K_3, L_1, L_2, N, T$ and ω are related by the relations $a\epsilon\frac{1}{2\omega}MTq$ and $b\epsilon\frac{1}{2}MTq$, and the condition $\omega Tq = (2p-1)\pi$, $(\omega q, 2p-1) = 1$, holds for every function $x(t)$ such that $x(t+T) = -x(t)$.

$$f\left(t+T, x(t+T), \dot{x}(t+T), \int_0^{h(t+T)} \varphi(t+T, s, x(s), \dot{x}(s))ds\right) =$$

$$= -f\left(t, x(t), \dot{x}(t), \int_0^{h(t)} \varphi(t, s, x(s), \dot{x}(s))ds\right), \quad \omega Tq = r\pi.$$

Here p, r and q are fixed positive integers, r is even and q is odd, $(\omega q, r) = 1$; $Tq\left[\frac{1}{\omega}(K_1 + K_3NL_1) + (K_2 + K_3NL_2)\right] < 2$.

Starting from the integral equation

$$x(t) = \frac{1}{2\omega}\int_0^t f\left(\tau, x(\tau), \dot{x}(\tau), \int_0^{h(\tau)} \varphi(\tau, s, x(s), \dot{x}(s))ds\right) \sin\omega(t-\tau)d\tau -$$

$$-\frac{1}{2\omega}\int_0^{Tq} f\left(\tau, x(\tau), \dot{x}(\tau), \int_0^{h(\tau)} \varphi(\tau, s, x(s), \dot{x}(s))ds\right) \sin\omega(t-\tau)d\tau \equiv$$

$$\equiv \frac{1}{2\omega}\int_0^{Tq} Q(\tau) f\left(\tau, x(\tau), \dot{x}(\tau), \int_0^{h(\tau)} \varphi(\tau, s, x(s), \dot{x}(s))ds\right) \sin\omega(t-\tau)d\tau, \tag{5.6}$$

where $Q(\tau)$ is given by the relation (see §2.1, Chapter 2)

$$Q(\tau) = \begin{cases} 1, & 0 \le \tau \le \tau, \\ -1, & t < \tau \le Tq, \end{cases} \tag{5.7}$$

and taking account of the conditions (5.3)–(5.5) and condition in Definition 5.1, one can prove, by using the method of successive approximations, the existence of T- and $2T$-periodic solutions of equation (5.1) [33] in a fashion similar to that used in the proof of Theorem 2.5 in Chapter 2. Following this line and starting from the integral equation

$$u(x,t) = \frac{1}{4}\int_0^t d\tau \int_\tau^t \{F[u, u_t, u_x](x+\theta-\tau, \tau)+$$

$$+F[u, u_t, u_x](x-\theta+\tau, \tau)\}d\theta - \frac{1}{4}\int_t^{Tq} d\tau \int_\tau^t \{F[u, u_t, u_x](x+\theta-\tau, \tau)+$$

$$+F[u, u_t, u_x](x-\theta+\tau, \tau)\}d\theta \equiv$$

$$\equiv \frac{1}{4}\int_0^{Tq} Q(\tau)d\tau \int_\tau^t \sum_{i=o}^{1} F[u,u_t,u_x](x+(-1)^i(\theta-\tau),\tau)d\theta, \tag{5.8}$$

where

$$F[u,u_t,u_x](x,t) = f\Bigg(x,t,u(x,t),u_t(x,t),u_x(x,t),$$

$$\int_0^{h(x,t)} \varphi(x,t,s,u(x,s),u_t(x,s),u_x(x,s))ds\Bigg); \tag{5.9}$$

T and q are reals and the function $Q(\tau)$ is given by (5.7), we shall establish a number of results concerning the hyperbolic boundary value problem of the form

$$u_{tt} - u_{xx} =$$

$$= f\left(x,t,u,u_t,u_x, \int_0^{h(x,t)} \varphi(x,t,s,u(x,s),u_t(x,s),u_x(x,s))ds\right),$$

$$u(0,t) = u(\pi,t) = 0. \tag{5.10}$$

in the class of functions satisfying Lipschitz condition. These results will be presented in the subsequent sections.

5.2. Classical and smooth periodic solutions

This section deals with the existence of T-periodic solutions of the boundary value problem

$$u_{tt} - u_{xx} = F[u,u_t,u_x]; \tag{5.11}$$

$$u(0,t) = u(\pi,t) = 0,\ t \in \mathbb{R}; \tag{5.12}$$

$$u(x,t+T) = u(x,t),\ t \in \mathbb{R},\ 0 \le x \le \pi, \tag{5.13}$$

where the operator $F[u,u_t,u_x]$ is given by (5.9).

1. T-systems of the first class. Suppose that the operator $F[u,u_t,u_x]$ admits an odd 2π-periodic continuation with respect to variable $x \in [0,\pi]$ to the whole real line $\mathbb{R} = (-\infty,+\infty)$. Investigating the existence of solutions, we restrict ourselves to the functional space G_x (see Chapter 3).

Consider the system of integral equations

$$u(x,t) = \frac{1}{4}\int_0^t d\tau \int_\tau^t [F[u,u_t,u_x](x+\theta-\tau,\tau)+$$

$$+F[u,u_t,u_x](x-\theta+\tau,\tau)]d\theta-$$

$$-\frac{1}{4}\int_t^{Tq} d\tau \int_\tau^t [F[u,u_t,u_x](x+\theta-\tau,\tau)+$$

$$+F[u,u_t,u_x](x-\theta+\tau,\tau)]d\theta,$$

$$u_t(x,t) = \frac{1}{4}\int_0^t [F[u,u_t,u_x](x+t-\tau,\tau)+$$

$$+F[u,u_t,u_x](x-t+\tau,\tau)]d\tau-$$

$$-\frac{1}{4}\int_t^{Tq}[F[u,u_t,u_x](x+t-\tau,\tau)+F[u,u_t,u_x](x-t+\tau,\tau)]d\tau, \tag{5.14}$$

$$u_x(x,t) = \frac{1}{4}\int_0^t [f[u,u_t,u_x](x+t-\tau,\tau)-$$

$$-F[u,u_t,u_x](x-t+\tau,\tau)]d\tau-$$

$$-\frac{1}{4}\int_t^{Tq}[F[u,u_t,u_x](x+t-\tau,\tau)-F[u,u_t,u_x](x-t+\tau,\tau)]d\tau,$$

or, in short form,

$$u(x,t) = \frac{1}{4}\int_0^{Tq} Q(\tau)d\tau \int_\tau^t \sum_{i=0}^{1} F[u,u_t,u_x](x+(-1)^i(\theta-\tau),\tau)d\theta,$$

$$u_t(x,t) = \frac{1}{4}\int_0^{Tq} Q(\tau)\sum_{i=0}^{1} F[u,u_t,u_x](x+(-1)^i(t-\tau),\tau)d\tau, \tag{5.15}$$

$$u_x(x,t) = \frac{1}{4}\int_0^{Tq} Q(\tau)\sum_{i=0}^{1}(-1)^i F[u,u_t,u_x](x+(-1)^i(t-\tau),\tau)d\tau,$$

where $F[u,u_t,u_x](x,t)$ is given by (5.9), T and q are reals and the function $Q(\tau)$ is given by (5.7).

We assume that the functions $f(x,t,u,u_t,u_x,v)$, where $v(x,t) = \int_0^{h(x,t)} \varphi(x,t,s,u(x,s),u_t(x,s),u_x(x,s))ds$, $\varphi(x,t,s,u,u_t,u_x)$ and $h(x,t)$ are defined in the region

$$x \in \mathbb{R}_x,\ t \in [0,Tq] = I,\ s \in [0,Tq] = I, \tag{5.16}$$

$$u \in [-b,b] = I^1;\ u_t \in [-c,c] = I^2;\ u_x \in [-c,c] = I^2;\ v \in \mathbb{R}_v^1,$$

are continuous jointly with respect to variables x,t,s,u,u_t,u_x and v and satisfy the conditions

$$|f(x,t,u,u_t,u_x,v)| \le M, \tag{5.17}$$

$$f(x,t,u,u_t,u_x,v) \in \mathrm{Lip}_{u,u_t,u_x,v}(K;\mathbb{R}_x \times I \times I^1 \times I^2 \times I^2 \times \mathbb{R}_v^1),$$

that is,

$$|f(x,t,u'',u_t'',u_x'',v'') - f(x,t,u',u_t',u_x',v')| \le$$

$$\le K\{|u''-u'|+|u_t''-u_t'|+|u_x''-u_x'|+|v''-v'|\}; \tag{5.18}$$

$$|\varphi(x,t,s,u,u_t,u_x)| \le M_1;$$

$$\varphi(x,t,s,u,u_t,u_x) \in \mathrm{Lip}_{u,u_t,u_x}(L;\mathbb{R}_x \times I \times I \times I^1 \times I^2 \times I^2), \tag{5.19}$$

that is,

$$|\varphi(x,t,s,u'',u_t'',u_x'') - \varphi(x,t,s,u',u_t',u_x')| \leq$$
$$\leq L\{|u''-u'| + |u_t''-u_t'| + |u_x''-u_x'|\}; \tag{5.20}$$
$$|h(x,t)| \leq N, \tag{5.21}$$

Here M, M_1, K, L, and N are positive constants.

Definition 5.2. The second order equation (5.1) is called a T-system of the first class in the region (5.6) if the constants b, c, M, K, L, N, T and q are related by the following inequalities

$$b\epsilon\frac{1}{4}M(Tq)^2,\ c\epsilon\frac{1}{2}MTq,\ KTq(1+LN)\left(2+\frac{Tq}{2}\right) < 1 \tag{5.22}$$

for certain fixed real T and q.

Theorem 5.1. *Let the functions* $f\left(x,t,u,u_t,u_x \int\limits_0^{h(x,t)} \varphi(x,t,s,u(x,s),u_t(x,s)u_x(x,s))ds\right), \varphi(x,t,s,u,u_t,u_x)$ *and* $h(x,t)$ *be defined in the region (5.16), continuous with respect to variables* x,t,s,u,u_t,u_x *and* v *and satisfy inequalities (5.17)–(5.22). Then, for any continuous function* $\mu = \mu(x,t)$ *such that*

$$|\mu(x,t)| \leq \frac{1}{4}M(Tq)^2,$$

$$|\mu_t(x,t)| \leq \frac{1}{2}MTq,\ |\mu_x(x,t)| \leq \frac{1}{2}MTq, \tag{5.23}$$

the sequences of functions

$$u^0(x,t) = \mu(x,t),\ u_t^0 = \mu_t(x,t),\ u_x^0(x,t) = \mu_x(x,t),$$

$$u^{m+1}(x,t) = \frac{1}{4}\int\limits_0^{Tq} Q(\tau)d\tau \int\limits_\tau^t \sum_{i=0}^{1} F[u^m,u_t^m,u_x^m](x+(-1)^i(\theta-\tau),\tau)d\theta, \tag{5.24}$$

$$u_t^{m+1}(x,t) = \frac{1}{4}\int\limits_0^{Tq} Q(\tau)\sum_{i=0}^{1} F[u^m,u_t^m,u_x^m](x+(-1)^i(t-\tau),\tau)d\tau,$$

$$u_x^{m+1}(x,t) = \frac{1}{4}\int\limits_0^{Tq} Q(\tau)\sum_{i=0}^{1}(-1)^i F[u^m,u_t^m,u_x^m](x+(-1)^i(t-\tau),\tau)d\tau,$$

$$m = 0,1,2,\ldots,$$

converge, as $m \to \infty$, *uniformly with respect to* $(x,t) \in \mathbb{R}_x \times I$ *to continuous functions* $u_0(x,t), u_{0t}(x,t)$ *and* $u_{0x}(x,t)$ *defined for all* $(x,t) \in \mathbb{R}_x \times I$ *and satisfying the system of integral equations (5.15).*

Proof. Proceeding by induction and using (5.23) and (5.24) one can show that the functions $u^m(x,t), u_t^m(x,t)$ and $u_x^m(x,t)$ take their values in the region (5.16) for all $m\epsilon 0$ and $(x,t) \in \mathbb{R}_x \times I$.

We shall now prove that the sequences (5.24) converge uniformly for all $(x,t) \in \mathbb{R}_x \times I$ and hence, the limit functions are continuous on this set. To do this, we observe that convergence of the sequences (5.24) is equivalent to convergence of the series

$$u^0 + (u^1-u^0) + (u^2-u^1) + \ldots + (u^m-u^{m-1}) + \ldots,$$

$$u_k^0 + (u_k^1 - u_k^0) + (u_k^2 - u_k^1) + \ldots + (u_k^m - u_k^{m-1}) + \ldots, \quad k = t, x, \tag{5.25}$$

since the functions $u^m(x,t), u_t^m(x,t), u_x^m(x,t)$ are just partial sums of these series.

We shall estimate the terms of the series (5.25). From inequalities (5.17) and (5.23) and representation (5.24) it follows that, for $m = 0$, the following estimates hold

$$|u^1 - u^0| \le |u^1| + |u^0| \le \frac{1}{2} M (Tq)^2, \tag{5.26}$$

$$|u_t^1 - u_t^0| \le |u_t^1| + |u_t^0| \le MTq, \ |u_x^1 - u_x^0| \le |u_x^1| + |u_x^0| \le MTq.$$

By virtue of inequalities (5.18) and (5.20) and representation (5.24), we now obtain, for $m = 0, 1$,

$$|u^2 - u^1| \le \frac{K}{4} \int_0^{Tq} d\tau \left| \int_\tau^t \sum_{j=0}^{2} \sum_{i=0}^{1} \left\{ |u_j^1(x + (-1)^i(\theta - \tau), \tau) - \right.\right.$$

$$- u_j^0(x + (-1)^i(\theta - \tau), \tau)| + L \int_0^{h(x+(-1)^i(\theta-\tau),\tau)} |u_j^1(x + (-1)^i(\theta - \tau), s) -$$

$$\left.\left. - u_j^0(x + (-1)^i(\theta - \tau), s)| ds \right\} d\theta \right|, \tag{5.27}$$

$$|u_t^2 - u_t^1| \le \frac{K}{4} \int_0^{Tq} \sum_{j=0}^{2} \sum_{i=0}^{1} \left\{ |u_j^1(x + (-1)^i(t - \tau), \tau) - \right.$$

$$- u_j^0(x + (-1)^i(t - \tau), \tau)| +$$

$$+ L \int_0^{h(x+(-1)^i(t-\tau),\tau)} |u_j^1(x + (-1)^i(t - \tau), s) -$$

$$\left. - u_j^0(x + (-1)^i(t - \tau), s)| ds \right\} d\tau,$$

$$|u_t^2 - u_x^1| \le \frac{K}{4} \int_0^{Tq} \sum_{j=0}^{2} \sum_{i=0}^{1} \left\{ |u_j^1(x + (-1)^i(t - \tau), \tau) - \right.$$

$$- u_j^0(x + (-1)^i(t - \tau), \tau)| +$$

$$+ L \int_0^{h(x+(-1)^i(t-\tau),\tau)} |u_j^1(x + (-1)^i(t - \tau), s) -$$

$$\left. - u_j^0(x + (-1)^i(t - \tau), s)| ds \right\} d\tau,$$

here $u_0(x,t) \equiv u(x,t)$; $u_1(x,t) \equiv u_t(x,t)$ and $u_2(x,t) \equiv u_x(x,t)$.

Taking account of the inequality

$$\int_0^{Tq} d\tau \left| \int_\tau^t d\theta \right| \le (Tq)^2/2 \tag{5.28}$$

along with inequalities (5.21), (5.266) and (5.27), we obtain

$$|u^2 - u^1| \le K \int_0^{Tq} (1 + LN) \left(\frac{1}{2}M(Tq)^2 + MTq + MTq\right) \left| \int_\tau^t d\theta \right| d\tau \le$$

$$\le \frac{1}{2}MTqK(1 + LN)\left(2 + \frac{Tq}{2}\right)\frac{(Tq)^2}{2} = \frac{1}{4}M(Tq)^2\gamma_1, \tag{5.29}$$

$$\gamma_1 = KTq(1 + LN)\left(2 + \frac{Tq}{2}\right),$$

$$|u_t^2 - u_t^1| \le \frac{1}{2}MTq\gamma_1, \; |u_x^2 - u_x^1| \le \frac{1}{2}MTq\gamma_1.$$

Similarly, we have $|u^3 - u^2| \le \frac{1}{4}M(Tq)^2\gamma_1^2$,

$$|u_t^3 - u_t^2| \le \frac{1}{2}MTq\gamma_1^2, \; |u_x^2 - u_x^1| \le \frac{1}{2}MTq\gamma_1^2,$$

....

$$|u^m - u^{m-1}| \le \frac{1}{4}M(Tq)^2\gamma_1^{m-1},$$

$$|u_t^m - u_t^{m-1}| \le \frac{1}{2}MTq\gamma_1^{m-1}, \; |u_x^m - u_x^{m-1}| \le \frac{1}{2}MTq\gamma_1^{m-1},$$

$$\gamma_1 = KTq(1 + LN)\left(2 + \frac{Tq}{2}\right),$$

.... (5.30)

By virtue of the estimates (5.23), (5.26), (5.29) and (5.30), the series in (5.25) are dominated, for all $(x, t) \in \mathbb{R}_x \times I$, by the following converging series with positive terms

$$\frac{1}{4}M(Tq)^2 + \frac{1}{2}M(Tq)^2 + \frac{1}{4}M(Tq)^2\gamma_1 +$$

$$+ \frac{1}{4}M(Tq)^2\gamma_1^2 + \ldots + \frac{1}{4}M(Tq)^2\gamma_1^{m-1} + \ldots,$$

$$\frac{1}{2}MTq + MTq + \frac{1}{2}MTq\gamma_1 + \frac{1}{2}MTq\gamma_1^2 + \ldots + \frac{1}{2}MTq\gamma_1^{m-1} + \ldots,$$

$$\gamma_1 = KTq(1 + LN)(2 + Tq/2).$$

By the Weierstrass criterion, each of the series in (5.25) converges uniformly for all $(x, t) \in \mathbb{R}_x \times I$.

Denoting the sums of the series in (5.25) or, which is the same, the limit functions of sequences (5.24) by $u_0(x,t)$, $u_{0t}(x,t)$ and $u_{0x}(x,t)$ and passing to the limit in the recurrence relations (5.24) as $m \to \infty$, we see that the limit functions $u_0(x,t)$, $u_{0t}(x,t)$ and $u_{0x}(x,t)$ satisfy the system of integral equations (5.15). The proof is complete.

By using Theorem 5.1 one can readily see that the following theorem is true.

Theorem 5.2. *Suppose the condition of Theorem 5.1 is satisfied. If* $f(x, t, u, u_t, u_x, v) \in C^1$ *with respect to variable* x, *then every continuously differentiable solution* (u, u_t, u_x) *of system (5.15) (or (5.14)) is a classical solution of equation (5.11).*

P r o o f. Differentiating the first equality in (5.14) yields $(u)'_t = u_t$, $(u)'_x = u_x$. Next, using the second and third equations of system (5.14) we obtain

$$u_{tt} \equiv (u_t)' = F\left(x, t, u(x,t), u_t(x,t), u_x(x,t), \int_0^{h(x,t)} \varphi(x,t,s,u(x,s),u_t(x,s),\right.$$

$$\left. u_x(x,s))ds\right) + \frac{1}{4}\int_0^{Tq} Q(\tau)\sum_{i=0}^{1}(-1)^i \frac{\partial F[u,u_t,u_x](x+(-1)^i(t-\tau),\tau)}{\partial x}d\tau,$$

$$u_{xx} \equiv (u_x)'_x = \frac{1}{4}\int_0^{Tq} Q(\tau)\sum_{i=0}^{1}(-1)^i \frac{\partial F[u,u_t,u_x](x+(-1)^i(t-\tau),\tau)}{\partial x}d\tau.$$

Therefore $u_{tt} - u_{xx} =$ $f\left(x,t,u(x,t),u_t(x,t),u_x(x,t), \int_0^{h(x,t)} \varphi(x,t,s,u(x,s),u_t(x,s),u_x(x,s))ds\right)$, as required to prove.

Theorem 5.3. *Suppose that the condition of Theorem 5.1 is satisfied and that a certain solution $(u(x,t), u_t(x,t), u_x(x,t))$ of system (5.15) (or (5.14)) satisfies the conditions*

$$1) \qquad F[u,u_t,u_x](-x,t) = -F[u,u_t,u_x](x,t);$$

$$2) \qquad F[u,u_t,u_x](x+2\pi,t) = F[u,u_t,u_x](x,t),$$

that is, the operator $F[u,u_t,u_x]$ admits an odd 2π-periodic continuation with respect to variable x.

Then the function $u(x,t)$ satisfies the boundary condition (5.12).

The proof proceeds in the same manner as the proof of Theorem 3.9 in Chapter 3.

The assertion below can also be proved in the similar fashion.

Theorem 5.4. *Suppose that the condition of Theorem 5.1 is satisfied and that a certain solution $(u(x,t), u_t(x,t), u_x(x,t))$ of system (5.15) (or (5.14)) satisfies the condition (1) in Theorem 5.3 and the condition*

$$1) \qquad F[u,u_t,u_x](\pi - x,t) = F[u,u_t,u_x](x,t).$$

Then the function $u(x,t)$ satisfies the boundary condition (5.12).

Remark 5.1. From the condition (2) in Theorem 5.3 it follows that if the condition (5.12) ($u(0,t) = u(\pi,t) = 0$) holds, then both the right-hand side of equation (2.1) and the solution $u(x,t)$ of the boundary value problem (5.11), (5.12) are 2π-periodic in x. The following theorem therefore holds.

Theorem 5.5. *Let the condition of Theorem 5.1 be fulfilled. If for a certain solution $(u(x,t), u_t(x,t), u_x(x,t))$ of system (5.14) the function*

$$f\left(x,t,u(x,t),u_t(x,t),u_x(x,t), \int_0^{h(x,t)} \varphi(x,t,s,u(x,s),u_t(x,s),u_x(x,s))ds\right)$$

can be expanded into a Fourier series of the form

$$f\left(x,t,u(x,t),u_t(x,t),u_x(x,t), \int_0^{h(x,t)} \varphi(x,t,s,u(x,s),\right.$$

$$u_t(x,s), u_x(x,s))ds) = \sum_{k=1}^{\infty} f_k(t)\sin kx, \tag{5.31}$$

where the functions $f_k(t)$ are continuous on the interval $[0, Tq]$ and satisfy the condition

$$|f_k(t)| \leq \frac{C}{k^\gamma}, \; C > 0, \; \gamma > 2, \tag{5.32}$$

then the function $u(x,t)$ is a classical solution of the boundary value problem (5.11), (5.12).

This theorem can be proved by the argument used in proving Theorem 3.10 in Chapter 3.

Definition 5.3. If there exists a continuous solution $(u(x,t), u_t(x,t), u_x(x,t))$ of the system of integral equations (5.14) (or (5.15)), then the finction $u(x,t) \in C^1$ satisfying the boundary condition $u(0,t) = u(\pi,t) = 0$ is called a smooth generalized or simply smooth solution of the boundary value problem (5.11), (5.12).

The conclusions similar to those drawn in Remark 2.2, Chapter 3 can be drawn for a solution $u(x,t)$ of the boundary value problem (5.11), (5.12) (or (5.10)).

2. **T- and $2T$-periodic systems of the first class.**

Definition 5.4. The boundary value problem (5.11), (5.12) is called a T-periodic system of the first class if the functions

$$f(x,t,u,u_t,u_x,v), \; \varphi(x,t,s,u,u_t,u_x), \; h(x,t)$$

are T-periodic in t, equation (5.11) is a T-system of the first class (see Definition 5.2) and the period T satisfies the relation $Tq = (2p-1)\pi$, $(q, 2p-1) = 1$, with p and q being certain positive integers.

The following theorem can be proved in the manner similar to that used in the proof of Theorem 3.11 in Chapter 3.

Theorem 5.6. *Suppose that the condition of Theorem 5.1 is satisfied and that*

1) the functions $f(x,t,u,u_t,u_x,v)$, $\varphi(x,t,s,u,u_t,u_x)$ and $h(x,t)$– are T-periodic in variable t,

2) the period T satisfies the realtion $Tq = (2p-1)\pi$, $(q, 2p-1) = 1$ with p and q fixed positive integers.

If for a certain smooth $(2p-1)\pi/q$-periodic initial function

$$|u^0(x,t)| \equiv |\mu(x,t)| = \left|\sum_{k=1}^{\infty} \mu_{2k-1}(t)\sin(2k-1)x\right| \leq \frac{1}{4}M(Tq)^2,$$

$$|\mu_t(x,t)| \leq \frac{1}{2}MTq, \; |\mu_x(x,t)| \leq \frac{1}{2}MTq, \tag{5.33}$$

the right-hand side of equation (5.11) can be expanded into a Fourier series

$$f\left(x,t,\mu(x,t),\mu_t(x,t),\mu_x(x,t), \int_0^{h(x,t)} \varphi(x,t,s,\mu(x,s),\right.$$

$$\left.\mu_t(x,s),\mu_x(x,s))ds\right) = \sum_{k=1}^{\infty} f^0_{2k-1}(t)\sin(2k-1)x, \tag{5.34}$$

the coefficients $f^0_{2k-1}(t)$ of which are continuous T-periodic functions satisfying the condition

$$|f^0_{2k-1}(t)| \leq \frac{C}{(2k-1)\gamma}, \; C > 0, \; \gamma > 1, \tag{5.35}$$

then the boundary value problem (5.11), (5.12) has a smooth T-periodic solution $u(x,t) \in C^1$.

Definition 5.5. The boundary value problem (5.11), (5.12) is called a $2T$-periodic system of the first class if the functions $f(x,t,u,u_t,u_x,v), \varphi(x,t,s,u,u_t,u_x)$ and $h(x,t)$ are $2T$-periodic in variable t, equation (5.11) is a T-system of the first class (see Definition 5.2), the number T satisfies the relation $Tq = r\pi, r = 2k, q = 2s-1, (r,q) = 1$ and, for every function $\mu(x,t) \in C^1$ satisfying $\mu(x,t+T) = -\mu(x,t)$, the following equality holds

$$f(x,t+T,\mu(x,t+T),\mu_t(x,t+T),\mu_x(x,t+T),v(x,t+T)) =$$

$$= -f(x,t,\mu(x,t),\mu_t(x,t),\mu_x(x,t),v(x,t)),$$

r and q being fixed positive integers.

Theorem 5.7. *Let the boundary value problem (5.11), (5.12) be a $2T$-periodic system of the first class and let the condition of Theorem 5.1 be fulfilled.*

If for a certain smooth $2T$-periodic initial function

$$|u^0(x,t)| \equiv |\mu(x,t)| = \left|\sum_{k=1}^{\infty} \mu_k(t)\sin kx\right| \le \frac{1}{4}M(Tq)^2,$$

$$|\mu_t(x,t)| \le \frac{1}{2}MTq, \; |\mu_x(x,t)| \le \frac{1}{2}MTq, \; \mu(x,t+T) = -\mu(x,t),$$

the right-hand side of equation (5.11) can be expanded into a Fourier series

$$f(x,t,\mu(x,t),\mu_t(x,t),\mu_x(x,t),v(x,t)) = \sum_{k=1}^{\infty} f_k^0(t)\sin kx,$$

the Fourier coefficients $f_k^0(t)$ of which are continuous $2T$-periodic functions satisfying

$$|f_k^0(t)| \le \frac{C}{k\gamma}, \; C > 0, \; \gamma > 1,$$

then the boundary value problem (5.11), (5.12) has a smooth $2r\pi/q$-periodic solution $u(x,t) \in C^1$.

The proof of Theorem 5.7 is similar to that of Theorem 5.6.

3. T-periodic systems of the first class in the functional space $G_{\pi t}$. The notion of nonlinear T-systems of the first class in the functional space $u(x,t)$, i. e. in the space of functions $G_{\pi k}$ which are continuous and bounded on $[0,\pi] \times \mathbb{R}$ together with their derivative with respect to t, can be introduced in the same way as it was done in §3.1, Chapter 3 by using the operators $P_k, R_k, k = 1,2,3$.

Consider a nonlinear periodic boundary value problem that is described by the simple integro-differential equation

$$u_{tt} - u_{xx} = f\left(x,t,u,\int_0^{h(x,t)} \varphi(x,t,s,u(x,s))ds\right), 0 \le x \le \pi, \; t \in \mathbb{R}, \; s \in \mathbb{R}, \qquad (5.36)$$

and the conditions (5.12) and (5.13).

By using the Schauder principle stated as Theorem 5.9 in Chapter 3, we shall investigate the existence of continuous T_k-periodic solutions of the nonlinear problem (5.12), (5.13) and (5.36) in the functional spaces $A_k^0, k = 1,2,3$ under sufficiently weak assumptions about the right-hand side of equation (5.36).

Theorem 5.8. *Let k be one of the numbers $1, 2, 3$ and let*

1) the functions $f(x,t,u,v)(v(x,t) = \int_0^{h(x,t)} \varphi(x,t,s,u(x,s))ds), \varphi(x,t,s,u)$ and $h(x,t)$ be defined in the region $\{(x,t) \in \mathbb{R}^2; s \in \mathbb{R}; u \in [-b,b]; v \in \mathbb{R}^1_v\}$, be jointly continuous with respect to variables x, t, s, u and v, T_k-periodic in t, 2π-periodic in x and, in addition, the function $f(x,t,u,v)$ satisfy the condition

$$|f(x,t,u,v)| \leq M, \ M = \text{const},$$

and, for every $(x,t) \in \mathbb{R}^2$, be continuous (with respect to u and v) uniformly with respect to $(x,t) \in \mathbb{R} \times [0, T_k q/2]$;

(2) the constants b, M, T_k and q are related by the inequality

$$b > \frac{1}{16} M (T_k q)^2;$$

(3) for any function $F[u](x,t) = f\left(x,t,u, \int_0^{h(x,t)} \varphi(x,t,s,u(x,s))ds\right)$, the function $u \in C \bigcap A^0_k$ belongs to the space $C \bigcap \tilde{A}^0_k$.

Then the boundary value problem (5.12), (5.13) and (5.36) has at least one continuous T_k-periodic solution.

P r o o f. In Theorem 3.15 of Chapter 3, set

$$X = C^{T_k, 2\pi}(\mathbb{R}^2),$$

where $C^{T_k,2\pi}(\mathbb{R}^2)$ is a set of continuous functions $u(x,t)$ that are 2π-periodic in variable x and T_k-periodic in variable t, and take the set of those $u(x,t) \in C^{T_k,2\pi}(\mathbb{R}^2)$ which satisfy

$$\|u\| \leq b, \ \|u\| = \max_{[0,2\pi]\times[0,T_k]} |u(x,t)|.$$

as Ω. Let the operator $\tilde{P}_k$ be defined as follows

$$z = \tilde{P}_k(u), \ z(x,t) = \frac{1}{4} \int_0^{c_k} Q(\tau) d\tau \int_{x-t+\tau}^{x+t-\tau} f(\eta, \tau, u(\eta,\tau),$$

$$\left. \int_0^{h(\eta,\tau)} \varphi(\eta, \tau, s, u(\eta, s)) ds \right) d\eta, \tag{5.37}$$

where

$$Q(\tau) = \begin{cases} 1, & 0 \leq \tau \leq t, \\ -1, & t < \tau \leq c_k, \end{cases} \quad c_k = T_k q/2, \quad k = 1,2,3.$$

The conditions (1)–(3) in Theorem 5.8 ensure the condition $\tilde{P}_k(u) \in \Omega$ holds. Indeed, from Lemma 3.4 of Chapter 3 and the condition (3) in Theorem 5.8 it follows that the operator $\tilde{P}_k$ sends every T_k-periodic function

$$u(x,t) \in \Omega \subset C^{T_k,2\pi}$$

into a T_k-periodic function $z(x,t)$ and

$$\|z\| = \|\tilde{P}_k(u)\| \leq \frac{1}{4} \left| \int_0^{c_k} Q(\tau) d\tau \int_{x-t+\tau}^{x+t-\tau} \right| f\left(\eta, \tau, u(\eta,\tau),\right.$$

$$\int_0^{h(\eta,\tau)} \varphi(\eta,\tau,s,u(\eta,s))ds\Bigg)\Bigg|d\eta\Bigg| \le \frac{1}{16}M(T_kq)^2 < b,\ u\in\Omega. \tag{5.38}$$

We shall now check that the operator $\tilde{P}_k$ is continuous. Let

$$u_n \to u_0,\ v_n \to v_0,\ u_n \in \Omega,\ v_n \in \Omega,\ z_n = \tilde{P}_k(u_n),\ n = 0,1,2,\dots$$

Since the function $f(x,t,u,v)$ is uniformly continuous with respect to u and v, for any $\varepsilon > 0$ one can find $\eta > 0, \eta = \min(\eta_1,\eta_2)$ such that the inequality

$$|f(x,t,u,v) - f(x,t,u',v')| \le |f(x,t,u,v) - f(x,t,u',v)| +$$

$$+|f(x,t,u',v) - f(x,t,u',v')| < \frac{\varepsilon}{2} + \frac{\varepsilon}{2} = \varepsilon,$$

$$(x,t) \in [0,2\pi]\times[0,T_k]. \tag{5.39}$$

holds for $|u-u'| < \eta_1$ and $|v-v'| < \eta_2$.

Since $\|u_n - u_0\| \to 0, \|v_n - v_0\| \to 0$, we have $\|u_n - u_0\| < \eta, \|v_n - v_0\| < \eta$ for sufficiently large $n(n\varepsilon n_0)$, and therefore $|u_n(x,t) - u_0(x,t)| < \eta, |v_n(x,t) - v_0(x,t)| < \eta, (x,t) \in [0,2\pi]\times[0,T_k]$. In view of (5.39), we can write down

$$|f(x,t,u_n(x,t),v_n(x,t)) - f(x,t,u_0)(x,t),v_0(x,t))| < \varepsilon,$$

$$(x,t) \in [0,2\pi]\times[0,T_k].$$

for these n. Hence, for $n\varepsilon n_0$ we have

$$\|z_n - z_0\| = \max_{[0,2\pi]\times[0,T_k]} |z_n(x,t) - z_0(x,t)| \le$$

$$\le \frac{1}{4}\max_{[0,2\pi]\times[0,T_k]} \int_0^{T_kq/2} d\tau \Bigg|\int_{x-t+\tau}^{x+t-\tau}\Bigg| f(\eta,\tau,u_n(\eta,\tau),v_n(\eta,\tau)) -$$

$$-f(\eta,\tau,u_0(\eta,\tau),v_0(\eta,\tau))\Bigg|d\eta\Bigg| \le \frac{1}{16}(T_kq)^2\varepsilon,$$

and therefore $z_n = \tilde{P}_k(u_n) \to z_0 = \tilde{P}_k(u_0)$.

In order to be able to use the Schauder principle we need only to establish relative compactness of the set $\tilde{P}_k(\Omega)$ because Ω is a convex closed set. To do this, we shall show that the functions of the family $\tilde{P}_k(\Omega)$ are equicontinuous (the fact that they are uniformly bounded follows from the inclusion $\tilde{P}_k(\Omega) \subset \Omega$, Ω being a bounded set).

In view of (5.37) we have

$$|z(x_2,t_2) - z(x_1,t_1)| = \Bigg|\frac{1}{4}\int_0^{c_k} Q(\tau)d\tau \int_{x_2-t_2+\tau}^{x_2+t_2-\tau} f(\eta,\tau,u(\eta,\tau),v(\eta,\tau))d\eta -$$

$$-\frac{1}{4}\int_0^{c_k} Q(\tau)d\tau \int_{x_1-t_1+\tau}^{x_1+t_1-\tau} f(\eta,\tau,u(\eta,\tau),v(\eta,\tau))d\eta\Bigg| \le$$

$$\le \frac{1}{4}\Bigg|\int_0^{t_2} d\tau \int_{x_2-t_2+\tau}^{x_2+t_2-\tau} f(\eta,\tau,u(\eta,\tau),v(\eta,\tau))d\eta -$$

$$- \int_0^{t_1} d\tau \int_{x_1-t_1+\tau}^{x_1+t_1-\tau} f(\eta,\tau,u(\eta,\tau),v(\eta,\tau))d\eta\Bigg| +$$

$$+\frac{1}{4}\Bigg|\int_{c_k}^{t_2} d\tau \int_{x_2-t_2+\tau}^{x_2+t_2-\tau} f(\eta,\tau,u(\eta,\tau),v(\eta,\tau))d\eta -$$

$$- \int_{c_k}^{t_1} d\tau \int_{x_1-t_1+\tau}^{x_1+t_1-\tau} f(\eta,\tau,u(\eta,\tau),v(\eta,\tau))d\eta\Bigg| = I_1 + I_2. \tag{5.40}$$

The expression I_1 can be estimated as follows

$$I_1 = \frac{1}{4}\Bigg|\int_0^{t_2} d\tau \int_{x_2-t_2+\tau}^{x_2+t_2-\tau} f(\eta,\tau,u(\eta,\tau),v(\eta,\tau))d\eta -$$

$$- \int_0^{t_1} d\tau \int_{x_2-t_2+\tau}^{x_2+t_2-\tau} f(\eta,\tau,u(\eta,\tau),v(\eta,\tau))d\eta +$$

$$+ \int_0^{t_1} d\tau \int_{x_2-t_2+\tau}^{x_2+t_2-\tau} f(\eta,\tau,u(\eta,\tau),v(\eta,\tau))d\eta -$$

$$- \int_0^{t_1} d\tau \int_{x_1-t_1+\tau}^{x_1+t_1-\tau} f(\eta,\tau,u(\eta,\tau),v(\eta,\tau))d\eta\Bigg| \leq$$

$$\leq \frac{1}{4}\int_{t_1}^{t_2}\Bigg|\int_{x_2-t_2+\tau}^{x_2+t_2-\tau} f(\eta,\tau,u(\eta,\tau),v(\eta,\tau))d\eta\Bigg| d\tau +$$

$$+\frac{1}{4}\int_0^{t_1}\Bigg|\int_{x_1-t_1+\tau}^{x_2-t_2+\tau} f(\eta,\tau,u(\eta,\tau),v(\eta,\tau))d\eta\Bigg| d\tau +$$

$$+\frac{1}{4}\int_0^{t_1}\Bigg|\int_{x_1+t_1-\tau}^{x_2+t_2-\tau} f(\eta,\tau,u(\eta,\tau),v(\eta,\tau))d\eta\Bigg| d\tau \leq$$

$$\leq \frac{1}{4}MT_kq|x_2-x_1| + \frac{3}{8}MT_kq|t_2-t_1|. \tag{5.41}$$

The estimate for I_2 can be obtained in the similar manner

$$I_2 \leq \frac{1}{4}MT_kq|x_2-x_1| + \frac{3}{8}MT_kq|t_2-t_1|. \tag{5.42}$$

Combining inequalities (5.40)–(5.42) yields

$$|z(x_2,t_2) - z(x_1,t_1)| \leq \frac{1}{2}MT_kq|x_2-x_1| + \frac{3}{4}MT_kq|t_2-t_1|,$$

whence it follows that the set $\tilde{P}_k(\Omega)$ is equicontinuous. All the conditions of Theorem 3.15 in Chapter 3 are therefore satisfied and it follows that the integral equation

$$u(x,t) = \frac{1}{4}\int_0^{c_k} Q(\tau)d\tau \int_{x-t+\tau}^{x+t-\tau} f\left(\eta, \tau, u(\eta,\tau), \int_0^{h(\eta,\tau)} \varphi(\eta,\tau,s,u(\eta,s))ds\right)d\eta$$

has at least one continuous T_k-periodic solution, which means that there exists a continuous T_k-periodic solution of the boundary value problem (5.12), (5.13) and (5.36). This completes the proof.

Remark 5.2. By using the same operators R_k, similar results can be obtained for the boundary value problem (5.12), (5.13) and (5.36) in the space $C_\pi \bigcap A_k$. For $k = 2$, such results were obtained in [34].

Remark 5.3. By using the theorems proved above along with the properties of the operators P_k and R_k, $k = 1, 2, 3$, one can deduce a statement analogous to Theorem 3.14 in Chapter 3 for equation (5.11) with small parameter ε.

Remark 5.4. The Schauder principle together with the results established in the present section enable the existence of continuous T_k-periodic solutions $k = 1, 2, 3$ of the boundary value problem (5.11)–(5.13) to be proved under weaker assumptions about the right-hand side of equation (5.11). The proof goes similarly to the proof of Theorem 5.8 by using the functional spaces A_k^0, A_k, $k = 1, 2, 3$ and operators P_k, R_k, $k = 1, 2, 3$.

5.3. The existence of generalized periodic solutions of hyperbolic integro-differential equations

This section is devoted to the investigation into the questions concerning the structure of generalized periodic solutions of the second order hyperbolic integro-differential equations of the firts and second kind. We make use of the theorem on existence and uniqueness of continuously generalized solutions of hyperbolic systems of integro-differential equations, the proof of which can be found in [31]. We start with the reduction of the wave nonlinear integro-differential equation to systems of integral equations (see also §1.4 and 1.5 in Chapter 1). This reduction will be needed later.

Consider a periodic boundary value problem

$$u_{tt} - u_{xx} = f\left(x, t, u, u_t, u_x, \int_0^{h(x,t)} \varphi(x,t,s,u(x,s),u_t(x,s),u_x(x,s))ds\right),$$

$$u(0,t) = u(\pi,t) = 0, \quad u(x,t+2\pi) = u(x,t). \tag{5.43}$$

Suppose that the functions $f(x,t,u,u_t,u_x,v)$, where

$$v(x,t) = \int_0^{h(x,t)} \varphi(x,t,s,u(x,s),u_t(x,s),u_x(x,s))ds,$$

$\varphi(x,t,s,u,u_t,u_0)$ and $h(x,t)$ are defined in the region

$$x \in [0,\pi] = I, \quad t \in \mathbb{R}_t, \quad s \in \mathbb{R}_s, \quad u \in [a,b] = I^1,$$

$$u_t \in [-c,c] = I^2, \quad u_x \in [-c,c] = I^2, \quad v \in \mathbb{R}_v^1, \tag{5.44}$$

are continous with respect to variables x, t and s, 2π-periodic in t and satisfy the condition

$$|f(x,t,u,u_t,u_x,v)| \le M, \tag{5.45}$$

$$f(x,t,u,u_t,u_x,v) \in \mathrm{Lip}_{u,u_t,u_x,v}(K; I \times \mathbb{R}_t \times I^1 \times I^2 \times I^2 \times \mathbb{R}^1_v),$$

that is,

$$|f(x,t,u'',u_t'',u_x'',v'') - f(x,t,u',u_t',u_x',v')| \leq$$
$$\leq K\{|u'' - u'| + |u_t'' - u_t'| + |u_x'' - u_x'| + |v'' - v'|\}; \tag{5.46}$$

$$|\varphi(x,t,s,u,u_t,u_x)| \leq M_1; \tag{5.47}$$

$$\varphi(x,t,s,u,u_t,u_x) \in \mathrm{Lip}_{u,u_t,u_x}(L; I \times \mathbb{R}_t \times \mathbb{R}_s \times I^1 \times I^2 \times I^2),$$

i. e.

$$|\varphi(x,t,s,u'',u_t'',u_x'') - \varphi(x,t,s,u',u_t',u_x')| \leq$$
$$\leq L\{|u'' - u'| + |u_t'' - u_t'| + |u_x'' - u_x'|\}; \tag{5.48}$$

$$|h(x,t)| \leq N, \tag{5.49}$$

M, M_1, K, L and N being positive constants.

1. 2π-hyperbolic integro-differential systems of the first kind. We first introduce 2π-hyperbolic integro-differential systems (called HID systems in the sequel) of the first kind for the boundary value problem (5.43). The introduction of these systems does not involve much difficulty when the integral term of the equation is of Fredholm type or when it is of the form as in (5.43). Before giving the definition we consider a class of continuous 2π-periodic functions $\mu(t)$ expandable into a Fourier series

$$\mu(t) = \frac{\alpha_0}{2} + \sum_{k=1}^{\infty}(\alpha_k'\cos kt + \alpha_k''\sin kt).$$

Let $\alpha = \{\alpha_0, \alpha_1', \alpha_1'', \alpha_2', \alpha_2'', \ldots\}$ denote the collection of the Fourier coefficients defining a function $\mu(t)$ such that

$$|\mu(t)| \leq c - M\pi, \ c > M\pi. \tag{5.50}$$

Let $\mathcal{D}_\mu = \{\alpha\}$ denote the set of the Fourier coefficients which satisfy (5.50).

Definition 5.6. An integro-differential equation of second order (5.43) is called a 2π-HID system of the first kind in the region (5.44) if the constants a, b, c and M are related by the inequalities

$$c > M\pi, \ a \leq -\left(c\pi - \frac{M\pi^2}{2}\right) < c\pi - \frac{M\pi^2}{2} \leq b \tag{5.51}$$

and the following condition is satisfied

$$f\left(x,t,0,0,0,\int_0^{h(x,t)}\varphi(x,t,s,0,0,0)ds\right) \equiv 0.$$

Lemma 5.1. *Suppose that the functions $f(x,t,u,u_t,u_x,v), \varphi(x,t,s,u,u_t,u_x)$ and $h(x,t)$ are defined in the region (5.44), are continuous with respect to variables x, t, and s, 2π-periodic in t and satisfy inequalities (5.45)-(5.49), (5.51). Then, for any continuous 2π-periodic function $|\mu(t)| \leq c - M\pi$, the following sequences of functions 2π-periodic in t*

$$u_1^0(x,t,\alpha) = \mu(t+x), \ u_2^0(x,t,\alpha) = -\mu(t-x),$$

$$u^0(x,t,\alpha) = \int_0^x \frac{\mu(t+\xi) + \mu(t-\xi)}{2}d\xi,$$

$$u_1^{m+1}(x,t,\alpha) = \mu(t+x) - \int_0^x \tilde{F}[u^m, u_1^m, u_2^m](\eta, t+x-\eta)d\eta, \tag{5.52}$$

$$u_2^{m+1}(x,t,\alpha) = -\mu(t-x) + \int_0^x \tilde{F}[u^m, u_1^m, u_2^m](\eta, t-x+\eta)d\eta,$$

$$u^{m+1}(x,t,\alpha) = \int_0^x \frac{u_1^{m+1}(\xi,t,\alpha) - u_2^{m+1}(\xi,t,\alpha)}{2} d\xi \equiv$$

$$\equiv \int_0^x \frac{\mu(t+\xi) + \mu(t-\xi)}{2} d\xi -$$

$$-\frac{1}{2}\int_0^x d\xi \int_0^\xi \{\tilde{F}[u^m, u_1^m, u_2^m](\eta, t+\xi-\eta) + \tilde{F}[u^m, u_1^m, u_2^m] \times$$

$$\times (\eta, t-\xi+\eta)\} d\eta, \; m = 0, 1, 2, \ldots; \tag{5.53}$$

$$\tilde{F}[u, u_1, u_2](x,t) = f\Bigg(x, t, u(x,t), u_t(x,t), u_x(x,t),$$

$$\int_0^{h(x,t)} \varphi(x,t,s,u(x,s),u_t(x,s),u_x(x,s))ds\Bigg) \equiv$$

$$\equiv f\Bigg(x, t, u(x,t), \frac{u_1(x,t)+u_2(x,t)}{2}, \frac{u_1(x,t)-u_2(x,t)}{2},$$

$$\int_0^{h(x,t)} \varphi\left(x,t,s,u(x,s), \frac{u_1(x,s)+u_2(x,s)}{2}, \frac{u_1(x,s)-u_2(x,s)}{2}\right) ds\Bigg)$$

$$u_1 = u_t + u_x, \; u_2 = u_t - u_x \Leftrightarrow u_t = \frac{u_1+u_2}{2}, \; u_x = \frac{u_1-u_2}{2}$$

converege uniformly with respect to

$$(x,t,\alpha) \in I \times \mathbb{R}_t \times \mathcal{D}_\mu \tag{5.54}$$

as $m \to \infty$ to functions $u_{01}(x,t,\alpha)$, $u_{02}(x,t,\alpha)$ and $u_0(x,t,\alpha)$ defined in the region (5.54), 2π-periodic in t and satisfying the system of integral equations

$$u_1(x,t,\alpha) = \mu(t+x) - \int_0^x \tilde{F}[u, u_1, u_2](\eta, t+x-\eta)d\eta,$$

$$u_2(x,t,\alpha) = -\mu(t-x) + \int_0^x \tilde{F}[u, u_1, u_2](\eta, t-x+\eta)d\eta, \tag{5.55}$$

$$u(x,t,\alpha) = \int_0^x \frac{u_1(\xi,t,\alpha) - u_2(\xi,t,\alpha)}{2} d\xi \equiv \int_0^x \frac{\mu(t+\xi)+\mu(t-\xi)}{2} d\xi -$$

$$-\frac{1}{2}\int_0^x d\xi \int_0^\xi \{\tilde{F}[u,u_1,u_2](\eta,t+\xi-\eta)+\tilde{F}[u,u_1,u_2]\times$$
$$\times(\eta,t,-\xi+\eta)\}d\eta.$$

Furthermore,

$$|u_1(x,t,\alpha)-u_0(x,t,\alpha)|\leq KM(1+LN)\left(2+\frac{\pi}{2}\right)\frac{\pi^3}{3!},$$

$$|u^m(x,t,\alpha)-u_0(x,t,\alpha)|\leq (3K)^{m-2}MK^2(1+LN)^m\beta_1\frac{\pi^{m+2}}{(m+2)!}, \tag{5.56}$$

$$\beta_1=\left(2+\frac{\pi}{2}\right)\left(2+\frac{\pi}{3}\right),\quad m=2,3,\ldots$$

The proof of Lemma 5.1 proceeds in the way similar to that of Lemma 4.1, Chapter 4 and can be found in [31]. In this case, the error arising because of the replacement of the exact solution $u_0(x,t,\alpha)$ of system (5.55) by the approximate solution $u^m(x,t,\alpha)$ is estimated by (5.56).

We now consider the Cauchy problem

$$u_{tt}-u_{xx}=f\left(x,t,u,u_t,u_x,\int_0^{h(x,t)}\varphi(x,t,s,u(x,s),u_t(x,s),u_x(x,s))ds\right),$$

$$u(0,t)=0,\quad u_x(0,t)=\mu(t). \tag{5.57}$$

In the class of smooth functions, (5.57) is equivalent, in the triangle $\overline{\Delta}=\{0\leq x\leq \pi, x\leq t\leq 2\pi-x\}$, to system (5.55). By using Lemma 3.1, one can readily see that the following theorem is true.

Theorem 5.9. *Suppose the condition of Lemma 3.1 is satisfied. Then, for any continuous 2π-periodic function $|\mu(t)|\leq c-M\pi$, the Cauchy problem (3.57) has a unique continuous 2π-periodic solution $u_{\overline{\Delta}}(x,t,\alpha)$ in the region*

$$\Delta_\infty=\{0\leq x\leq\pi,\ 2\pi n+x\leq t\leq 2\pi(n+1)-x,\ n=0,\pm1,\pm2,\ldots\}.$$

Remark 5.5. The solution $u_{\overline{\Delta}}(x,t,\alpha)$ of the Cauchy problem (5.57) always satisfies the first boundary condition $u_{\overline{\Delta}}(0,t,\alpha)=0$. The second boundary condition $u_{\overline{\Delta}}(\pi,\pi,\alpha)=0$ is obviously satisfied not for every initial function $\mu(t)$ and not for every function $f(x,t,u,u_t,u_x,v)$.

Theorem 5.10. *Suppose that a 2π-HID system of the first kind satisfies the condition of Lemma 5.1. Then, for any continuous 2π-periodic function $|\mu(t)|\leq c-M\pi$, the boundary value problem (5.43) has, in the rectangle $\prod_{2\pi}=\{0\leq x\leq\pi,\ 0\leq t\leq 2\pi\}$, a solution $\tilde{u}(x,t,\alpha)\in L^\infty$ of the form*

$$\tilde{u}(x,t,\alpha)=\begin{cases}u_{\overline{\Delta}}(x,t,\alpha), & (x,t)\in\overline{\Delta},\\ 0, & (x,t)\in\prod_{2\pi}/\overline{\Delta},\end{cases} \tag{5.58}$$

if and only if the following condition is satisfied

$$\tilde{\Delta}(\alpha)\equiv\int_0^{2\pi}\mu(\xi)d\xi+\int_0^\pi d\xi\int_0^\xi\{F[u,u_t,u_x](\eta,\pi+\xi-\eta)+$$

$$+F[u,u_t,u_x](\eta,\pi_\xi+\eta))\}d\eta=0, \tag{5.59}$$

where

$$F[u,u_t,u_x](x,t)=f\Bigg(x,t,u(x,t),u_t(x,t),u_x(x,t),$$

$$\int_0^{h(x,t)}\varphi(x,t,s,u(x,s),u_t(x,s),u_x(x,s))ds\Bigg). \tag{5.60}$$

P r o o f. Let the boundary value problem (5.43) have a solution of the form (5.58) in $\prod_{2\pi}$. Then the boundary conditions are fufilled $\tilde{u}(0,t,\alpha)=0$ and $\tilde{u}(\pi,t,\alpha)=0$ or, equivalently, $u_{\overline{\Delta}}(0,t,\alpha)$ and $u_{\overline{\Delta}(\pi,\pi,\alpha)=0}$. Since the function $u_{\overline{\Delta}}(x,t,\alpha)$ satisfies the system of integral equations (5.55), from the relations $u_{\overline{\Delta}}(\pi,\pi,\alpha)=0$ it follows that (5.59) holds.

Let now the condition (5.59) be satisfied with an initial function $|u_x(0,t)|\equiv|\mu(t)|\leq c-M\pi$ and a solution $u_{\overline{\Delta}}(x,t,\alpha)$ of the Cauchy problem (5.57). The solution of (5.58) is then a solution of the boundary value problem (5.43), which proves the theorem.

Remark 1.1. Suppose that the condition of Theorem 5.10 is satisfied and the free term α_0 in the collection of the Fourier coefficients $\alpha=(\alpha_0,\alpha_1',\alpha_1'',\alpha_2',\alpha_2'',\ldots)$, determining a 2π-periodic function $|\mu(t)|\leq c-M\pi$ equals zero. The boundary value problem (5.43) has a solution $\tilde{u}(x,t,\alpha)\in L^\infty$ of the form (5.58) in $\prod_{2\pi}$ if and only if the following equality holds

$$\Delta(\alpha)=\int_0^\pi d\xi\int_0^\xi\{F[u,u_t,u_x](\eta,\pi+\xi-\eta)+$$

$$+F[u,u_t,u_x](\eta,\pi-\xi+\eta)\}d\eta=0. \tag{5.61}$$

Thus the condition (5.61) (or (5.59)) is a condition on computing a collection of the Fourier coefficients $\alpha=(\alpha_0,\alpha_1',\alpha_1'',\alpha_2',\alpha_2'',\ldots)$, determining a 2π-periodic function $|\mu(t)|\leq c-M\pi$ which is used in the construction of a 2π-periodic solution $\tilde{u}\in L^\infty$ of a 2π-HID system of the first kind. Therefore, the problem on the existence and construction of generalized 2π-periodic solutions ($\tilde{u}\in L^\infty$) of 2π-HID systems of the first kind (hyperbolic integro-differential equations (5.43) of the special form $f\left(x,t,0,0,0,\int_0^{h(x,t)}\varphi(x,t,s,0,0,0)ds\equiv 0\right)$) is equivalent to the existence of points α (a collection of Fourier coefficients) determining an initial function $|\mu(t)|\leq c-M\pi$ satisfying $\tilde{\Delta}(\alpha)=0$ ($\Delta(\alpha)=0$).

The points α satisfying $\tilde{\Delta}(\alpha)=0$ are the singular points of the mapping [103]

$$\tilde{\Delta}:\mathcal{D}_\mu\to\mathbb{R},\quad \tilde{\Delta}(\alpha)=\int_0^{2\pi}\mu(z)dz+$$

$$+\int_0^{2\pi}d\xi\int_0^\xi\{F[u,u_t,u_x](\eta,\pi+\xi-\eta)+F[u,u_t,u_x](\eta,\pi-\xi+\eta)\}d\eta. \tag{5.62}$$

This mapping can only be computed approximately by computing, say, the following functions

$$\tilde{\Delta}_m(\alpha)=\int_0^{2\pi}\mu(z)dz+\int_0^\pi d\xi\int_0^\xi\{F[u^m,u_t^m,u_x^m](\eta,\pi+\xi-\eta)+$$

$$+F[u^m,u_t^m,u_x^m](\eta,\pi-\xi+\eta)\}d\eta. \tag{5.63}$$

In this connection a problem appears how, proceeding from the mapping (5.63), one can make a conclusion about zeros of the mapping $\tilde{\Delta}(\alpha)$, and hence about the existence of generalized periodic solutions of 2π-systems of the first kind. For so-called T-systems of ordinary differential equations, this problem was solved in [103], as equation (5.61) is scalar, the function $\mu(t)$ can be sought in the form of $\mu(t) = \alpha_k^{'}\cos kt$ or $\mu(t) = \alpha_r^{''}\sin kt$. Then (5.61) is an equation to determine the desired Fourier coefficients $\alpha_k^{'}$ or $\alpha_k^{''}$.

Since the functions $\Delta(\alpha_k^{'})$ and $\Delta(\alpha_k^{''})$ can be computed only approximately starting from the sequences

$$\Delta_m(\alpha_k^{\nu}) = \int_0^{\pi} d\xi \int_0^{\xi} \{F[u^m, u_t^m, u_x^m](\eta, \pi + \xi - \eta) +$$

$$+F[u^m, u_t^m, u_x^m](\eta, \pi - \xi + \eta)\}d\eta, \tag{5.64}$$

where ν stands for a prime or two primes, a problem arises how one can judge about zeros of the function $\Delta_m(\alpha_k^{\nu})$ proceeding from the zeros of the function $|\delta(\alpha_k^{\nu})$. To solve it, one should take account of the fact that each function $\Delta_m(\alpha_k^{\nu})$ (as the limit function $\Delta(\alpha_k^{\nu})$) is defined and continuous for α_k^{ν} in the interval and satisfies the inequality

$$||\Delta_m(\alpha_k^{\nu}) - \Delta(\alpha_k^{\nu})| \leq (3K)^{m-1} M K^2 (1 + LN)^m \beta_1 \frac{\pi^{m+3}}{(m+3)!} = d_m,$$

$$\beta_1 = \left(2 + \frac{\pi}{2}\right)\left(2 + \frac{\pi}{3}\right), \quad m = 2, 3, \dots. \tag{5.65}$$

The last estimate results in the following assertion hold.

Theorem 5.11. *Suppose that equation (5.43) is a 2π-HID system of the first kind and that, for a certain m, the function $\Delta_m(\alpha_k^{\nu})$ given by (5.64) satisfies the inequalities*

$$\min_{\alpha_k^{\nu} \in \mathcal{D}_\mu} \Delta_m(\alpha_k^{\nu}) \leq -d_m, \quad \max_{\alpha_k^{\nu} \in \mathcal{D}_\mu} \Delta_m(\alpha_k^{\nu}) \mathrm{e} d_m, \tag{5.66}$$

where $\mathcal{D}_\mu = [-c + M\pi, c - M\pi]$ and d_m is the constant from (5.65). Then the boundary value problem (5.43) has a solution $\tilde{u}(x, t, \alpha_k^{\nu}) \in L^\infty$ of the form (5.58).

The proof of this theorem is similar to that of Theorem 4.3, Chapter 4.

Finally, the following theorem can be proved similarly to Theorem 4.5 in Chapter 4.

Theorem 5.12. *Let for a 2π-HID system of the first kind the condition of Lemma 5.1 and the following condition be satisfied*

(1) for any function $u(x, t) \in C^1$ odd in variable t, the function $F[u, u_t, u_x](x, t)$ is odd in t.

Then for every continuous odd 2π-periodic function $|u_x(0, t)| \equiv |\mu(t)| \leq c - M\pi$, the boundary value problem (3.1) has a 2π-periodic solution $\tilde{u}(x, t, \alpha) \in L^\infty$ of the form (5.58).

2. 2π-hyperbolic integro-differential systems of the second kind.

Definition 5.7. A 2π-HID system in the region (5.44) is an integro-differential equation of second order (5.43) provided the constants a, b, c, M, K, L and N are related by the inequalities

$$c > 2M\pi, \quad a \leq -c\pi < c\pi \leq b, \tag{5.67}$$

$$2\pi K(1 + LN)(2 + \pi) < 1. \tag{5.68}$$

Let the rectangle $\Pi_{2\pi} = \{0 \leq x \leq \pi, \ 0 \leq t \leq 2\pi\}$ be partitioned into the three parts

$$\overline{\Delta} = \{(x, t) \in \Pi_{2\pi} : 0 \leq x \leq \pi, \quad x \leq t \leq 2\pi - x\},$$

$$\overline{\Delta}_1 = \{(x,t) \in \Pi_{2\pi} : 0 \le x \le \pi, \quad 0 \le t \le x\},$$

$$\overline{\Delta}_2 = \{(x,t) \in \Pi_{2\pi} : 0 \le x \le \pi, \quad 2\pi - x \le t \le 2\pi\}.$$

by the lines $t = x$ and $t = 2\pi - x$. As shown in §1.9, Chapter 1 and will be shown later (see Theorem 3.5), in the class of smooth functions, equation (5.43) is equivalent to the system of integral equations (5.55) for all $(x,t) \in \overline{\Delta}$ and to the systems

$$u_1(x,t,\alpha) = \mu(t+x) - \int_0^x \tilde{F}[u,u_1,u_2](\eta, t+x-\eta)d\eta,$$

$$u_2(x,t,\alpha) = -\mu(t-x) + \int_0^\pi \tilde{F}[u,u_1,u_2](\eta, 2\pi+t-x-\eta)d\eta -$$

$$-\int_0^\pi \tilde{F}[u,u_1,u_2](\eta, t-x+\eta)d\eta, \tag{5.69}$$

$$u(x,t,\alpha) = \int_\pi^x \frac{u_1(\xi,t) - u_2(\xi,t)}{2} d\xi \;\; \forall (x,t) \in \overline{\Delta}_1;$$

$$u_1(x,t,\alpha) = \mu(t+x) - \int_0^\pi \tilde{F}[u,u_1,u_2](\eta, 2\pi+t+x+\eta)d\eta +$$

$$+\int_x^\pi \tilde{F}[u,u_1,u_2](\eta, t+x-\eta)d\eta, \tag{5.70}$$

$$u_2(x,t,\alpha) = -\mu(t-x) + \int_0^x \tilde{F}[u,u_1,u_2](\eta, t-x+\eta)d\eta,$$

$$u(x,t,\alpha) = \int_\pi^x \frac{u_1(\xi,t) - u_2(\xi,t)}{2} d\xi \;\; \forall (x,t) \in \overline{\Delta}_2,$$

where $\mu(t) \in C^1$ is an arbitrary function and $\tilde{F}[u,u_1,u_2](x,t)$ is given by (5.53).

Lemma 5.2. *Suppose that the functions $f(x,t,u,u_t,u_x,v)$, $\varphi(x,t,s,u,u_t,u_x)$ and $h(x,t)$ are defined in the region (5.44), are continuous with respect to variables x,t and s and satisfy the inequalities (5.45)–(5.49), (5.67) and (5.68). Then, for any continuous 2π-periodic function*

$$|\mu(t)| \equiv \left| \frac{\alpha_0}{2} + \sum_{k=1}^{\infty} (\alpha_k^{'} \cos kt + \alpha_k^{''} \sin kt) \right| \le C - 2M\pi \tag{5.71}$$

the sequences of 2π-periodic in t functions (5.52) and the sequences

$$\tilde{u}_1^0(x,t,\alpha) = \mu(t+x), \;\; \tilde{u}_2(x,t,\alpha) = -\mu(t-x),$$

$$\tilde{u}^0(x,t,\alpha) = \int_\pi^x \frac{\mu(t+\xi) + \mu(t-\xi)}{2} d\xi,$$

$$\tilde{u}_1^{m+1}(x,t,\alpha) = \mu(t+x) - \int_0^x \tilde{F}[\tilde{u}^m, \tilde{u}_1^m, \tilde{u}_2^m](\eta, t+x-\eta)d\eta,$$

$$\tilde{u}_2^{m+1}(x,t,\alpha) = -\mu(t-x) + \int_0^\pi \tilde{F}[\tilde{u}^m, \tilde{u}_1^m, \tilde{u}_2^m](\eta, 2\pi+t-x-\eta)d\eta -$$

$$- \int_x^\pi \tilde{F}[\tilde{u}^m, \tilde{u}_1^m, \tilde{u}_2^m](\eta, t-x+\eta)d\eta, \tag{5.72}$$

$$\tilde{u}^{m+1}(x,t,\alpha) = \frac{1}{2}\int_\pi^x [\tilde{u}_1^{m+1}(\xi,t,\alpha) - \tilde{u}_2^{m+1}(\xi,t,\alpha)]d\xi,$$

$$m = 0,1,2,\ldots;$$

$$\tilde{\tilde{u}}_1^0(x,t,\alpha) = \mu(t+x),\ \tilde{\tilde{u}}_2^0(x,t,\alpha) = -\mu(t-x),$$

$$\tilde{\tilde{u}}^0(x,t,\alpha) = \int_\pi^x \frac{\mu(t+\xi)+\mu(t-\xi)}{2}d\xi,$$

$$\tilde{\tilde{u}}_1^{m+1}(x,t,\alpha) = \mu(t+x) - \int_0^\pi \tilde{F}[\tilde{\tilde{u}}^m, \tilde{\tilde{u}}_1^m, \tilde{\tilde{u}}_2^m](\eta, 2\pi+t+x+\eta)d\eta +$$

$$+ \int_x^\pi \tilde{F}[\tilde{\tilde{u}}^m, \tilde{\tilde{u}}_1^m, \tilde{\tilde{u}}_2^m](\eta, t+x-\eta)d\eta, \tag{5.73}$$

$$\tilde{\tilde{u}}_2^{m+1}(x,t,\alpha) = -\mu(t-x) + \int_0^x \tilde{F}[\tilde{\tilde{u}}^m, \tilde{\tilde{u}}_1^m, \tilde{\tilde{u}}_2^m](\eta, t-x+\eta)d\eta,$$

$$\tilde{\tilde{u}}^{m+1}(x,t,\alpha) = \frac{1}{2}\int_\pi^x [\tilde{\tilde{u}}_1^{m+1}(\xi,t,\alpha) - \tilde{\tilde{u}}_2^{m+1}(\xi,t,\alpha)]d\xi,$$

$$m = 0,1,2,\ldots,$$

converge, as $m \to \infty$, *uniformly with respect to* $(x,t,\alpha) \in I \times \mathbb{R}_t \times \mathcal{D}_\mu$ *to functions* $(u_{01}(x,t,\alpha), u_{02}(x,t,\alpha), u_0(x,t,\alpha))$ *and* $(\tilde{u}_{01}(x,t,\alpha), \tilde{u}_{02}(x,t,\alpha), \tilde{u}_0(x,t,\alpha))$, $(\tilde{\tilde{u}}_{01}(x,t,\alpha), \tilde{\tilde{u}}_{02}(x,t,\alpha), \tilde{\tilde{u}}_0(x,t,\alpha))$ *defined in the region (5.54), -periodic in t and satisfying the systems of integral equations (5.55), (5.69) and (5.70), respectively. Furthermore,*

$$|u^m(x,t,\alpha) - u_0(x,t,\alpha)| \le (3K)^{m-2}MK^2(1+LN)^m\beta_1\frac{\pi^{m+2}}{(m+2)!},$$

$$\beta_1 = \left(2+\frac{\pi}{2}\right)\left(2+\frac{\pi}{3}\right),\quad m = 2,3,\ldots,$$

$$|\tilde{u}^m(x,t,\alpha) - \tilde{u}_0(x,t,\alpha)| \le 2M\pi^2\beta_2^m, \tag{5.74}$$

$$|\tilde{\tilde{u}}^m(x,t,\alpha) - \tilde{\tilde{u}}_0(x,t,\alpha)| \le 2M\pi^2\beta_2^m,$$

$$\beta_2 = 2\pi K(1+LN)(2+\pi) < 1,\quad m = 1,2,\ldots$$

The proof of Lemma 5.2 is similar to that of Lemma 4.2 in Chapter 4.

Following the results established in §4.3, Chapter 4 one might study the existence of continuous solutions of the boundary value problem (5.43) in the rectangle $\Pi_{2\pi} = \{0 \le x \le \pi,\ 0 \le t \le 2\pi\}$ in the similar manner. The following statements can readily be proved.

Theorem 5.13. *Suppose that the condition of Lemma 5.2 is satisfied. Then, for any continuous 2π-periodic function $|u_x(o,t)| \equiv |\mu(t)| \le C - 2M\pi$ satisfying (5.59), which is equivalent to the equation*

$$u(\pi,\pi,\alpha) \equiv \int_0^{2\pi} \mu(z)dz - \int_0^{\pi} d\eta \int_{\eta}^{2\pi-\eta} F[u,u_t,u_x](\eta,\xi)d\xi = 0, \tag{5.75}$$

the boundary value problem (5.43) has a unique solution that is continuous in the rectangle $\Pi_{2\pi}$ and has the form

$$u(x,t,\alpha) = \begin{cases} u_0(x,t,\alpha), & (x,t) \in \overline{\Delta}, \\ \tilde{u}_0(x,t,\alpha), & (x,t) \in \Pi_{2\pi}/\{\overline{\Delta} \bigcup \overline{\Delta}_2\}, \\ \tilde{\tilde{u}}_0(x,t,\alpha), & (x,t) \in \Pi_{2\pi}/\{\overline{\Delta} \bigcup \overline{\Delta}_1\}, \end{cases}$$

where $u_0(x,t,\alpha)$ is a solution of system (5.55), $\tilde{u}_0(x,t,\alpha)$ a solution of (5.69) and $\tilde{\tilde{u}}_0(x,t,\alpha)$ a solution of (5.70).

Theorem 5.14. *Let the condition of Lemma 5.2 and condtion (1) in Theorem 5.12 be fulfilled. Then, for any continuous odd 2π-periodic function $|u_x(0,t)| = |\mu(t)| \le C - 2M\pi$, the boundary value problem (5.43) has a unique continuous solution $u(x,t,\alpha)$ defined in $\Pi_{2\pi}$.*

Theorem 5.15. *Suppose that the condition of Lemma 5.2 and the following condition are satisfied*

1) for every function $u(x,t) \in C^1$ even with respect to t, the function $F[u,u_t,u_x](x,t)$ is even with respect to t.

Then, for any continuous even 2π-periodic function $|u_x(0,t)| \equiv |\mu(t)| \le C - 2M\pi$, the boundary value problem (5.43) has a 2π-periodic solution $\overline{u} \in L^\infty$ of the form

$$\overline{u}(x,t,\alpha) = \begin{cases} u_0(x,t,\alpha), & (x,t) \in \overline{\Delta}/\{(x,t) : t = x,\ 0 < x \le \pi\}, \\ \tilde{u}_0(x,t,\alpha), & (x,t) \in \overline{\Delta}_1/\{(0,0)\}, \\ \tilde{\tilde{u}}_0(x,t,\alpha), & (x,t) \in \overline{\Delta}_2/\{(x,t) : t = 2\pi - x,\ 0 \le x \le \pi\}. \end{cases} \tag{5.76}$$

Note that constructing solutions of the boundary value problem (5.43) and proving their existence, the properties of 2π-HID systems of the second kind allow us to verify the following statement being true for 2π-HID systems of the first kind.

Theorem 5.16. *Let a 2π-HID system of the firts kind satisfies the condition of Lemma 3.1. Then, for any continuous 2π-periodic function $|\mu(t)| \le C - M\pi$, the boundary value problem (5.43) has a 2π-periodic solution $\tilde{\tilde{u}}(x,t,\alpha) \in L^\infty$ of the form*

$$\tilde{\tilde{u}}(x,t,\alpha) = \begin{cases} u_{\overline{\Delta}}(x,t,\alpha) \equiv u_0(x,t,\alpha), & (x,t) \in \overline{\Delta}/\{(\pi,\pi)\}, \\ 0, & (x,t) \in \Pi_{2\pi}/\{\overline{\Delta}\{(\pi,\pi)\}\}. \end{cases} \tag{5.77}$$

5.4. Periodic solutions of nonlinear wave equations with small parameter

In this section we investigate the problems concerning the existence of 2π-periodic solutions of boundary value problems with a small parameter ε, which are of a general form

$$u_{tt} - u_{xx} =$$

$$= \varepsilon f\left(x,t,u,u_t,u_x,\int\limits_0^{h(x,t)}\varphi(x,t,s,u(x,s),u_t(x,s),u_x(x,s))ds\right), \qquad (5.78)$$

$$u(0,t)=u(\pi,t)=0, \quad u(x,t+2\pi)=u(x,t),$$

The results obtained are then illustrated by some particular examples.

Theorem 5.17. *Suppose that the functions $f(x,t,u,u_t,u_x,v)$ and $\varphi(x,t,s,u,u_t,u_x)$ are defined in the region $\Omega_\infty=\{0\le x\le\pi,\ t\in\mathbb{R}_t, s\in\mathbb{R}_s, |u|\le a, |u_t|\le a, |u_x|\le a, v\in\mathbb{R}^1_v\}$, are continuous with respect to varibales x,t and s, 2π-periodic in t and $f(x,t,u,u_t,u_x,v)\in \mathrm{Lip}_{u,u_t,u_x,v}(K;\Omega_\infty), \varphi(x,t,s,u,u_t,u_x)\in\mathrm{Lip}_{u,u_t,u_x}(L;\Omega_\infty)$, K and L being positive constants. Suppose also that the function $h(x,t)$ is defined in the region $\Pi_\infty=\{0\le x\le\pi, t\in\mathbb{R}_t\}$, is continuous with respect to x and t, 2π-periodic in t and $|h(x,t)|\le M$, M being positive constant. Then, for any continuous and 2π-periodic function $|u_x(0,t)|=|\tilde\mu(t)|<\frac{a}{2\pi}$, the Cauchy* problem

$$u_{tt}-u_{xx}=$$

$$= \varepsilon f\left(x,t,u,u_t,u_x,\int\limits_0^{h(x,t)}\varphi(x,t,s,u(x,s),u_t(x,s),u_x(x,s))ds\right), \qquad (5.79)$$

$$u(0,t)=0, \quad u_x(0,t)=\tilde\mu(t)$$

has a unique continuous 2π-periodic solution in the region $\Delta_\infty=\{0\le x\le\pi, 2\pi n+x\le t\le 2\pi(n+1)-x, n=0,\pm1,\pm2,\ldots\}$ for all values of $\varepsilon, 0<|\varepsilon|<\infty$ for which $|u|\le a, |u_t|\le a, |u_x|\le a$ hold for $(x,t)\in\overline{\Delta}=\{0\le x\le\pi, x\le t\le 2\pi-x\}$.

P r o o f. In the class of smooth functions, the Cauchy problem (4.2) is equivalent to the following system of integral equations (see §1.5, Chapter 1)

$$u_1(x,t)=\tilde\mu(t+x)-\varepsilon\int\limits_0^x \tilde F[u,u_1,u_2](\eta,t+x-\eta)d\eta,$$

$$u_2(x,t)=-\tilde\mu(t-x)+\varepsilon\int\limits_0^x \tilde F[u,u_1,u_2](\eta,t-x+\eta)d\eta,$$

$$u(x,t)=\frac{1}{2}\int\limits_0^x\{u_1(\xi,t)-u_2(\xi,t)\}d\xi \quad \forall (x,t)\in\overline{\Delta}=$$

$$=\{0\le x\le\pi,\ x\le t\le 2\pi-x\}, \qquad (5.80)$$

where u_1, u_2 and $\tilde F[u,u_1,u_2](x,t)$ are given by (5.53).

Solving (5.80) by the method of successive approximations, we see Theorem 4.1 hold.

Remark 5.6. The solution $u_{\overline{\Delta}}(x,t,\alpha)$ thus constructed evidently satisfies the first boundary condition $u(0,t)=0$, and the second boundary condition $u(\pi,t)=0$ is satisfied, generally speaking, not for every initial function $\tilde\mu(t)$.

Theorem 5.18. *Suppose that in the regions Ω_∞ and Π_∞, the condition of Theorem 4.1 and the following conditions are fulfilled*

1) $\quad f\left(x,t,0,0,0,\int\limits_0^{h(x,t)}\varphi(x,t,s,0,0,0)ds\right)\not\equiv 0,$

2) $$f\left(x,-t,0,0,0,\int\limits_0^{h(x,-t)}\varphi(x,-t,s,0,0,0)ds\right)=,$$

$$=-f\left(x,t,0,0,0,\int\limits_0^{h(x,t)}\varphi(x,t,s,0,0,0)ds\right);$$

3) for any function $u(x,t)\in C^1$ odd in variable t, the function $\tilde{F}[u,u_1,u_2](x,t)$ is odd in variable t.

Then, for $u_x(0,t)=0$, the solution of the Cauchy problem (5.79) in the region $\overline{\Delta}=\{0\le x\le\pi,\ x\le t\le 2\pi-x\}$ satisfies the equality $u(\pi,\pi)=0$.

P r o o f. From system (5.80) it follows that, for $u_x(0,t)=\tilde{\mu}(t)\equiv 0$, the solution of the Cauchy problem (5.79) satisfies the system of integral equations

$$u_1(x,t)=-\varepsilon\int\limits_0^x\tilde{F}[u,u_1,u_2](\eta,t+x-\eta)d\eta,$$

$$u_2(x,t)=\varepsilon\int\limits_0^x\tilde{F}[u,u_1,u_2](\eta,t-x+\eta)d\eta,$$

$$u(x,t)=-\frac{\varepsilon}{2}\int\limits_0^x d\xi\int\limits_0^\xi\{\tilde{F}[u,u_1,u_2](\eta,t+\xi-\eta)+$$
$$+\tilde{F}[u,u_1,u_2](\eta,t-\xi+\eta)\}d\eta. \tag{5.81}$$

Choosing now $u_1^0(x,t)=0$, $u_2^0(x,t)=0$, $u^0(x,t)=0$ as a zero approximation and taking account of (5.81), we construct the sequences of functions $(u_1^n(x,t))$, $(u_2^n(x,t))$, $(u^n(x,t))$. We shall show all the members of the sequence $(u^n(x,t))$ to be odd functions of variable t. We have

$$u^1(x,t)=-\frac{\varepsilon}{2}\int\limits_0^x d\xi\int\limits_0^\xi\{\tilde{F}[0,0,0](\eta,t+\xi-\eta)+$$
$$+\tilde{F}[0,0,0](\eta,t-\xi+\eta)\}d\eta\equiv$$
$$\equiv-\frac{\varepsilon}{2}\int\limits_0^x d\xi\int\limits_0^\xi\left\{f\left(\eta,t+\xi-\eta,0,0,0,\int\limits_0^{h(\eta,t+\xi-\eta)}\varphi(\eta,t+\xi-\right.\right.$$
$$\left.-\eta,s,0,0,0)ds\right)+f\left(\eta,t-\xi+\eta,0,0,0,\int\limits_0^{h(\eta,t-\xi+\eta)}\varphi(\eta,t-\xi+\right.$$
$$\left.\left.+\eta,s,0,0,0)ds\right)\right\}d\eta\equiv-u_{11}(x,t). \tag{5.82}$$

It is easy to show that $u_{11}(x,-t)=-u_{11}(x,t)$, provided the condition (2) in Theorem 5.18 is fulfilled. From the last equality and (5.82) it follows that $u^1(x,-t)=-u^1(x,t)$.

Next, proceeding by induction and using the conditions (2) and (3) in Theorem 5.18 and the third equation in (5.81), we see that all the members of the sequence $(u^n(x,t))$ are odd functions of variable t and $u^n(\pi,\pi)=0$, $n=0,1,2,\ldots$. Hence, the limit function

$u_{\overline{\Delta}}(x,t) = \lim_{n\to\infty} u^n(x,t)$ (a solution of the Cauchy problem (5.79) in the region $\overline{\Delta}$) satisfies the boundary condition $u(0,t) = 0, u(\pi,\pi) = 0$. This completes the proof.

Theorem 5.19. *Suppose that the condition of Theorem 4.1 and the condition (3) in Theorem 5.18 are satisfied in the regions Ω_∞ and Π_∞. Then, for any continuous odd and 2π-periodic function $|u_x(0,t)| = |\tilde{\mu}(t)| < \frac{a}{2\pi}$, the solution of the Cauchy problem in the region $\overline{\Delta}$ satisfies $u(\pi,\pi) = 0$.*

P r o o f. Take the odd function $u^0(x,t)$, which is 2π-periodic in t, as the zero approximation $u^0(x,t) = \frac{1}{2}\int\limits_0^x(\tilde{\mu}(t+\xi) + \tilde{\mu}(t-\xi))d\xi$ in the system of integral eqautions (5.80), set $u_1^0(x,t) = \tilde{\mu}(t, u_2^0(x,t) = -\tilde{\mu}(t))$ and take account of the condition of Theorem 5.19. The theorem then follows.

Theorem 5.20. *Let the condition of Theorem 4.1 be fulfilled in the regions Ω_∞ and Π_∞. Then, for any continuous 2π-periodic function $|u_x(0,t)| = |\tilde{\mu}(t)| < \frac{a}{2\pi}$ that satisfies the equation*

$$u(\pi,\pi) \equiv \int\limits_o^{2\pi} u_x(0,\alpha)d\alpha - \varepsilon \int\limits_0^{\pi} d\eta \int\limits_\eta^{2\pi-\eta} F[u,u_t,u_x](\eta,\xi)d\xi = 0, \tag{5.83}$$

where $F[u,u_t,u_x](x,t)$ is given by (5.60), the boundary value problem (5.78) has a unique continuous solution $u(x,t)$ in the rectangle $\Pi_{2\pi} = \{0 \le x \le \pi, 0 \le t \le 2\pi\}$, provided $|\varepsilon|$ is sufficiently small.

This theorem can be proved in the same manner as Theorem 4.6 in Chapter 4.

Remark 5.7. Note that, proving the existence of a continuous solution of the boundary value problem (4.1), we assumed the condition (5.83) to hold. If this condition does not hold, the solution of the boundary value problem (5.78) will usually be discontinuous along the characteristics OC and AC (see Fig. 2).

Proceeding as in the proofs of Theorems 5.18 and 5.19, one can readily show the following statements to be true.

Theorem 5.21. *Let the condition of Theorem 5.17 and conditions (1) and (2) in Theorem 5.18 be satisfied in the regions Ω_∞ and Π_∞. Let the following condition also hold in these regions*

(1) for every odd function $u(x,t) \in C^1$ odd in variable t, the function $F[u,u_t,u_x](x,t)$ is odd in t.

Then the boundary value problem (5.78) has a unique continuous solution $u(x,t)$ defined in $\Pi_{2\pi}$, provided $|\varepsilon|$ is small enough.

Theorem 5.22. *Suppose that the condition of Theorem 5.17 and the condition (1) in Theorem 5.21 are fuifilled in the regions Ω_∞ and Π_∞. Then, for any continuous odd 2π-periodic function $|u_x(0,t)| = |\tilde{\mu}(t)| < \frac{a}{2\pi}$, the boundary value problem (5.78) has a unique continuous solution $u(x,t)$ defined in $\Pi_{2\pi}$, provided $|\varepsilon|$ is small enough.*

Note that the continuous solution $u(x,t)$ of (5.78) constructed in the rectangle $\Pi_{2\pi}$ according to Theorems 5.20–5.22 does not satisfy, in general, the periodicity condition, i. e. $u(x,0) \neq u(x,2\pi)$ for all $x \in [0,\pi]$. This condition reqiures an addition condition to be held. The latter can easily be found as it was described in Remark 4.2, Chapter 4, which leads us to the following statement.

Theorem 5.23. *Let the condition of Theorem 5.17 together with the following condition*

1) for every function $u(x,t)$ even in variable t, the function $Fu,u_t,u_x[(x,t)$ is even in t.

be fulfilled in the regions Ω_∞ and Π_∞.

Then, for any continuous even 2π-periodic function $|u_x(0,t)| = |\tilde{\mu}(t)| < \frac{a}{2\pi}$, the boundary value problem (5.78) has a unique weak solution $u(x,t) \in L^\infty$, provided $|\varepsilon|$ is small enough.

The proof of this theorem is similar to that of Theorem 4.8 in Chapter 4.

Finally, assume that $f(x,t,0,0,0,\int_0^{h(x,t)} \varphi(x)t,s,0,0,0)ds \equiv 0$. Then, setting $u \equiv 0$ in the regions $\overline{\Delta}_1$ and $\overline{\Delta}_2$ and using Theorem 5.22, we arrive at the statement.

Theorem 5.24. *Suppose that the condition of Theorem 5.22 is satisfied and that*

$$f\left(x,t,0,0,0,\int_0^{h(x,t)} \varphi(x,t,s,0,0,0)ds\right) \equiv 0.$$

holds.

Then, for any continuous odd 2π-periodic function $|u_x(0,t)| = |\tilde{\mu}(t)| < \frac{a}{2\pi}$, the boundary value problem (5.78) has a unique solution $u(x,t) \in L^\infty$ for all $|\varepsilon|$.

P r o o f. Let the rectangle $\Pi_{2\pi}$ be partitioned into the three parts $\overline{\Delta}, \overline{\Delta}_1$ and $\overline{\Delta}_2$ by the lines $t = x$ and $t = 2\pi - x$, $0 \leq x \leq \pi$ (see Fig. 2). Setting $u(x,t) \equiv 0$ in the regions $\overline{\Delta}_1$ and $\overline{\Delta}_2$ and $u(x,t) = u_{\overline{\Delta}}(x,t)$ (the solution of the Cauchy problem (5.79) in $\overline{\Delta}$) in the region $\overline{\Delta}$ yields a bounded function that is 2π-periodic in t. Being defined in the rectangle $\Pi_{2\pi}$, this function is a solution of the boundary value problem (4.1) in $\Pi_{2\pi}$ and belongs to the space L^∞, as required.

Examples. Note that the problems described by the equations with small parameter such as (5.78), where the functions $f(x,t,u,u_t,u_x,v), \varphi(x,t,s,u,u_t,u_x)$ and $h(x,t)$ satisfy, in the region (5.44), the conditions (5.45)–(5.49) and $f(x,t,0,0,0,\int_0^{h(x,t)} \varphi(x,t,s,0,0,0)ds) \equiv 0$, are always 2π-HID periodic systems of the first kind, provided ε is small enough.

For instance, the equation

$$u_{tt} - u_{xx} = \varepsilon(\alpha u + \beta u^3) \int_0^{\sin t} u(x,s)ds, \quad 0 < \alpha,\ \beta < \infty, \tag{5.84}$$

the right-hand side of which $(\alpha u + \beta u^3) \int_0^{\sin t} u(x,s)ds$ satisfies the condition (1) in Theorem 5.12, has infinitely many 2π-periodic solutions of the form (5.58) for sufficiently small ε, i. e. for every odd function $\mu(t) = \alpha_k'' \sin kt$, $|\alpha_k''| < c - \varepsilon M\pi$, the boundary value problem (5.78) has a 2π-periodic solution of the form (5.58) if $f(x,t,u,u_t,u_x,v) \equiv (\alpha u + \beta u^3) \int_0^{\sin t} u(x,s)ds$. Such a solution obviously depends on the real parameter α_k''.

On the other hand, in the class of smooth functions, the system of integral eqautions (5.55) is also equivalent to the Cauchy problem (5.57) if $\mu(t) = 0$ and $f(x,t,0,0,0,\int_0^{h(x,t)} \varphi(x,t,s,0,0,0)ds) \equiv 0$. Therefore, depending on the choice of the initial approximation $u^0(x,t,0)$, the solution $u(x,t,0)$ of (5.55) for $\mu(t) \equiv 0$ may satisfy or may not satisfy the equality $u(\pi,\pi,0) = 0$. Note that there is no additional condition for the equality $u(\pi,\pi,0) = 0$ to hold. However, for $\mu(t) \equiv 0$, there are infinitely many 2π-periodic continuous solutions of the boundary value problem (5.78). These solutions are of another kind than the solution (5.58), which can easily be explained by the following example.

Consider the boundary value problem given as

$$u_{tt} - u_{xx} = \varepsilon(\alpha u_t + \beta u_t^3) \int\limits_0^{\sin^2 t} u(x, t-s)ds, \quad 0 < \alpha, \ \beta < \infty,$$

$$u(0,t) = u(\pi,t) = 0, \quad u(x, t+2\pi) = u(x,t). \tag{5.85}$$

If we pick the function $u^0(x,t) = \alpha \cos t \cos x$ as the zero approximation this function evidently belongs to the class $A_2 = \{u^0 : u^0(x,t) = u^0(x,t+2\pi) = u^0(\pi - x, t+\pi)\}$, (see §3.1, Chapter 3), then the function $f(x,t,u^0,u_t^0,u_x^0,v^0) = (\alpha u_t^0 + \beta(u_t^0)^3) \int\limits_0^{\sin^2 t} u^0(x,t-s)ds = -\alpha^3(\cos x \sin t + \alpha\beta \cos^3 x \sin^3 t) \int\limits_0^{\sin^2 t} \cos x \cos(t-s)ds$, according to which the first approximation $u^1(x,t)$ for (5.85) is defined, also belongs to the class A_2. As shown in Chapter 3 and the present chapter, in this case there are other kinds of operator (distinct from (5.55)) transforming A_2 into itself. Note that in the space A_2 there are also classical 2π-periodic solutions of the boundary value problem (5.78). The results of §5.2 show that for certain values of the parameter α in the initial approximation $u^0(x,t) = \alpha \cos t \cos x$ and the parameter $\varepsilon, \varepsilon \ll 1$, the boundary value problem (5.85) has infinitely many smooth 2π-periodic solutions. The same statement holds for the boundary value problem

$$u_{tt} - u_{xx} = \varepsilon(1 - u^2)u_t \int\limits_0^{\cos x \sin t} u(x,t-s)ds,$$

$$u(0,t) = u(\pi,t) = 0, \quad u(x,t+2\pi) = u(x,t). \tag{5.86}$$

However, if the right-hand sides of equations (5.85) and (5.86) are of the form

$$f(x,t,u,u_t,u_x,v) \equiv g(x,t) +$$

$$+\varepsilon(\alpha u_t + \beta u_t^3) \int\limits_0^{\sin^2 t} u(x,t-s)ds, \quad 0 < \alpha, \ \beta < \infty,$$

$$f(x,t,u,u_t,u_x,v) \equiv g(x,t) +$$

$$+\varepsilon(1 - u^2)u_t \int\limits_0^{\cos x \sin t} u(x,t-s)ds, \quad g(x,t) \in A_2, \tag{5.87}$$

then, following [21, 76], one can readily show the solution to be unique in both cases.

It is noteworthy that the solutions indicated above are solutions of the second order wave equations of the first class. An essential point here is that the periodicity condition (4.68) (see §4.3, Chapter 4) holds in these cases, i. e. in the space A_2 this condition is satisfied automatically for $\tilde{\mu}(t) \equiv 0$.

A number of other problems can be treated in the similar fashion

One can similarly deal with a number of other problems for periodic systems of the first class, solutions of which exist not only in A_2 but also in the spaces $A_1, A_3, A_k^0, k = 1,2,3$ (see §3.1, Chapter 3). It is also possible to construct examples for systems of the second class [92, 93, 97].

CHAPTER 6

HYPERBOLIC SYSTEMS WITH FAST AND SLOW VARIABLES AND ASYMPTOTIC METHODS FOR SOLVING THEM

The present chapter deals with the construction and justification of general asymptotic expansions of solutions of hyperbolic equations not subjected to boundary and initial conditions.

6.1. The first approximation of asymptotic solutions of the second order equations

1. The first approximation for equations without fast variables. In [24] a new nonlinear approach, resulted from extending asymtotic methods to the partial differential equations of elliptic type, was developed to study flat waves in a stratified medium. We shall consider its natural generalisation to hyperbolic equations [134].

Consider a wave solution of the quasilinear equation

$$u_{tt} - a^2 u_{xx} = \lambda u + \varepsilon f(\eta, \tau, u, p, q, \varepsilon), \tag{6.1}$$

where ε is a small parameter; $\eta = \varepsilon x$, $\tau = \varepsilon t$; $\varepsilon f = \varepsilon f_1 + \varepsilon^2 f_2 + \ldots$, $f_\alpha = f_\alpha(\eta, \tau, u, p, q)$, $\alpha = 1, 2, \ldots$; $p = u_t$; $q = u_x$; a and λ are constants.

Suppose the function f is sufficiently smooth in the considered region. The expansion $\varepsilon f = \varepsilon f_1 + \varepsilon f_2 + \ldots$ is generally thought of as asymptotic, i. e.

$$\left| f - \sum_{\alpha=1}^{k} \varepsilon^{\alpha-1} f_a \right| < C_k \varepsilon^k, \quad C_k = \text{const.}$$

for small ε.

In the sequel we shall investigate wave solutions of equation (6.1) corresponding to a simple harmonic flat wave in a region of large extension (equivalent to $1/\varepsilon$) in all the coordinates x and t.

Letting $\varepsilon = 0$ in (6.1), we arrive at nonperturbed equation

$$u_{0tt} - a^2 u_{0xx} = \lambda u_0. \tag{6.2}$$

Assume that equation (6.2) has a wave solution

$$u_0 = u_0(\omega t - kx), \tag{6.3}$$

$(\omega, -k)$ being a nonzero wave vector and u_o a function 2π-periodic with respect to wave phase $\xi = \omega t - kx$.

Substituting (6.3) into (6.2) gives an ordinary differential equation $(\omega^2 - a^2k^2)u_{0\xi\xi} = \lambda u_0$, which has a solution $\exp\{\pm i\sqrt{\lambda/(a^2k^2 - \omega^2)}\xi\}$, provided $\lambda(\omega^2 - a^2k^2) < 0$ and $\omega^2 - a^2k^2 \neq 0$. Hence, u_0 is a 2π-periodic function of ξ if and only if

$$a^2k^2 - \omega^2 = \lambda, \quad \lambda(\omega^2 - a^2k^2) < 0, \quad \omega^2 - a^2k^2 \neq 0. \tag{6.4}$$

In that case the general solution of (6.3) is of the form

$$u_0 = c\cos(\omega t - kx + h), \tag{6.5}$$

where c and h are arbitrary constants and ω and k satisfy (6.4).

To build up asymptotic approximations to the solutions of perturbed equation (6.1), we make use of the modifying parameter method together with the asymptotic change of variables and set

$$\begin{aligned} u(x,t,\varepsilon) &= c\,(x,t,\varepsilon)\cos[\xi(x,t,\varepsilon)],\\ p(x,t,\varepsilon) &= -\omega c\,(x,t,\varepsilon)\sin[\xi(x,t,\varepsilon)],\\ q(x,t,\varepsilon) &= kc\,(x,t,\varepsilon)\sin[\xi(x,t,\varepsilon)], \end{aligned} \tag{6.6}$$

where the functions $c(x,t,\varepsilon)$ and $\xi(x,t,\varepsilon)$ are defined by the expansions

$$\begin{aligned} c &= \bar{c} + \varepsilon u_1(\eta,\tau,\bar{c},\bar{\xi}) + \varepsilon^2 u_2(\eta,\tau,\bar{c},\bar{\xi}) + \cdots,\\ \xi &= \bar{\xi} + \varepsilon v_1(\eta,\tau,\bar{c},\bar{\xi}) + \varepsilon^2 v_2(\eta,\tau,\bar{c},\bar{\xi}) + \cdots. \end{aligned} \tag{6.7}$$

Now, substituting (6.7) into (6.6), we obtain asymptotic expansions for u, p and q. Next, finding the second derivatives u_{tt} and u_{xx}, substituting them and functions u, p and q into equation (6.1) and equating the coefficients at equal powers in the resulted expression, we obtain a chain system of equations. From this system one can successively find u_i and v_i, $i = 1, 2, \ldots$ and construct amplitude-phase equations with respect to $\bar{c}$ and $\bar{\xi}$ for corresponding approximations. We shall construct the first improved approximation of the solution.

Using (6.7), the expansions of u, p and q into powers of ε (ε^2 and ε) can be written down as

$$u = \bar{c}\cos\bar{\xi} + \varepsilon[u_1\cos\bar{\xi} - \bar{c}v_1\sin\bar{\xi}]; \tag{6.8}$$

$$p = -\omega(\bar{c}\sin\bar{\xi} + \varepsilon[u_1\sin\bar{\xi} + \bar{c}v_1\cos\bar{\xi}]); \tag{6.9}$$

$$q = k(\bar{c}\sin\bar{\xi} + \varepsilon[u_1\sin\bar{\xi} + \bar{c}v_1\cos\bar{\xi}]). \tag{6.10}$$

Suppose that $\bar{c} = \bar{c}(\xi_0)$, $\bar{\xi} = \omega t - kx + \bar{h}(\xi_0)$ in the first approximation, where $\xi_0 = \varepsilon(\omega t - kx)$.

Differentiating (6.8) with respect to t and equating the resulted expression to (6.9) yield

$$\frac{d\bar{c}}{d\xi_0}\cos\bar{\xi} - \frac{d\bar{h}}{d\xi_0}\bar{c}\sin\bar{\xi} + \frac{\partial u_1}{\partial\bar{\xi}}\cos\bar{\xi} - \frac{\partial v_1}{\partial\bar{\xi}}\bar{c}\sin\bar{\xi} = 0. \tag{6.11}$$

Substituting the second derivatives $u_t = p_t$ and $u_{xx} = q_x$ and relations (6.8)–(6.10) into equation (6.1), we obtain

$$(a^2k^2 - \omega^2)\left[\frac{d\bar{c}}{d\xi_0}\sin\bar{\xi} + \frac{d\bar{h}}{d\xi_0}\bar{c}\cos\bar{\xi} + \frac{\partial\bar{u}_1}{\partial\bar{\xi}}\sin\bar{\xi} + \frac{\partial\bar{v}_1}{\partial\bar{\xi}}\bar{c}\cos\bar{\xi}\right] = f_1; \tag{6.12}$$

$$f_1 = f_1(\eta,\tau,\bar{c}\cos\bar{\xi},\ -\omega\bar{c}\sin\bar{\xi},\ k\bar{c}\sin\bar{\xi}). \tag{6.13}$$

The system of equations (6.11) and (6.12) is always solvable with respect to $\partial u_1/\partial\bar{\xi}$ and $\partial v_1/\partial\bar{\xi}$, i. e.

$$\begin{aligned} (a^2k^2 - \omega^2)\frac{\partial u_1}{\partial\bar{\xi}} &= f_1\sin\bar{\xi} - \langle f_1\sin\bar{\xi}\rangle,\\ \bar{c}(a^2k^2 - \omega^2)\frac{\partial v_1}{\partial\bar{\xi}} &= f_1\cos\bar{\xi} - \langle f_1\cos\bar{\xi}\rangle; \end{aligned} \tag{6.14}$$

$$(a^2k^2-\omega^2)\frac{d\bar c}{d\xi_0}=\langle f_1\sin\bar\xi\rangle,$$

$$\bar c(a^2k^2-\omega^2)\frac{d\bar h}{d\xi_0}=\langle f_1\cos\bar\xi\rangle, \tag{6.15}$$

where

$$\langle f_1\sin\bar\xi\rangle=\frac{1}{2\pi}\int_0^{2\pi} f_1\sin\bar\xi\,d\bar\xi;\quad \langle f_1\cos\bar\xi\rangle=\frac{1}{2\pi}\int_0^{2\pi} f_1\cos\bar\xi\,d\bar\xi.$$

Taking into account the relations $\bar c=\bar c(\xi_0)$ and $\bar\xi=\omega t-kx+\bar h(\xi_0)$, $\xi_0=\varepsilon(\omega t-kx)$ we can write

$$\omega\frac{\partial\bar c}{\partial\bar t}+a^2k\frac{\partial\bar c}{\partial x}=\varepsilon(\omega^2-a^2k^2)\frac{d\bar c}{d\xi_0},$$

$$\omega\frac{\partial\bar h}{\partial\bar t}+a^2k\frac{\partial\bar h}{\partial x}=\varepsilon(\omega^2-a^2k^2)\frac{d\bar h}{d\xi_0}. \tag{6.16}$$

Combining (6.15) and (6.16) gives

$$\frac{\partial\bar c}{\partial\bar t}+\frac{a^2k}{\omega}\frac{\partial\bar c}{\partial x}=-\frac{\varepsilon}{\omega}\langle f_1\sin\xi\rangle,$$

$$\frac{\partial\bar h}{\partial\bar t}+\frac{a^2k}{\omega}\frac{\partial\bar h}{\partial x}=-\frac{\varepsilon}{\omega\bar c}\langle f_1\cos\xi\rangle. \tag{6.17}$$

Thus the first improved approximation of the asymptotic solution of equation (6.1) has the form

$$u=\bar c\cos\bar\xi+\varepsilon[u_1\cos\bar\xi-\bar c v_1\sin\bar\xi],$$

The amplitude $\bar c$ and phase $\bar h$ can be found from the system of partial differential equations of the first approximation (6.17). The functions u_1 and v_1 can be obtained by integrating system (6.14) with respect to $\bar\xi$ from zero to $\bar\xi$, $u_1|_{\bar\xi=0}$ and $v_1|_{\bar\xi=0}$ being arbitrary functions (for example, $u_1|_{\bar\xi=0}=0$, $v_1|_{\bar\xi=0}$).

Note that expression (6.10) becomes a consequence of expressions (6.8) and (6.9) if equality (6.11) is taken into account. We need this equality in order to be able to write the derivative $u_x=q$. In this case $u_{tx}=u_{xt}$ $(p_x=q_t)$.

2. The first approximation to an equation of the general form. Consider the equation

$$u_{tt}-a^2u_{xx}=\lambda u+\varepsilon f(x,t,\eta,\tau,u,p,q,\varepsilon), \tag{6.18}$$

where $p=u_y$; $q=u_x$; $\eta=\varepsilon x$; $\tau=\varepsilon t$ and $\varepsilon>0$ is a small parameter.

Letting $\varepsilon=0$ in (6.18) gives the nonperturbed equation (6.2). Assume that equation (6.2) has a wave solution $u_0=a\cos(\omega f-kx+h)$, where ω and k satisfy (6.4).

The solution of eqaution (6.18) is sought as $u=c(x,t,\varepsilon)\cos\beta$, $p=-\omega c(x,t,\varepsilon)\sin\beta$, $q=kc(x,t,\varepsilon)\sin\beta$ where $\beta=\omega t-kx+h(x,t,\varepsilon)$. Differentiating the function u with respect to u_t and u_x and taking into account the equalities $p=u_t$ and $q=u_x$, we obtain

$$\frac{\partial c}{\partial t}\cos\beta-\frac{\partial h}{\partial t}c\sin\beta=0, \tag{6.19}$$

$$\frac{\partial c}{\partial x}\cos\beta-\frac{\partial h}{\partial x}c\sin\beta=0. \tag{6.20}$$

Next, computing the second derivatives u_{tt} and u_{xx} and using equation (6.18), we have

$$-\omega\frac{\partial c}{\partial x}\sin\beta-\omega\frac{\partial h}{\partial t}c\cos\beta-a^2k\frac{\partial c}{\partial x}\sin\beta-a^2k\frac{\partial h}{\partial x}c\cos\beta=\varepsilon f^*(x,t,\eta,\tau,c,\beta), \tag{6.21}$$

where $f^*(x,t,\eta,\tau,c,\beta) = f(x,t,\eta,\tau,c\cos\beta,\ -\omega c\sin\beta,\ kc\sin\beta)$.

Combining equalities (6.19) and (6.20), we can write

$$\omega\frac{\partial c}{\partial x} + a^2k\omega\frac{\partial c}{\partial x}\sin^2\beta + a^2kc\sin\beta\cos\beta\omega\frac{\partial h}{\partial x} = -\varepsilon f^*(x,t,\eta,\tau,c,\beta)\sin\beta. \tag{6.22}$$

From equality (6.20) it follows that $\frac{\partial h}{\partial x}c\sin\beta = \omega\frac{\partial c}{\partial x}\cos\beta$. So, (6.22) can be rewritten as

$$\omega\frac{\partial c}{\partial t} + a^2k\omega\frac{\partial c}{\partial x} = -\varepsilon f^*\sin\beta. \tag{6.23}$$

From equalities (6.19)–(6.21) we obtain another equation

$$c\left(\omega\frac{\partial h}{\partial t} + a^2k\frac{\partial h}{\partial x}\right) = -\varepsilon f^*\cos\beta, \tag{6.24}$$

which constitutes, together with equation (6.23), the system

$$\begin{gathered}\omega\frac{\partial c}{\partial t} + a^2k\frac{\partial c}{\partial x} = -\varepsilon f^*\sin\beta,\\ c\left(\omega\frac{\partial h}{\partial t} + a^2k\frac{\partial h}{\partial x}\right) = -\varepsilon f^*\cos\beta\end{gathered} \tag{6.25}$$

which enable us to determine the desired functions c and h.

Applying the averaging methods (they will be presented in §6 of this chapter) to the hyperbolic type system (6.25), we can construct the first approximation to the solution u of equation (6.18). Note that this general method of determining single harmonic wave solutions of equation (6.18) leaves open the question about the fulfillement of equality condition for mixed derivatives p_x and q_t. Differentiating the expressions for p and q, $p = -\omega c\sin(\omega t - kx + h)$, $q = kc\sin(\omega t - kx + h)$ with respect to x and t, respectively, we come to the conclusion that for the equality $p_x = q_t$ to hold it is necessary that c and h be solutions of a linear homogeneous partial differential system of the first order

$$-\omega\frac{\partial c}{\partial x} = k\frac{\partial c}{\partial t},\quad -\omega\frac{\partial h}{\partial x} = k\frac{\partial h}{\partial t},$$

i. e. c and h are to be functions of the form $c = c(\omega t - kx)$ and $h = h(\omega t - kx)$. This property is used in constructing asymptotic expansions.

3. The first approximation to the asymptotic solution of a quasilinear equation. Consider an equation

$$u_{tt} - a^2u_{xx} = \lambda u + \varepsilon f(\eta,\tau,u,p,q,p_x,q_x,\varepsilon) \tag{6.26}$$

where $p = u_t$; $q = u_x$; $\eta = \varepsilon x$; $\tau = \varepsilon t$; $\varepsilon f = \varepsilon f_1 + \varepsilon^2 + \ldots$, $f_\alpha = f_\alpha(\eta,\tau,u,p,q,p_x,q_x)$, $\alpha = 1,2,\ldots$ and ε is a small parameter.

Setting $\varepsilon = 0$ in (6.26) gives the nonperturbed equation (6.2). Suppose that equation (6.2) possesses a wave solution $u_0 = c\cos(\omega t - kx + h)$, with ω and k satisfying the condition (6.4).

In order to construct asymptotic approximations to solutions of the perturbed equation (6.26) we use the asymptotic change of variables (6.6) and (6.7). It results in the first improved approximation $u = \bar{c}\cos\bar{\xi} + \varepsilon[u_1\cos\bar{\xi} - \bar{c}v_1\sin\bar{\xi}]$, where the function u_1 and v_1 are given by the relations

$$u_1(\eta,\tau,\bar{c},\bar{\xi}) = u_{10}(\eta,\tau,\bar{c}) + \frac{1}{a^2k^2-\omega^2}\int_0^{\bar{\xi}}(f_1^*\sin\bar{\xi} - \langle f_1^*\sin\bar{\xi}\rangle)d\bar{\xi},$$

$$v_1(\eta,\tau,\bar{c},\bar{\xi}) = v_{10}(\eta,\tau,\bar{c}) + \frac{1}{\bar{c}(a^2k^2-\omega^2)}\int_0^{\bar{\xi}} (f_1^*\cos\bar{\xi} - \langle f_1^*\cos\bar{\xi}\rangle)d\bar{\xi}. \tag{6.27}$$

The amplitude $\bar{c}$ and phase $\bar{h}$ are described by the partial differential system of the first approximation

$$\frac{\partial \bar{c}}{\partial t} + \frac{a^2k}{\omega}\frac{\partial \bar{c}}{\partial x} = -\frac{\varepsilon}{\omega}\langle f_1^*\sin\bar{\xi}\rangle,$$

$$\frac{\partial \bar{h}}{\partial t} + \frac{a^2k}{\omega}\frac{\partial \bar{h}}{\partial x} = -\frac{\varepsilon}{\bar{c}\omega}\langle f_1^*\cos\bar{\xi}\rangle,$$

$$(\bar{\xi} = \omega t - kx + \bar{h}(\varepsilon(\omega t - kx))),$$

$$f_1^* = f_1(\eta,\tau,\bar{c}\cos\bar{\xi},\ -\omega\bar{c}\sin\bar{\xi},\ k\bar{c}\sin\bar{\xi}, k\omega\bar{c}\cos\bar{\xi},\ -k^2c\cos\bar{\xi},0). \tag{6.28}$$

Remark 6.1. The presented method for constructing solutions of the second order quasilinear wave equation can also be applied to the equations of the form

$$u_{tt} - a^2(\eta,\tau)u_{xx} = \lambda(\eta,\tau)u + \varepsilon f(\eta,\tau,u,p,q,p_x,q_x,\varepsilon), \tag{6.29}$$

where $p = u_t$; $q = u_x$; $\eta = \varepsilon x$; $\tau = \varepsilon t$; $\varepsilon f = \varepsilon f_1 + \varepsilon^2 + \ldots$, $f_\alpha = f_\alpha(\eta,\tau,u,p,q,p_x,q_x)$, $\alpha = 1,2,\ldots$ and ε is a small parameter.

It should be noted that the first approximation expressions (6.17) and (6.28) generalises the Krylov — Bogolyubov — Mitropol'skii first approximation expressions for the second order ordinary differential equations

$$\ddot{u} + \omega^2 u = \varepsilon f(\tau,u,\dot{u}). \tag{6.30}$$

Indeed, if the wave solution $u(x,t\varepsilon)$ does not depend on x, equation (6.1) turns into equation (6.30) ($-\lambda = \omega^2$) and the expressions (6.17) turns into the Krylov — Bogolyubov — Mitropol'skii first approximation expressions [14, 63, 74].

4. Application of asymptotic methods to partial differential equations of second order. As already mentioned at the beginning of this section, a new nonlinear approach to investigating flat waves in a stratified medium was developed in [24]. We now turn our attention to this method and its generalisation to partial differential equations of second order.

In oceanography, one uses the Boussinesq approximation to hydrodynamic equations [24]. Disregarding the viscosity of water and Coriolis forces, one can write down the motion equation in vector form as follows

$$\rho_0 u_t + \rho_0(\vec{u}\cdot\vec{\nabla})\vec{u} = -\vec{\nabla}p + \rho\vec{g}, \tag{6.31}$$

where $p(x,y,z,t)$ and $\rho(x,y,z,t)$ are water pressure and density, respectively, $\rho_0 = \text{const}$ is water density at the ocean surface, $\vec{u}(x,y,z,t)$ is the local velocity of liquid particles, $\vec{\nabla} = \left\{\frac{\partial}{\partial x},\frac{\partial}{\partial y},\frac{\partial}{\partial x}\right\}$ Hamiltonian, and $\vec{g} = (0,0,g)$ the gravity force. This equation is supplemented by the liqiud incompressibily condition

$$\vec{\nabla}\cdot\vec{u} = 0 \tag{6.32}$$

and the equation of continuity

$$\frac{\partial \rho}{\partial t} + (\vec{u}\cdot\vec{\nabla})\rho = 0. \tag{6.33}$$

In studying periodic wave motion, the work [24] assumes the density ρ to be average with respect to time and depending on the coordinate z only (i.e., the medium is considered to be stratified). As shown in [24], system (6.31)–(6.33) reduces to the equation

$$\frac{\partial^2 T}{\partial z^2} + \frac{\partial^2 T}{\partial \theta^2} + \frac{1}{\rho_0}[\varphi'(T - vz) - \Phi'(T - vz)] = 0, \tag{6.34}$$

where $T(z, \theta)$ is the flow function, $\theta = x - vt$, $v > 0$; φ and Φ are arbitrary functions.

Considered in the general case of arbitrary φ and Φ, equation (6.34) is difficult to integrate and there is no general method to solve it. Therefore, asymptotic methods acquire their significance as the means for obtaining quantitative and qualitative results in a number of important particular cases. For instance, in the case of slowly changing vertical gradient of the average density (ocean "smooth startification"), equation (6.34) can be reduced, under certain assumptions concerning meduim unboundedness, to a quasilinear equation

$$\frac{\partial^2 T}{\partial z^2} + \frac{\partial^2 T}{\partial \theta^2} + \chi^2(\eta)T = \varepsilon f_1(\eta, T) + \varepsilon^2 f_2(\eta, T) + \cdots, \tag{6.35}$$

where $\varepsilon > 0$ is a small parameter, $\eta = \varepsilon z$ a "slow" vertical coordinate.

So, the problem on nonlinear internal waves reduces, under the conditions mentioned above, to the construction of the asymptotics of wave solutions of equation (6.35).

Besides the quasilinear problem on nonlinear internal waves in a stratified medium given above, more general problems can also be considered.

5. The general problem on the asymptotics of nonlinear periodic flat waves in a startified medium [24, 25]. We start with setting forth the problem that generalises possible problems concerning nonlinear internal waves, including the problem described by equation (6.35).

Suppose we are given a process described by a scalar function $u(x_1, x_2, \ldots, x_N)$. One of the coordinates x_i, say x_{N-1}, may be viewed as time t. Suppose that the properties of the medium where the process runs can be characterized by a parameter (or several parameters) μ. Assume that the medium is stratified along a one coordinate, say $x_N^* = z$. With this interpretation in effect, we write the variables as $x_1, x_2, \ldots, x_n, t, z, n = N - 2$. If time t is not singled out, we shall write $x_1, x_2, \ldots, x_m, z, m = N - 1$. In the sequel we shall be interested in a "smooth" stratification, which can be described by an additional small positive parameter $\varepsilon : \mu = \mu(\varepsilon z)$.

Suppose that wave properties of the medium can be described by the principal nonlinear term $B(\mu, u) \equiv Q(\eta, u)$, $\mu = \mu(\eta)$, $\eta = \varepsilon z$ and by a small perturbing term $\varepsilon K\left(\mu, u, \frac{\partial u}{\partial x_1}, \ldots, \frac{\partial u}{\partial x_N}, \varepsilon\right)$, which allows for the effects not captured by the function B. The process equation can be written down as follows

$$Lu + B(\mu, u) = \varepsilon K \Leftrightarrow Lu + Q(\eta, u) = \varepsilon f, \quad \frac{d\eta}{dt} = s, \tag{6.36}$$

where L is a linear differential operator with constant coefficients containing second derivatives, i. e.

$$Lu = \sum_{i,j=1}^{N} a_{ij} \frac{\partial^2 u}{\partial x_i \partial x_j}, \quad a_{ij} = a_{ji} = \text{const}. \tag{6.37}$$

Obviously, (6.36) and (6.37) generalises equation (6.35).

If $\varepsilon = 0$ (which means that there is no stratification), equation (6.36) turns into a nonperturbed (degenerated) equation

$$\sum_{i,j=1}^{N} a_{ij}\frac{\partial^2 u_0}{\partial x_i \partial x_j} + Q(\eta, u_0) = 0 \quad (\eta = \text{const}), \tag{6.38}$$

containing η as a constant parameter. The nonperturbed equation (6.38) may admit, under certain condition, a wave solution which depends on a parameter or it may admit a family of solutions that depend on a certain number of arbitrary constants corresponding to the periodic flat wave. Such solution can naturally be written as

$$u_0 = u_0(k_1x_1 + k_2x_2 + \cdots + k_Nx_N), \tag{6.39}$$

where the function u_0 is periodic of a constant period T_0 with respect to the wave phase $\psi = k_1x_1 + \ldots + k_Nx_N$. Let $T_0 = 2\pi$. If we single out the coordinate $x_N = z$ the medium may be stratified with respect to and do not single out time t, we can rewrite (6.39) as follows

$$u_0 = u_0(k_1x_1 + k_2x_2 + \cdots + k_mx_m + l_0z) \quad (m = N-1,\ l_0 = k_N),$$

where

$$\vec{k} = (k_1, k_2, \ldots, k_m) \quad (m = N-1) \tag{6.40}$$

is an m-dimensional vector comprising the wave numbers k_1 $k_2, \ldots, k_m$ regarded subsequently as arbitrary but fixed.

The problem in question consists in constructing, for $\varepsilon \neq 0$, an asymptotic wave solution of the perturbed equation (6.36) satisfying this equation with the accuracy up to the magnitude of order ε^k and turning into (6.39) as $\varepsilon = 0$. If u is such a solution, then

$$u|_{\varepsilon=0} = u_0, \quad |Lu + Q(\eta, u) - \varepsilon f| < C_{k+1}\varepsilon^{k+1}, \quad C_{k+1} = \text{const}.$$

The desired solution u can be written as $u = u(\varepsilon, \eta, \psi(\eta, \varepsilon, x_1, x_2, \ldots, x_m, z))$, where the function u is 2π-periodic with respect to the phase $\psi = \psi(\eta, \varepsilon, x_1, x_2, \ldots, x_m, z)$ of perturbed wave. The phase gradient is assumed to be as follows $\operatorname{grad}\psi = (k_1, k_2, \ldots, k_m, l(\eta,\varepsilon), \eta, \varepsilon)$, where $l(\eta, \varepsilon)$ is a new slowly changing function turning into $l_0 = \text{const}$ as $\varepsilon = 0$. The gradient serves a perturbed wave vector. With this description of the wave process in use, the influence of stratification amounts to the change of oscillation amplitude, which becomes a slow function of z (this is reflected in the fact that u, too, explicitly depends on the argument $\varepsilon z = \eta$), and to the change of module and direction of the wave vector $\vec{k}$ on account of the component l slowly changing along the axis z, the rest of the components in (6.40) being fixed.

The conditions given here hold true for specific problems related to the description of internal waves in ocean. Indeed, for equation (6.35), which is a particular case of equation (6.36) $\left(\text{for } B = \chi^2(\eta)T,\ \varepsilon f = \varepsilon f_1 + \varepsilon^2 f_2 + \ldots \text{ and } L = \frac{\partial^2}{\partial z^2} + \frac{\partial^2}{\partial \theta^2}\right)$, the nonperturbed equation is of the form

$$\frac{\partial^2 T}{\partial z^2} + \frac{\partial^2 T}{\partial \theta^2} + \chi^2(\eta)T = 0, \quad \eta = \text{const}$$

and has a wave solution of the type $T = R\cos(k\theta + lz)$ $(k^2 + l^2 = \chi^2)$.

For $\varepsilon \neq 0$, these wave solutions generate the same solutions of equation (6.35). The equation

$$\sum_{i,j=1}^{N} a_{ij}\frac{\partial^2 u}{\partial x_i \partial x_j} + \chi^2(\eta)u = \varepsilon f\left(\eta, u, \frac{\partial u}{\partial x_1}, \cdots, \frac{\partial u}{\partial x_N}\varepsilon\right). \tag{6.41}$$

is a particular case of equation (6.36) corresponding to $Q(\eta, u) = \chi^2(\eta)u$. (Some asymptotic methods (the averaging method [24] and others) were developed for such the equation.)

In [24] the reader can find various generalizations of the above general problem concerning the asymptotics of nonlinear periodic waves in stratified medium. We consider a natural generalization of equation (6.36), namely the one where (6.36) has slowly changing coefficients,

$$\sum_{i,j=1}^{N} a_{ij}(\eta)\frac{\partial^2 u}{\partial x_i \partial x_j} + Q(\eta, u) = \varepsilon f\left(\eta, u, \frac{\partial u}{\partial x_1}, \dots, \frac{\partial u}{\partial x_N}, \varepsilon\right). \tag{6.42}$$

Here $\eta = (\eta_1, \eta_2, \dots, \eta_N) \equiv (\varepsilon x_1, \varepsilon x_2, \dots, \varepsilon x_N)$ and $\dfrac{\partial \eta_i}{\partial x_i} = \varepsilon$, $i = 1, 2, \dots, N$; $\varepsilon > 0$ is a small parameter.

In particular, if $Q(\eta, u) = \lambda(\eta)u$, equation (6.42) reduces to a quasilinear equation

$$\sum_{i,j=1}^{N} a_{ij}(\eta)\frac{\partial^2 u}{\partial x_i \partial x_j} + \lambda(\eta)u = \varepsilon f\left(\eta, u, \frac{\partial u}{\partial x_1}, \dots, \frac{\partial u}{\partial x_N}, \varepsilon\right), \tag{6.43}$$

which is a generalization of equation (6.41) as well as the corresponding equation of elliptic type considered in [66] ($\lambda(\eta) = \chi^2(\eta)$). In equations (6.42) and (6.43), no coordinate (according to which a medium can be stratified) is singled out and the coefficients depend on all "slow" variables.

6. The first approximation to the asymptotic solution of a general quasilinear equation with slowly changing variables. We investigate wave solutions of the general quasilinear equation (6.43), with $\varepsilon > 0$ being a small parameter,

$$\eta = (\eta_1, \eta_2, \dots, \eta_N), \quad \eta_i = \varepsilon x_i, \quad i = 1, 2, \dots, N.$$

$$\varepsilon f = \varepsilon f_1 + \varepsilon^2 f_2 + \dots, \quad f_\nu = f_\nu\left(\eta, u, \frac{\partial u}{\partial x_1} \cdots, \frac{\partial u}{\partial x_N}\right), \quad \nu = 1, 2, \dots.$$

Suppose that the functions f and λ are sufficiently smooth in the region in question, i. e. all their derivatives exist and are bounded. Furthermore, we assume that $\lambda(\varepsilon x_1, \varepsilon x_2, \dots, \varepsilon x_N) \neq 0$ in the region. The expansion $\varepsilon f = \varepsilon f_1 + \varepsilon^2 f_2 + \dots$ are generally thought of as asymptotic, i. e.

$$\left| f - \sum_{\nu=1}^{k} \varepsilon^{\nu-1} f_\nu \right| \le C_k \varepsilon^k, \quad C_k = \text{const.}$$

holds for small $\varepsilon, \varepsilon \ll 1$.

We shall be interested in wave solutions of equation (6.43) corresponding to a simple harmonic flat wave in a region of large extension (equivalent to $1/\varepsilon$) in all the coordinates x_i.

Setting $\varepsilon = 0$, we obtain a nonpertubed equation

$$\sum_{i,j=1}^{N} a_{ij}(\eta)\frac{\partial^2 u_0}{\partial x_i \partial x_j} + \lambda(\eta)u_0 = 0, \quad \vec{\eta} = \overrightarrow{\text{const}}, \tag{6.44}$$

from (6.43), which contains η as a collection of constant parameters.

Suppose that equation (6.44) has a wave solution

$$u_0 = u_0(\omega_1 x_1 + \omega_2 x_2 + \cdots + \omega_N x_N, \eta), \tag{6.45}$$

which depends on η as a collection of parameters, $(\omega_1, \ldots, \omega_N)$ being a nonzero wave vector and u_0 being a function 2π-periodic with respect to the wave phase $\xi = \omega_1 x_1 + \omega_2 x_2 + \ldots + \omega_N x_N$.

Substituting (6.45) into equation (6.44) gives a differential equation

$$\left(\sum_{i,j=1}^{N} a_{ij}(\eta)\omega_i\omega_j\right) \frac{\partial^2 u_0}{\partial \xi^2} + \lambda(\eta) u_0 = 0, \quad \vec{\eta} = \overrightarrow{\text{const}},$$

which has a solution $\exp\{\pm\sqrt{\lambda(\eta)/s(\eta)}\,\xi\}$, provided the conditions

$$s(\eta) \equiv \sum_{i,j=1}^{N} a_{ij}(\eta)\omega_i\omega_j \neq 0, \quad s(\eta)\lambda(\eta) > 0 \tag{6.46}$$

hold.

From this one can readily see u_0 to be a 2π-periodic function if and only if

$$\sum_{i,j=1}^{N} a_{ij}(\eta)\omega_i\omega_j - \lambda(\eta) = 0, \quad \vec{\eta} = \overrightarrow{\text{const}}\,. \tag{6.47}$$

In that case, the solution (6.45) is of a general form

$$u_0 = a\cos(\omega_1(\eta)x_1 + \cdots + \omega_N(\eta)x_N + h), \tag{6.48}$$

where a and h are arbitrary constants and $\omega_i(\eta)$, $i = 1, 2, \ldots, N$ are to be chosen so that equality (6.47) and inequality (6.46) hold.

To construct asymptotic approximations to solutions of the perturbed equation (6.43) we use the method of variation of parameters and asymptotic change of variables, which are analogou to the known averaging method in connection with partial differential equations.

For $\varepsilon \neq 0$, the wave solution of equation (6.43) is sought in the form

$$u = a(x_1, \ldots, x_N, \varepsilon)\cos[\xi(x_1, x_2, \ldots, x_N, \varepsilon)], \tag{6.49}$$

where a and ξ are new unknown functions. For $\varepsilon = 0$, this solution has to turn into (6.48).

Generally speaking, if $\varepsilon \neq 0$, the amplitude a varies and the phase ξ changes differently. As the new functions a and ξ are introduced, additional relations are required to determine them uniquely. These relations can be introduced much in the same way as the variable change is introduced in the averaging method

$$\frac{\partial u}{\partial x_j} = -\omega_j(\eta)a(x_1, x_2, \ldots, x_N, \varepsilon)\sin[\xi(x_1, x_2, \ldots, x_N, \varepsilon)],\ i = 1, 2, \ldots, N. \tag{6.50}$$

To separate slow changes and oscillations, the following expansions are build up for the functions a and ξ

$$\begin{aligned} a &= \alpha + \varepsilon u_1(\eta, \alpha, \bar{\xi}) + \varepsilon^2 u_2(\eta, \alpha, \bar{\xi}) + \ldots, \\ \xi &= \bar{\xi} + \varepsilon v_1(\eta, \alpha, \bar{\xi}) + \varepsilon^2 v_2(\eta, \alpha, \bar{\xi}) + \ldots, \end{aligned} \tag{6.51}$$

Here α and $\bar{\xi}$ are the averaged amplitude and phase describing the regular motion (change); u_i and v_i, $i = 1, 2, \ldots$, are bounded smooth functions 2π-periodic in $\bar{\xi}$, corresponding to small oscillations of a and ξ near their first approximations α and $\bar{\xi}$.

The functions α and $\bar{\xi}$, u_i and v_i $i = 1, 2, \ldots$, are to be determined. Substituting expansions in (6.51) into expressions (6.49) and (6.50), we obtain asymptotic expansions for u and $\dfrac{\partial u}{\partial x_j}$, $i = 1, \ldots, N$. Next, differentiating u twice, substituting u and $\dfrac{\partial u}{\partial x_j}$, $i = 1, 2, \ldots, N$ into equation (6.43) and equating the coefficients at equal powers of ε, we obtain a chain system of equations, which enable us to find u_i, v_i, $i = 1, 2, \ldots$, one after another and to construct amplitude-phase equations with respect to α and $\bar{\xi}$ for the corresponding approximations. We restrict ourselves to constructing the first approximation of the solution.

Thus, according to the above scheme, we have

$$u = \alpha \cos \bar{\xi} + \varepsilon[u_1 \cos \bar{\xi} - \alpha v_1 \sin \bar{\xi}]; \tag{6.52}$$

$$\frac{\partial u}{\partial x_j} = -\alpha\omega_j(\eta) \sin \bar{\xi} + \varepsilon[-u_1 \sin \bar{\xi} - \alpha v_1 \cos \bar{\xi}]\omega_j(\eta), \ i = 1, 2, \ldots, N \tag{6.53}$$

as the first approximation. Using the above results, we set

$$\alpha = \alpha(\xi_0), \ \bar{\xi} = \omega_1 x_1 + \cdots + \omega_N x_N + h(\xi_0),$$

$$\xi_0 = \varepsilon(\omega_1 x_1 + \cdots + \omega_N x_N), \tag{6.54}$$

where $\varepsilon > 0$ is a small parameter.

Differentiating (6.52) with respect to x_j and comparing (6.53) with the expression obtained, we get

$$\frac{\partial u_1}{\partial \bar{\xi}} \cos \bar{\xi} - \alpha \frac{\partial v_1}{\partial \bar{\xi}} \sin \bar{\xi} + \frac{d\alpha}{d\xi_0} \cos \bar{\xi} - \alpha \frac{dh}{d\xi_0} \sin \bar{\xi} = 0. \tag{6.55}$$

By (6.51)–(6.53), we can write

$$\lambda(\eta)u = \lambda(\eta)\alpha \cos \bar{\xi} + \varepsilon\lambda(\eta)[u_1 \cos \bar{\xi} - \alpha v_1 \sin \bar{\xi}]; \tag{6.56}$$

$$\varepsilon f = \varepsilon f_1(\eta, \alpha \cos \bar{\xi}, \ -\alpha\omega_1 \sin \bar{\xi}, \ldots, \ -\alpha\omega_N \sin \bar{\xi}) \equiv \varepsilon f_1 \tag{6.57}$$

as the first approximation.

Differentiating (6.53) with respect to x_1, $i = 1, 2, \ldots, N$ substituting (6.56), (6.57) and the expression obtained in equation (6.43) and equating the coefficients at equal powers of $\varepsilon(\varepsilon^0$ and $\varepsilon)$, we obtain relations, which constitute a system of equations. Solving this system with respect to $\dfrac{\partial u_1}{\partial \bar{\xi}}$ and $\dfrac{\partial v_1}{\partial \bar{\xi}}$ yields

$$\frac{\partial u_1}{\partial \bar{\xi}} = -\frac{1}{s(\eta)} f_1 \sin \bar{\xi} - \frac{d\alpha}{d\xi_0} - \frac{\alpha}{s(\eta)} \sum_{i,j=1}^{N} a_{ij} \frac{\partial \omega_j}{\partial \eta_i} \sin^2 \bar{\xi},$$

$$\frac{\partial v_1}{\partial \bar{\xi}} = -\frac{1}{\alpha s(\eta)} f_1 \cos \bar{\xi} - \frac{dh}{d\xi_0} - \frac{1}{2s(\eta)} \sum_{i,j=1}^{N} a_{ij} \frac{\partial \omega_j}{\partial \eta_i} \sin 2\bar{\xi}, \tag{6.58}$$

where $s(\eta) = \sum_{i,j=1}^{N} a_{ij}\omega_i\omega_j \neq 0$.

Note that for the mixed derivatives to be equal, the following condition have to be satisfied

$$\frac{\partial \omega_i(\eta)}{\partial \eta_j} = \frac{\partial \omega_j(\eta)}{\partial \eta_i}, \quad i, j = 1, 2, \ldots, N. \tag{6.59}$$

Let us now introduce the notation of averaging

$$\langle \psi \rangle = \frac{1}{2\pi} \int_0^{2\pi} \psi(\bar{\xi})\, d\bar{\xi}. \tag{6.60}$$

Keeping in mind periodicity of the functions u_1 and v_1 and notation (6.60), we average, with respect to $\bar{\xi}$, the left- and right-hand sides of equations in (6.58). The following system of amplitude-phase equations result

$$\frac{d\alpha}{\partial \xi_0} = -\frac{1}{s(\eta)}\langle f_1 \sin \bar{\xi}\rangle - \frac{\alpha}{2s(\eta)} \sum_{i,j=1}^{N} a_{ij} \frac{\partial \omega_j}{\partial \eta_i},$$

$$\frac{dh}{\partial \xi_0} = -\frac{1}{\alpha s(\eta)}\langle f_1 \cos \bar{\xi}\rangle. \tag{6.61}$$

Integrating system (6.58) with respect to $\bar{\xi}$ and using (6.61), we obtain

$$u_1(\eta, \alpha, \bar{\xi}) = u_{10}(\eta, \alpha) + \frac{\alpha}{4s(\eta)} \sum_{i,j=1}^{N} a_{ij} \frac{\partial \omega_j}{\partial \eta_i} \sin 2\bar{\xi} -$$

$$-\frac{1}{s(\eta)} \int_0^{\bar{\xi}} (f_1 \sin \bar{\xi} - \langle f_1 \sin \bar{\xi}\rangle)\, d\bar{\xi},$$

$$v_1(\eta, \alpha, \bar{\xi}) = v_{10}(\eta, \alpha) + \frac{1}{4s(\eta)} \sum_{i,j=1}^{N} a_{ij} \frac{\partial \omega_j}{\partial \eta_i} (\cos 2\bar{\xi} - 1) -$$

$$-\frac{1}{\alpha s(\eta)} \int_0^{\bar{\xi}} (f_1 \cos \bar{\xi} - \langle f_1 \cos \bar{\xi}\rangle)\, d\bar{\xi}, \tag{6.62}$$

where the functions u_{10} and v_{10} can be chosen arbitrarily (for example, $v_{10} = u_{10} = 0$).

In view of (6.54) we can write

$$\sum_{i,j=1}^{N} a_{ij}\omega_j \frac{\partial \alpha}{\partial x_i} = \varepsilon \sum_{i,j=1}^{N} a_{ij}\omega_i\omega_j \frac{d\alpha}{d\xi_0} \equiv \varepsilon s(\eta) \frac{d\alpha}{d\xi_0},$$

$$\sum_{i,j=1}^{N} a_{ij}\omega_j \frac{\partial h}{\partial x_i} = \varepsilon \sum_{i,j=1}^{N} a_{ij}\omega_i\omega_j \frac{dh}{d\xi_0} \equiv \varepsilon s(\eta) \frac{dh}{d\xi_0}. \tag{6.63}$$

Combining (6.61) and (6.63) yields

$$\sum_{i,j=1}^{N} a_{ij}\omega_j \frac{\partial \alpha}{\partial x_i} = -\varepsilon\langle f_1 \sin \bar{\xi}\rangle - \frac{\varepsilon\alpha}{2} \sum_{i,j=1}^{N} a_{ij}(\eta) \frac{\partial \omega_j}{\partial \eta_i},$$

$$\sum_{i,j=1}^{N} a_{ij}\omega_j \frac{\partial h}{\partial x_i} = -\frac{\varepsilon}{\alpha}\langle f_1 \cos\bar{\xi}\rangle. \tag{6.64}$$

Note that the above asymptotic method for finding wave solutions is valid for both hyperbolic and elliptic equations of the form (6.43) [22, 41, 46, 47, 81, 82, 88] as well as for generalized equation (6.42). For the elliptic equation, the above result completely coincides with the result in [66]. And if the amplitude and phase of the first approximation slowly depend on the phase of nonperturbed equation (6.44) (see (6.54)), the additional conditions (due to Moseenkov) imposed on the vector $\vec{\omega}(\eta) = (\omega_1(\eta), \ldots, \omega_N(\eta))$ and functions α and h [66],

$$\frac{\partial \alpha}{\partial \eta_i}\omega_j(\eta) = \frac{\partial \alpha}{\partial \eta_j}\omega_i(\eta), \qquad \frac{\partial h}{\partial \eta_i}\omega_j(\eta) = \frac{\partial h}{\partial \eta_j}\omega_i(\eta) \tag{6.65}$$

hold, the first approximation equations (6.64) being the same.

The specific character of the way in which equation (6.43) depends on many slow variables is thus reflected in the amplitude-phase equations (6.64), which are partial differential equations, as one might expect. In the particular case of $a_{ij} = \text{const}\ i,j = 1,2,\ldots,N$, and $\eta = \eta_N = \varepsilon x_N$ [24] (x_N is a slow coordinate with respect to which a medium can be stratified), the perturbed system is descibed by equation (6.41); the amplitude-phase equations (6.64) turn into a system of ordinary differential equations

$$\frac{d\alpha}{d\eta_N} = -\left[\sum_{j=1}^{N-1} a_{jN}\omega_j + a_{NN}\omega_N(\eta)\right]^{-1} \left\{\langle f_1 \sin\bar{\xi}\rangle + \frac{a_{NN}\alpha}{2}\frac{d\omega_N}{d\eta_N}\right\},$$

$$\frac{dh}{d\eta_N} = -\frac{1}{\alpha}\left[\sum_{j=1}^{N-1} a_{jN}\omega_j + a_{NN}\omega_N(\eta)\right]^{-1} \langle f_1 \cos\bar{\xi}\rangle.$$

If equation (6.43) takes the form of equation (6.1), the amplitude-phase equations (6.64) go into the first order system of partial differential equations (6.17).

Thus (6.52) is the first approximation to the asymptotic solution of equation (6.43). The functions α and h in (6.52) are determined by the system of partial differential equations (6.64) of the first approximation, and $\bar{\xi} = \omega_1 x_1 + \ldots + \omega_N x_N + h$.

Example. Consider the elliptic equation (6.35). Note that this equation is a particular case of the hyperbolic equation

$$\frac{\partial^2 u}{\partial z^2} + 2\frac{\partial^2 u}{\partial x^2} - \frac{1}{v^2}\frac{\partial^2 u}{\partial t^2} + \chi^2(\eta)u = \varepsilon f_1(\eta, u) + \varepsilon^2 f_2(\eta, u) + \ldots,$$

if one sets $u(x,z,t) = T(z, x - vt) \equiv T(z,\theta)$. Therefore, the above asymptotic method for finding wave solutions of the second order partial differential equation (6.43) can be used if and only if the following conditions are fulfilled

$$\lambda(\eta) \neq 0, \quad s(\eta) \equiv \sum_{i,j=1}^{N} a_{ij}(\eta)\omega_i\omega_j \neq 0, \quad s(\eta)\lambda(\eta) > 0,$$

$$s(\eta) - \lambda(\eta) = 0, \quad \frac{\partial \omega_i}{\partial \eta_j} = \frac{\partial \omega_j}{\partial \eta_i}, \quad \vec{\eta} = \overrightarrow{\text{const}}, \quad i,j = 1,2,\ldots,N. \tag{6.66}$$

The first approximations of the amplitude α and phase h (see (6.64)) are slow functions of the total phase of the solution $u_0 = u_0(\omega_1 x_1 + \ldots) + \omega_N x_N$ of nonpertubed equation (6.44).

6.2. Analytical dependence of solutions of hyperbolic equations on parameter

The Poincaré theorem on analytical dependence of solution on parameter is one of the major results in the analytical theory of ordinary differential equations. In some cases, this theorem itself can be a source of computational algorithms [80]. The simplest proof of this theorem for a first order system of ordinary differential equations

$$\frac{dx}{dt} = f(x, t, \mu)$$

can be found in [112].

In this section we shall prove the similar theorem for hyperbolic partial differential equations. Consider a hyperbolic system of first order

$$D_i u_i \equiv \left(\frac{\partial}{\partial t} + \lambda_i(x,t)\frac{\partial}{\partial x}\right) u_i = F_i(x, t, u_1, \ldots, u_m, \mu), \quad i = 1, 2, \ldots, m, \tag{6.67}$$

on the set $\Pi_\infty = \{(x,t) \in \mathbb{R}^2_{x,t} : a \le x \le b,\, t \ge 0\}$, $-\infty < a < b < +\infty$, with the boundary condition

$$u_i(x, 0) = g_i(x), \qquad a \le x \le b, \tag{6.68}$$

where $\lambda_i(x,t)$ and $g_i(x)$ are known real continuous functions, μ a complex parameter.

Suppose that m real characteristics of system (6.67) pass through every point $(x_0, t_0) \in \Pi_\infty$ in the direction of decreasing t and that these characteristics can be represented with the help of functions $x = x_i(t; x_0, t_0)$, where $x_i(t; x_0, t_0)$ is a solution of the equation of characteristics

$$\frac{dx}{dt} = \lambda_i(x,t), \qquad x(t_0) = x_0, \qquad i = 1, 2, \ldots, m. \tag{6.69}$$

Let $\theta_i(x_0, t_0)$ denote the smallest value of t for such a solution $(0 \le \theta_i\ (x_0, t_0) \le t_0)$. As known [2, 52], the Cauchy problem (6.67), (6.68) has a unique solution on a closed bounded set of points $(x, t) \in \Pi_\infty$, for which $\theta_i(x,t) = 0$, $i = 1, 2, \ldots, m$. According to this we introduce the notation $\Pi_g = \{(x,t) \in \Pi_\infty : \theta_i(x,t) = 0\ \forall i = 1, 2, \ldots, m\}$.

Let $u(x,t,\mu_0) = (u_1(x,t,\mu_0),\ \ldots,\ u_m(x,t,\mu_0))$ be a solution of the Cauchy problem (6.67), (6.68) defined on the set Π_g and let the following conditions be fulfilled in the region $\Omega = \{(x, t, u_1, u_2,\ \ldots,\ u_m, \mu) : (x,t) \in \Pi_g,\ |u_i - u_i(x,t,\mu_0)| < b_i,\ i = 1, 2, \ldots, m,$ $|\mu - \mu_0| < \Delta\}$:

1) the functions $F_i(x, t, u_1, u_2,\ \ldots,\ u_m, \mu)$, $i = 1, 2, \ldots, m$ are analytic with respect to variables $u_1, u_2, \ldots, u_m$ for fixed x, t, μ and with respect to μ for fixed $x, t,\ u_1,\ u_2, \ldots, u_m$;

2) $\partial F_i/\partial u_j$ and $\partial F_i/\partial \mu$, $i, j = 1, 2, \ldots, m$ are continuous functions of variables $x, t,$ $u_1, u_2, \ldots, u_m, \mu$;

3) there is a constant M such that $|F_i(x, t, u_1, u_2, \ldots, u_m, \mu)| \le M$, $|\partial F_i/\partial u_j| \le M$, $|\partial F_i/\partial \mu| \le M$, $i, j = 1, 2, \ldots, m$, for all $(x, t,\ u_1, u_2, \ldots, u_m, \mu) \in \Omega$;

4) the functions $g_i(x)$ and $\lambda_i(x,t)$ satisfy the conditions ensuring the existence of a continuous solution of the Cauchy problem (6.67), (6.68).

These conditions ensure a continuous solution $u(x,t,\mu)$ of the Cauchy problem (6.1), (6.2) to exist and be unique and the solution $u(x,t,\mu)$ to depend continuously on the initial values and the parameter [2, 52].

We shall prove the solution $u(x,t,\mu)$ of the Cauchy problem (6.67), (6.68) to be an analytic function of the parameter μ over the whole set Π_g in a sufficiently small region $|\mu - \mu_0| < \varepsilon$. Indeed, there exists $\varepsilon < \Delta$ such that all the solutions $\{u(x,t,\mu)\} = \{(u(x,t,\mu)),$ $\ldots,\ u_m(x,t,\mu))\}$ of system (6.67) for μ satisfying $|\mu - \mu_0| < \varepsilon$ belong to the region in question $|u_i - u_i(x,t,\mu_0)| < b_i$, $i = 1, 2, \ldots, m$.

Let μ_1 be an arbitrary inner point of the circle $|\mu - \mu_0| < \varepsilon$. Consider the following difference relations

$$\frac{\Delta u_1}{\Delta\mu} = \frac{u_i(x,t,\mu) - u_j(x,t,\mu_1)}{\Delta\mu}, \quad \Delta\mu = \mu - \mu_1, \quad i = 1,2,\ldots,m.$$

These relations satisfy a linear system of equations

$$D_1\left(\frac{\Delta u_i}{\Delta\mu}\right) = \frac{1}{\Delta u_1}[F_i(x,t,u_1(x,t,\mu),u_2(x,t,\mu),\ldots,u_m(x,t,\mu),\mu)-$$

$$-F_i(x,t,u_1(x,t,\mu_1),u_2(x,t,\mu),\ldots,u_m(x,t,\mu),\mu)]\frac{\Delta u_1}{\Delta\mu}+\cdots$$

$$\cdots+\frac{1}{\Delta u_m}[F_i(x,t,u_1(x,t,\mu_1),\ldots,u_{m-1}(x,t,\mu_1),u_m(x,t,\mu),\mu)-$$

$$-F_i(x,t,u_1(x,t,\mu_1),\ldots,u_{m-1}(x,t,\mu_1),u_m(x,t,\mu_1),\mu)]\frac{\Delta u_m}{\Delta\mu}+$$

$$+\frac{1}{\Delta\mu}[F_i(x,t,u_1(x,t,\mu_1),\ldots,u_{m-1}(x,t,\mu_1),u_m(x,t,\mu_1),\mu)-$$

$$-F_i(x,t,u_1(x,t,\mu_1),\ldots,u_{m-1}(x,t,\mu_1),u_m(x,t,\mu_1),\mu_1)],$$

$$\left.\frac{\Delta u_i}{\Delta\mu}\right|_{t=0} = 0, \qquad i = 1,2,\ldots,m,$$

a continuous solution of which is unique and tends, as $\Delta\mu$ tends to zero in arbitrary way, to a unique continuous solution of the following linear system of equations

$$D_iU_i = \sum_{j=1}^{m}\frac{\partial F_i(x,t,u_1(x,t,\mu_1),\ldots,u_m(x,t,\mu_1),\mu_1)}{\partial u_j}U_j+$$

$$+\frac{\partial F_i(x,t,u_1(x,t,\mu_1),\ldots,u_m(x,t,\mu_1),\mu_1)}{\partial\mu}\quad U_i|_{t=0} = 0, \ i = 1,2,\ldots,m.$$

Hence

$$\lim_{\Delta\mu\to0}\frac{\Delta u_i}{\Delta\mu} = U_i, \ U_i(x,0,\mu_1) = 0, \ i = 1,2,\ldots,m.$$

Whence it follows that the continuous solution $u(x,t,\mu)$ of system (6.67) is an analytic function of the parameter μ by virtue of the definition due to Cauchy.

Thus we arrive at the following statement.

Theorem 6.1 [145]. *Suppose the conditions (1)-(4) are fulfilled. The continuous solution $u(x,t,\mu)$ of the Cauchy problem (6.67), (6.68) is then an analytic function of the parameter μ in a neighbourhood of the point $\mu = \mu_0$.*

Remark 6.2. Since quasilinear hyperbolic equations of order m [61] can be reduced to an equivalent system of first order, the similar result is valid for the quasilinear hyperbolic equations of order m too. Moreover, in [174] a theorem analogous to Theorem 6.1 was proved for the case where system (6.67) consists of two equations and the right-hand sides are polynomials.

6.3. Bounded solutions of a linear hyperbolic system of first order

As known, the condition of the existence of a periodic (almost periodic) solution of the standard system $dx/dt = \varepsilon X(t,x)$, where $\varepsilon > 0$ is a small parameter, is given by the second basic theorem of Bogolyubov [12, 14, 67, 74, 118, 119]. Note that this theorem is an easy consequence of the theorems due to Bohl [36], Biryuk [10] and the Bogolyubov transformations reducing the standard system to a special one $dh/dt = \varepsilon Hh + Q(t,h,\varepsilon)$, where H is a constant $n \times n$ matrix.

We shall prove that the hyperbolic system of first order

$$Du = \varepsilon F(x,t,u), \tag{6.70}$$

share the same properties. In (6.70) $D = \operatorname{diag}(D_1, D_2, \ldots, D_m)$ is a diagonal matrix-operator, $D_i = \frac{\partial}{\partial t} + a_i \frac{\partial}{\partial x}$, $i = 1, 2, \ldots, m$; u and F are m-dimensional vectors; x and t are scalar variables; $\varepsilon > 0$ a small parameter [135, 138].

We shall first establish several lemmas and theorems related to differential equation systems (6.70).

Lemma 6.1 [138]. *Let*

$$Du = Au + f(x,t), \tag{6.71}$$

where A is a constant $m \times m$-matrix and $D = \operatorname{diag}(D_1, D_2, \ldots, D_m)$, $D_i = \partial/\partial t + a_i \partial/\partial x$, $i = 1, 2, \ldots, m$, Suppose that the following conditions are fulfilled:

1) $a_i = \text{const}$, $|a_i| \le K$;

2) $\operatorname{Re} \lambda_j(A) \ne 0$, $j = 1, 2, \ldots, m$, where $\lambda_j(A)$ are the roots of the equation $|A - \lambda E| = 0$;

3) $f(x,t) \in C^{1,0}_{x,t}(\mathbb{R}^2)$, $\sup\limits_{(x,t)\in\mathbb{R}^2} \|f(x,t)\| = \sup\limits_{x,t} \max\limits_i |f_i(x,t)| = K_1 < \infty$, $\sup\limits_{(x,t)\in\mathbb{R}^2} \|\partial f/\partial x\| = K_2 < \infty$.

Then there is a matrix $G(t) \in C^\infty$ $(0 < |t| < \infty)$ with the following properties.

1. *$G(+\infty) - G(-\infty) = E_m$, where E_m is a unit $m \times m$-matrix.*
2. *$\|G(t)\| \le ce^{-\alpha|t|}$, where c and α are positive constants.*
3. *$D_s G_{sk} = \sum\limits_{r=1}^{m} G_{sr} A_{rk} = \sum\limits_{r=1}^{m} A_{sr} G_{rk}$ $s = 1, 2, \ldots, m$ for $t \ne 0$.*
4. *$w_s(x,t) = \int\limits_{-\infty}^{\infty} \sum\limits_{k=1}^{m} G_{sk}(t-\tau) f_k(\varphi_s(\tau; x,t), \tau) d\tau$,*

$$s = 1, 2, \ldots, m, \tag{6.72}$$

is a solution of (6.71) bounded on $\mathbb{R}^2$, $\varphi_s(\tau; x,t) = a_s(\tau - t) + x$ is a solution of the Cauchy problem $d\xi/dt = a_s$, $\xi|_{\tau=t} = x$, which is defined and continuously differentiable with respect to x and t for all $(x,t) \in \mathbb{R}^2$, and $\left\|\frac{\partial \varphi}{\partial x}\right\| = 1$, $\left\|\frac{\partial \varphi}{\partial t}\right\| \le K$.

Proof. A proof of the properties 1)–3) and the construction of the matrix $G(t)$ itself (with regard to the fact that $D_s \dot{G}_{sk} \equiv \dot{G}$) is given in [36].

From the condition (2) of Lemma 3.1 it follows that

$$\int\limits_{-\infty}^{\infty} \|G(t)\| dt \le 2C \int\limits_{0}^{\infty} e^{-\alpha t} dt = \frac{2c}{\alpha}. \tag{6.73}$$

Here $\|G\| = \max\limits_s \sum\limits_{k=1}^{m} |G|$ is one of the three norms of the matrix $G(t)$. Therefore, the integral in (6.3) converges for any $(x,t) \in \mathbb{R}^2$ and the convergence is uniform over any finite interval $a < t < b$ and $x \in \mathbb{R}$. Differentiating the following equality

$$w_s(x,t) = \int\limits_{-\infty}^{t} \sum_{k=1}^{m} G_{sk}(t-\tau) f_k(\varphi_s, \tau) d\tau + \int\limits_{t}^{\infty} \sum_{k=1}^{m} G_{sk}(t-\tau) f_k(\varphi_s, \tau) d\tau,$$

with respect to the parameters t and x gives

$$\frac{\partial w_s}{\partial t} = \sum_{k=1}^{m}[G_{sk}(+0) - G_{sk}(-0)]f_k(x,\tau)d\tau +$$

$$+\int_{-\infty}^{t}\sum_{k=1}^{m}\dot{G}_{sk}(t-\tau)f_k(\varphi_s,\tau)d\tau + \int_{t}^{\infty}\sum_{k=1}^{m}\dot{G}_{sk}(t-\tau)f_k(\varphi_s,\tau)d\tau +$$

$$+\int_{-\infty}^{t}\sum_{k=1}^{m}G_{sk}(t-\tau)\frac{\partial f_k}{\partial\varphi_s}\frac{\partial\varphi_s}{\partial t}d\tau + \int_{t}^{\infty}\sum_{k=1}^{m}G_{sk}(t-\tau)\frac{\partial f_k}{\partial\varphi_s}\frac{\partial\varphi_s}{\partial t}d\tau; \tag{6.74}$$

$$\frac{\partial w_s}{\partial x} = \int_{-\infty}^{t}\sum_{k=1}^{m}G_{sk}(t-\tau)\frac{\partial f_k}{\partial\varphi_s}\frac{\partial\varphi_s}{\partial x} + \int_{t}^{\infty}\sum_{k=1}^{m}G_{sk}(t-\tau)\frac{\partial f_k}{\partial\varphi_s}\frac{\partial\varphi_s}{\partial x}d\tau. \tag{6.75}$$

Differentiation is valid here because the improper integrals resulted after formal differentiation uniformly converge on any finite interval $a < t < b$ and $x \in \mathbb{R}$. The equality $\frac{\partial\varphi_s}{\partial t} + a_s\frac{\partial\varphi_s}{\partial x} = 0$ and condition 3) of Lemma 6.1 combined with (6.74) and (6.75) imply the equality

$$D_s w_s(x,t) = \sum_{r=1}^{m}A_{sr}w_r + f_s(x,t), \quad s = 1,2,\ldots,m,$$

where $w(x,t)$ given by (6.72) is a solution of system (6.71). Estimating the norm of $w(x,t)$ and taking account of (6.73) we obtain

$$\|w(x,t)\| \le \sup_{x,t}\max_i|f_i(x,t)|\int_{-\infty}^{\infty}\|G(t-\tau)\|d\tau \le K_1\frac{2c}{\alpha} < \infty. \tag{6.76}$$

Hence the solution $w(x,t)$ is bounded, which proves the lemma.

Corollary 6.1. The bounded solution $w(x,t)$ satisfies the inequality $\sup_{x,t}\|w(x,t)\| \le K_3 \sup_{x,t}\|f(x,t)\|$, with the constant K_3 depending on the matrix A (see inequality (6.76)).

Corollary 6.2. If the free term $f(x,t)$ of system (6.71) is a vector-function ω-periodic in $t : f(x,t+\omega) = f(x,t)$, then the bounded solution $w(x,t)$ is also ω-periodic.

6.4. Almost periodic solutions of an almost linear hyperbolic system of first order

Suppose we are given a real system

$$Du = Au + \psi(x,t,u), \quad D = \operatorname{diag}(D_1, D_2, \ldots, D_m),$$

$$D_1 = \partial/\partial t + a_i\partial/\partial x, \qquad i = 1,2,\ldots,m, \tag{6.77}$$

where $A = [a_{ij}]$ is a constant $m \times m$-matrix and $a_i = \text{const}$. Assuming $\psi(x,t,u)$ to be a free term (similarly to (6.72)), consider a system of integral equations

$$u_s(x,t) = \int_{-\infty}^{\infty}\sum_{k=1}^{m}G_{sk}(t-\tau)\psi_k(\varphi_s,\tau,u(\varphi_s,\tau))d\tau, \quad s = 1,2,\ldots,m, \tag{6.78}$$

where G_{sk} and φ_k are the functions defined in Lemma 6.1.

Definition 6.1. A continuous solution $u_s(x,t)$, $s = 1, 2, \ldots, m$, of the integral equation system (6.78) is referred to as a continuous solution of the differential equation system (6.77).

Theorem 6.2 (an analogue to Bohl's theorem [138]). *Suppose that the following conditions are fulfilled for system (6.77):*

1) $a_i = \text{const}$ *and* $|a_i| \leq K$;

2) $\operatorname{Re} \lambda_j(A) \neq 0$, $j = 1, \ldots, m$, where $\lambda_j(A)$ *are the roots of the equation* $|A - \lambda E| = 0$;

3) $\sup_{(x,t)\in\mathbb{R}^2} \|\psi(x,t,0)\| = \Gamma < \infty$;

4) $\psi(x,t,u) \in C_{x,t,u}(\mathbb{R}^2 \times \|u\| < \infty)$ *and a Lipschitz condition* $\|\psi(x,t,\tilde{\tilde{u}}) - \psi(x,t,\tilde{u})\| \leq N\|\tilde{\tilde{u}} - \tilde{u}\|$ *holds for* $\tilde{\tilde{u}} \in \mathbb{R}^m_u$ *and* $\tilde{u} \in \mathbb{R}^m_u$.

Then, for a sufficiently small constant N, there is a continuous solution $w = w(x,t)$ *of system (6.77) defined and bounded on* $\mathbb{R}^2$.

The proof of Theorem 6.2 goes similarly to that of Bohl's theorem [36].

Remark 6.3. The bounded continuous solution $w = w(x,t)$ of system (6.77) is a classical one if the function $\psi(x,t,u)$ has bounded partial derivatives with respect to x and u.

Corollary 6.3. If the function $\psi(x,t,u)$ is ω-periodic in t, then a bounded extension of $w(x,t)$ is ω-periodic too.

Theorem 6.3 (an analogue to the theorem of Biryuk [138]). *Suppose we are given a real almost linear system*

$$Du = Au + f(x,t) + \varepsilon\psi(x,t,u), \quad u = (u_1, u_2, \ldots, u_m), \tag{6.79}$$

where $D = \operatorname{diag}(D_1, D_2, \ldots, D_m)$, $D_i = \partial/\partial t + a_i\partial/\partial x$ $i = 1, 2, \ldots, m$; $A = [a_{ij}]$ *is a constant* $m \times m$*-matrix and* ε *a small parameter. Assume that*

1) $a_i = \text{const}$ *and* $\|a_i\| \leq K$;

2) $\operatorname{Re} \lambda_j(A) \neq 0$, $j = 1, 2, \ldots, m$;

3) $f(x,t) \in C(\mathbb{R}^2)$, $\|f(x,t)\| \leq K_1$, *and* $f(x,t)$ *is almost periodic in t uniformly with respect to* $x \in \mathbb{R}$;

4) $\psi(x,t,u) \in C(\mathbb{R}^2 \times \mathbb{R}^m_u)$ *and* $\psi(x,t,u)$ *is almost periodic in t with respect to* $x \in \mathbb{R}$, $u \in U \subset \mathbb{R}^m_u$, *where U is a compact, and a Lipschitz condition* $\|\psi(x,t,\tilde{\tilde{u}}) - \psi(x,t,\tilde{u})\| \leq N \|\tilde{\tilde{u}} - \tilde{u}\|$ *holds. Then, for sufficiently small* ε, $|\varepsilon| < \varepsilon_0$ *system (6.79) admits a solution* $w = w(x,t,\varepsilon)$ *almost periodic in t.*

The proof of Theorem 6.3 is similar to that of the theorem of Biryuk [10, 36].

6.5. Mathematical justification of the Bogolyubov averaging method over the infinite time interval for hyperbolic systems of first order

1. Suppose that the limit

$$\lim_{(X,T)\to\infty} \frac{1}{XT} \int_0^T \int_0^X F(x,t,u)\,dt\,dx = \overline{F}(u). \tag{6.80}$$

exists. Besides system (6.70) we shall consider an averaged system

$$Dv = \varepsilon\overline{F}(v) \tag{6.81}$$

and shall investigate the case where system (6.8) has a "uasistatic" solution, corresponding to the equilibrium point $v = v_0$, $\overline{F}(v_0) = 0$. All the eigenvalues $\lambda_j(\partial\overline{F}(v)/\partial v|_{v=v_0})$ of the matrix $(\partial\overline{F}/\partial v)_{v=v_0}$ are assumed to have a nonzero real part. The right-hand side in (6.70) are assumed to satisfy the following conditions:

a) The vector-function $F(x,t,u)$ and its partial derivatives with respect to u are continuous in variable t, bounded and uniformly continuous with respect to x and u in the region $\Omega = \{(x,t) \in \mathbb{R}^2,\ u \in S_r \subset \mathbb{R}_u^m\}$, where $S_r = S(v_0; r)$ is an r-neighbourhood of the point v_0;

b) There exists a bounded function $F_1(x,u)$ with bounded first derivatives with respect to u such that, for every point of the region Ω,

$$\frac{1}{T}\int_t^{t+T} [F_i(\varphi_i,\tau,u) - F_{1i}(\varphi_i,u)]d\tau \to 0 \qquad (T\to\infty) \tag{6.82}$$

uniformly with respect to $(x,t) \in \mathbb{R}^2$, $u \in S_r$ and

$$\frac{1}{X}\int_x^{x+X} F_1(x,u)\,dx \to \overline{F}(u) \qquad (X\to\infty) \tag{6.83}$$

uniformly with respect to $x \in \mathbb{R}$, $u \in S_r$ $(\varphi_i = a_i(\tau - t) + x)$.

Let $u = v_0 + z$ in system (6.70), where v_0 is a "quasistatic" solution of equation (6.81) and z a new variable. System (6.70) can now be written as

$$Dz = \varepsilon Hz + \varepsilon B(x,t,z) + \varepsilon C(x,z). \tag{6.84}$$

Here we use the notation

$$H = (\partial\overline{F}(v)/\partial v)_{v=v_0}; \tag{6.85}$$

$$B(x,t,z) = F(x,t,v_0+z) - F_1(x,v_0+z); \tag{6.86}$$

$$C(x,z) = (F_1(x,v_0+z) - \overline{F}(v_0+z) + \overline{F}(v_0+z) - Hz). \tag{6.87}$$

From the conditions a) and b) it follows that, in the r-neighbourhood S_r^* of the point $z=0$, the functions $B(x,t,z)$ and $C(x,z)$ and their first partial derivatives with respect to z are bounded and uniformly continuous with respect to z in the region $(x,t) \in \mathbb{R}^2$, $\|z\| \le r$. Moreover, by the conditions (6.82), (6.83) and notation (6.86), (6.87),

$$\frac{1}{T}\int_t^{t+T} B_i(\varphi_i,\tau,z)\,d\tau \to 0 \qquad (T\to\infty),$$

$$\frac{1}{T}\int_t^{t+T} \frac{\partial B_i(\varphi_i,\tau,z)}{\partial z_j}\,d\tau \to 0 \qquad (T\to\infty), \tag{6.88}$$

uniformly with respect to $(x,t) \in \mathbb{R}^2$, $\|z\| \le r$ and

$$\frac{1}{X}\int_x^{x+X} C_1(x,z)\,dx \to 0 \qquad (X\to\infty),$$

$$\frac{1}{X}\int_x^{x+X} \frac{\partial C_1(x,z)}{\partial z}\,dx \to 0 \qquad (X\to\infty) \tag{6.89}$$

uniformly with respect to $x \in \mathbb{R}$, $\|z\| \leq r$ at every point z of the r-neighbourhood of the point $z = 0$. Here $C_1(x,z) = C(x,z) - C_2(z)$, where $C_2(z) = \overline{F}(v_) + z) - Hz$.

We now apply the following change of variables

$$z = h(x,t) + \varepsilon v_1(x,t,h) + \varepsilon v_2(x,h) \tag{6.90}$$

to system (6.84). In (6.90), $h(x,t)$, $v_1(x,t,h)$ and $v_2(x,h)$ are unknown functions.

Applying the operator D to (6.90) and substituting the resulted expression and the value of z into system (6.84), we obtain

$$Dh + \varepsilon\frac{\partial(v_1+v_2)}{\partial h}Dh = -\varepsilon Dv_1(x,t,h)|_{h=\text{const}} - \varepsilon Dv_2(x,h)|_{h=\text{const}} +$$
$$+\varepsilon Hh + \varepsilon^2 H(v_1+v_2) + \varepsilon B(x,t,h+\varepsilon(v_1+v_2)) +$$
$$+\varepsilon C(x, h+\varepsilon(v_1+v_2)). \tag{6.91}$$

To determine functions $v_1(x,t,h)$ and $v_2(x,h)$, $h = \text{const}$ let us consider the differential equations'

$$Dv_1 = -\eta_1 v_1(x,t,h) + B(x,t,h), \tag{6.92}$$

$$\Lambda(\partial v_2/\partial x) = -\eta_2 v_2(x,h) + C(x,h), \qquad h = \text{const}, \tag{6.93}$$

where η_1 and η_2 are certain positive parameters, $\Lambda = \operatorname{diag}(a_1, a_2, \ldots, a_m)$.

Lemma 6.1 now shows that

$$v_{1s}(x,t,h) = \int_{-\infty}^{t} e^{-\eta_1(t-\tau)} B_s(\varphi_s, \tau, h), \quad s = 1,2,\ldots,m; \tag{6.94}$$

$$v_{2s}(x,h) = \frac{1}{a_s}\int_{-\infty}^{t} e^{-\eta_2(x-\theta)} C_{1s}(\theta,h)d\theta, \quad s = 1,2,\ldots,m; \tag{6.95}$$

are bounded solutions of systems (6.92) and (6.93) in the region $(x,t) \in \mathbb{R}^2$, $\|h + \varepsilon(v_1 + v_2)\| \leq r$, $\eta_1 > 0$, $\eta_2 > 0$ provided $|a_i| > 0$, $i = 1,2,\ldots,m$.

Taking account of the conditions a), b), (6.88) and (6.89) and proceeding as in the Bogolyubov transformations [63, 74], we obtain

$$\|v_1(x,t,h)\| \leq \frac{\zeta_1(\eta_1)}{\eta_1}, \qquad \|v_2(x,t,h)\| \leq \frac{\zeta_2(\eta_2)}{\eta_2}; \tag{6.96}$$

$$\left\|\frac{\partial v_1}{\partial h}\right\| \leq \frac{\zeta_3(\eta_1)}{\eta_1}, \qquad \left\|\frac{\partial v_2}{\partial h}\right\| \leq \frac{\zeta_4(\eta_2)}{\eta_2}, \tag{6.97}$$

in the region $(x,t) \in \mathbb{R}^2$, $\|h + \varepsilon v_1(x,t,h) + \varepsilon v_2(x,h)\| \leq r$. Moreover, we have $\eta_i(\varepsilon) \to 0$, $j = 1,2,3,4$; $i = 1,2$ as $\zeta_j(\eta_j) \to 0$ in this region.

Successively using the Bogolyubov transformations from the proof of the second basic theorem [63, 74], we get $\varepsilon_0 > 0$ such that, for $\varepsilon < \varepsilon_0$, system (6.91) can be reduced to the form

$$Dh = \left(E + \varepsilon\frac{\partial(v_1+v_2)}{\partial h}\right)^{-1} [\varepsilon Hh + \varepsilon L_1(x,t,h,\varepsilon) + \varepsilon C_2(h)]$$

or

$$Dh = \varepsilon Hh + \varepsilon Q(x,t,h,\varepsilon),$$

where

$$L_1(x,t,h,\varepsilon) = \eta_1 v_1(x,t,h) + \Lambda\eta_2 v_2(x,h) + \varepsilon H(v_1+v_2) + B(x,t,h+\varepsilon v_1+\varepsilon v_2) -$$

$$-B(x,t,h) + C(x+h+\varepsilon v_1+\varepsilon v_2) - C(x,h),$$

E being the m-dimensional unit matrix. In the region $\Omega_2 = \{(x,t) \in \mathbb{R}^2, \|h\| \leq r_1$ $(0 < r - r_1)\}$ and its first partial derivatives with respect to h are bounded by a function of ε, tending to zero as $\varepsilon \to 0$. By using the properties of the functions $L_1(x,t,h,\varepsilon)$ and $C_0(h)$ one can readily see that the function $Q(x,t,h,\varepsilon)$ satisfies the following conditions:

1) the vector-function $Q(x,t,h,\varepsilon)$ is defined in the region $\Omega_2 \times I = [0,\varepsilon_0]$;

2) for all $(x,t) \in \mathbb{R}^2$ and $0 < \varepsilon < \varepsilon_0$, the inequality $\|Q(x,t,h,0,\varepsilon)\| \leq M(\varepsilon)$ holds with $M(\varepsilon) \to 0$ as $\varepsilon \to 0$;

3) for any positive $\sigma < r_1$ and $\|h'\| \leq \sigma$ $\|h''\| \leq \sigma$, the inequality

$$\|Q(x,t,h',\varepsilon) - Q(x,t,h'',\varepsilon)\| \leq \lambda(\varepsilon,\sigma)\|h' - h''\|$$

holds in the region $\Omega_2 \times I$ with $\lambda(\varepsilon,\sigma) \to 0$ as $\varepsilon \to 0$, $\sigma \to 0$.

Summarising the above and returning to the previous variable u, we arrive at the following theorem.

Theorem 6.4 [138]. *Suppose that*

1) the vector-function $F(x,t,u)$ (the right-hand side of system (6.70)) satisfies the conditions

a) $F(x,t,u)$ and its first partial derivatives with respect to u are continuous with respect to x and t, bounded and uniformly continuous with respect to u in the region $\Omega = \{\mathbb{R}^2_{x,t} \times S_r \subset \mathbb{R}^m_u\}$;

b) the limit

$$\lim_{(TX)\to\infty} \frac{1}{TX} \int_0^T \int_0^X F(x,t,u)\, dt dx = \overline{F}(u)$$

exists and there is a bounded function $F_1(x,u)$ possessing bounded first derivatives with respect to u and such that, at any point of the region Ω, we have

$$\frac{1}{T} \int_t^{t+T} [F_i(\varphi_i,\tau,u) - F_{1i}(\varphi_i,u)]d\tau \to 0 \qquad (T\to\infty) \tag{6.98}$$

uniformly with respect to $(x,t) \in \mathbb{R}^2$, $u \in S_r$ and

$$\frac{1}{X} \int_x^{x+X} F_1(x,u)dx \to \overline{F}(u) \qquad (X\to\infty) \tag{6.99}$$

uniformly with respect to $x \in \mathbb{R}$, $u \in S_r$ $(\varphi_i = a_i(\tau - t) + x)$;

2) the averaged system (6.81) $(Dv = \varepsilon\overline{F}(v))$ has a "quasistatic" solution $v = v_0$, $\overline{F}(v_0) = 0$;

3) the real parts of all m roots of the characteristic equation $|(\partial\overline{F}(v)/\partial v)_{v=v_0} - \lambda E| = 0$ are nonzero;

4) it is possible to indicate an r-neighbourhood $S_r = \{u \in dbR^m : \rho(u,\, v_0) < r\}$ of the point v_0 such that the function $F(x,t,u)$ is almost periodic in t uniformly with respect to $x \in \mathbb{R}$, $u \in S_r$ in this neighbourhood;

5) $0 < |a_i| \leq K$, $i = 1, 2, \ldots, m$.

Then for all ($\forall\sigma_0 : \sigma_0 < r$ *there exists* ($\exists\varepsilon_0^* > 0$) *such that for any* ε $0 < \varepsilon < \varepsilon_0^*$ *the following statements hold.*

A. System (6.70) has a bounded continuous solution $\tilde{u}(x,t)$ *defined for all* $(x,t) \in \mathbb{R}^2$.

Б. $\|\tilde{u}(x,t) - v_0\| < \sigma_0$ $\forall(x,t) \in \mathbb{R}^2$ *for all* .

B. The solution $\tilde{u}(x,t)$ *is almost periodic in t with the same frequency basis as the right-hand side of system (6.70).*

Г. If the function $F(x,t,u)$ *is periodic in t of period* ω *in the neighbourhood* S_r, *then the solution* $\tilde{u}(x,t)$ *is periodic in t of period* ω.

Remark 6.4. The condition "Б" in the Theorem 6.4 requires the existence of a function $F_1(x,u)$ such that the conditions (6.82), (6.83) and the conditions (6.98) and (6.99) in Theorem 6.4 are satisfied. In certain cases $F_1(x,u)$ can easily be determined. For example, let the limit

$$\lim_{T\to\infty} \frac{1}{T} \int_0^T F(x,t,u)dt = \underset{t}{M}\{F(x,t,u)\} \tag{6.100}$$

exist. Then

$$\overline{F}(u) = \lim_{X\to\infty} \int_0^X \underset{t}{M}\{F(x,t,u)\}\,dx.$$

If the difference $F(x,t,u) - \underset{t}{M}\{F(x,t,u)\}$ is independent of x and

$$\frac{1}{T} \int_t^{t+T} [F(x,t,u) - \underset{t}{M}\{F(x,t,u)\}]dt \to 0 \qquad (T \to \infty) \tag{6.101}$$

uniformly with respect to $t \in \mathbb{R}$, then, for $F_1(x,u) = \underset{t}{M}\{F(x,t,u)\}$, (6.101) implies (6.98). If, in addition, the function $F_1(x,u) = \underset{t}{M}\{F(x,t,u)\}$ satisfies the condition (6.99), then $F_1(x,u)$ exists and is defined by (6.100) as the mean value with respect to explicitly involved time t of the right-hand side of the first order hyperbolic system (6.70).

Theorem 6.4 coincides with the second basic theorem of Bogolyubov if the right-hand side $F(x,t,u)$ in (6.70) does not depend on x.

It is worth to note that a theorem analogous to Theorem 6.4 was proved by another method in [27] for the case where the difference $F(x,t,u) - \underset{t}{M}\{F(x,t,u)\}$ is independent of x (in particular, this takes place when $F(x,t,u) = F_2(t,u) + F_3(x,u)$ and the mean values of $\underset{t}{M}\{F_2(t,u)\}$ and $\underset{x}{M}\{F_3(x,u)\}$ with respect to t and x exist).

2. Suppose that the mean values

$$\lim_{T\to\infty} \frac{1}{T} \int_0^T F_i(\varphi_i,\tau,u)d\tau = F_i^{(0)}(u), \quad i = 1,2,\ldots,m$$

exist and are independent of x and t ($\varphi_i = a_i(\tau - t) + x$).

Besides system (6.70) we consider the averaged system

$$Dv = \varepsilon F^{(0)}(v) \tag{6.102}$$

and assume that (6.102) has a "quasistatic" solution $v = v_0$, $F^{(0)}(v) = 0$. Then, by using only the Bogolyubov transformations given in the proof of his second basic theorem [63, 74] and Lemma 6.1, we see that the following theorem is true.

Theorem 6.5 [138]. *Suppose the function $F(x,t,u)$ (the right-hand side in (6.70)) satisfies the conditions*

1a) *$F(x,t,u)$ and its first partial derivatives with respect to u are continuous with respect to x and t, bounded and uniformly continuous with respect to u in the region $\Omega = \{(x,t) \in \mathbb{R}^2,\ u \in S_r \subset \mathbb{R}^m_u$;*

1б) *the limits*

$$\lim_{T\to\infty} \frac{1}{T}\int_0^T F_i(\varphi_i,\tau,u)d\tau = F_i^{(0)}(u), \quad i = 1,2,\dots,m$$

$$\varphi_i = a_i(\tau - t) + x,$$

exist and are independent of x and t and, at every point of Ω,

$$\frac{1}{T}\int_0^T F_i(\varphi_i,\tau,u)d\tau \to F_i^{(0)}(u) \quad (T\to\infty)$$

uniformly with respect to $(x,t) \in \mathbb{R}^2$; $u \in S_r$;

2) *the averaged system (6.102) has a quasistatic solution $v = v_0$, $F^{(0)}(v_0) = 0$;*

3) *the real parts of all m roots of the characteristic equation $|(\partial F(v)/\partial v)_{v=v_0} - \lambda E| = 0$ are nonzero;*

4) *it is possible to indicate an r-neighbourhood $S_r = \{u \in \mathbb{R}^m_u : \rho(u,v_0) < r\}$ of the point $v = v_0$ such that in this neighbourhood $F(x,t,u)$ is almost periodic in t uniformly with respetc to $x \in \mathbb{R}$ and $u \in S_r$.*

Then for all $(\forall\sigma : \sigma < r_0)$ $(\exists\varepsilon_0 > 0)$ there exists such that for any ε, $0 < \varepsilon < \varepsilon_0$ the following statements are true.

А) *System (6.70) has a bounded continuous solution $\tilde{u}(x,t)$ defined for all $(x,t) \in \mathbb{R}^2$.*

Б) $\|\tilde{u}(x,t) - v_0\| < \sigma\ \forall(x,t) \in \mathbb{R}^2$.

В) *The solution $\tilde{u}(x,t)$ is almost periodic in t with the same frequency basis as the right-hand side in (6.70).*

Г) *If S_r is ω-periodic in t in the neighbourhood $F(x,t,u)$, then the solution $\tilde{u}(x,t)$ is ω-periodic in t.*

Corollary 6.4. If in the condition 4) of Theorem 6.4 and Theorem 6.5 we required the function $F(x,t,u)$ to be almost periodic (periodic) in x in an r-neighbourhood $S_r = \{u \in \mathbb{R}^m_u : \rho(u,v_0) < r\}$ of the point v_0, then the solution $\tilde{u}(x,t)$ would be almost periodic (periodic) in x with the same frequency basis as the function $F(x,t,u)$. If $F(x,t,u)$ is almost periodic (periodic) in x and t uniformly with respect to u in the neighbourhood S, then so is the solution $\tilde{u}(x,t)$.

We shall show that the above results can be used in the investigation into integral manifolds of standard systems of ordinary differential equations

$$\frac{dx}{dt} = \varepsilon X(t,x), \tag{6.103}$$

when the averaged system $\dfrac{dy}{dt} = \varepsilon Y(y) = \varepsilon M\{X(t,y)\}$ has a periodic solution $y = y(\varepsilon\omega t)$, $y(\theta+2\pi) = y(\theta)$ and the right-hand side $X(t,x)$ satisfies the conditions of the second basic theorem of Bogolyubov (see [12, 14, 63, 74]). Indeed, replacing x by $y(\theta) + z$ we reduce equation (6.103) to the equation

$$\frac{dz}{dt} = \varepsilon H(\theta)z + \varepsilon B(t,\theta,z), \tag{6.104}$$

where $H(\theta) = Y_y'(y(\theta))$; $B(t,\theta,z) = X(t,y+z) - Y(y+z) + B_1(\theta,z)$ and

$$B_1(\theta,z) \equiv Y(y+z) - Y(y) - Hz.$$

Now, setting $z = v(t,\theta) + \varepsilon w(t,\theta,v)$ and

$$w(t,\theta,v) = \int_{-\infty}^{t} e^{-\eta(t-\tau)}[B(\tau,\varepsilon\omega(\tau-t)+\theta,v) - B_1(\varepsilon\omega(\tau-t)+\theta,v)]d\tau$$

we obtain, by analogy with the Bogolyubov transformations,

$$\frac{\partial v}{\partial t} + \varepsilon\omega\frac{\partial v}{\partial \theta} = \varepsilon\omega H(\theta)v + \varepsilon Q(t,\theta,v,\varepsilon), \tag{6.105}$$

where the function $Q(t,\theta,v,\varepsilon)$ is almost periodic in t, 2π-periodic in θ and satisfies the conditions of Theorem 6.2 (the analogue of Bohl's theorem), $\eta = \eta(\varepsilon) > 0$ is a function of ε.

Next, taking account of the identity $\omega X'(\theta) = H(\theta)X(\theta)$, where $X(\theta) = \Phi(\theta)e^{\Lambda\theta}$ is a real fundamental matrix of the system $d\xi/dt = \varepsilon H(\theta)\xi$, $\Phi(\theta)$ is a bounded 2π-periodic nonsingular matrix, $\Lambda = \text{diag}\,(0,C)$, C is a constant $(n-1)\times(n-1)$-matrix, and changing the variable $v = \Phi(\theta)u(t,\theta)$, from (6.105) we obtain

$$\frac{\partial u}{\partial t} + \varepsilon\omega\frac{\partial u}{\partial \theta} = \varepsilon\omega\Lambda u + \varepsilon\Phi(\theta)Q(t,\theta,\Phi u,\varepsilon).$$

Now, letting $u = (g,h_1,h_2,\ldots,h_{n-1})$ and keeping in mind that $\Lambda = \text{diag}\,(0,C)$, from the last equation we obtain

$$\frac{\partial g}{\partial t} + \varepsilon\omega\frac{\partial g}{\partial \theta} = \varepsilon R_1(t,\theta,g,h,\varepsilon),$$

$$\frac{\partial h}{\partial t} + \varepsilon\omega\frac{\partial h}{\partial \theta} = \varepsilon\omega Ch + \varepsilon R_2(t,\theta,g,h,\varepsilon). \tag{6.106}$$

Thus, if the first equation in (6.106) has a constant solution $g = g_0$ or periodic solution $g = g(t,\theta)$, then the second equation of this system has an integral manifold periodic in t and θ, provided the function $R_2(t,\theta,g,h,\varepsilon)$ is periodic in t and θ. In the general case the standard system (6.103) has a local integral manifold $h = f(t,\theta,g,\varepsilon)$, i. e. one can always indicate a positive ε_0 such that for any ε, $0 < \varepsilon < \varepsilon_0$ equations (6.106) have a one-parameter local integral manifold representable by the relation $h = f(t,\theta,g,\varepsilon)$ where the vector-function $f(t,\theta,g,\varepsilon)$ is defined in the region $\mathbb{R}_t \times I_\theta \times U \times E_{\varepsilon_0}$, is 2π-periodic in θ and almost periodic or periodic in t [65].

Remark 6.5. Since the hyperbolic equation of second order $u_{tt} - a^2u_{xx} = \varepsilon f(x,t,u,u_t,u_x)$ can be reduced to hyperbolic systems of first order having the form (6.1) [61], the investigation into the existence of periodic (almost periodic) solutions of this equation can be done with the help of Theorem 6.4 and Theorem 6.5.

6.6. The averaging methods for hyperbolic systems with fast and slow variables

The present section is concerned with the questions of justification of various averaging schemes for hyperbolic systems of first order with fast and slow variables [72, 73, 127–132].

1. Averaging along characteristics. Consider, on the set $\Pi = \{(x, t) \in \mathbb{R}^2 : x \in \mathbb{R},\ t \geq 0\} \times I = [0, \varepsilon^*]$, the hyperbolic system

$$Du = \varepsilon F(x, t, u), \tag{6.107}$$

with the initial condition

$$u(x, 0) = g(x, \varepsilon), \qquad x \in \mathbb{R}, \quad \varepsilon \in I, \tag{6.108}$$

where D is a diagonal matrix having $D_i = \dfrac{\partial}{\partial t} + \lambda_i(x, t, \varepsilon)\dfrac{\partial}{\partial x}$, $i = 1, 2, \ldots, m$ as its entries, u and F are m-dimensional vectors, $\lambda_i(x, t, \varepsilon)$ and $g(x, \varepsilon)$ are known real vector-functions, $\varepsilon > 0$ a small parameter. The functions $\lambda_i(x, t, \varepsilon)$ are considered to be continuous in $\Pi \times I$ and be chosen to guarantee that the initial problem $x = x_i(t; x_0, t_0, \varepsilon)$, $0 \leq t < \infty$ for the equations of characteristics $dx/dt = \lambda_i(x, t, \varepsilon)$, $i = 1, 2, \ldots, m$ has a unique solution $(x(t_0) = x_0)$.

Let the limit

$$\lim_{T \to \infty} \frac{1}{T} \int_0^T F_i(x_i(t; x_0, t_0, \varepsilon), t, u)dt = F_i^{(0)}(u), \quad i = 1, 2, \ldots, m, \tag{6.109}$$

exist and be independent of the initial data (x_0, t_0).

With system (6.107) we associate the averaged system

$$Dv = \varepsilon F^0(v) \tag{6.110}$$

with the initial condition

$$v(x, 0) = g(x, \varepsilon), \qquad x \in \mathbb{R}, \quad \varepsilon \in I. \tag{6.111}$$

We shall state and prove a more general theorem on averaging along the characteristics (such a theorem was proved, for the first time, in [71] using stronger restrictions).

Theorem 6.6 [72]. *Let a vector-function $\lambda(x, t, \varepsilon)$ be defined on the set $\Pi \times I$ and a vector-function $F(x, t, u)$ on the set $\Omega = \{(x, t, u) \in \mathbb{R}^{m+2} : (x, t) \in \Pi,\ u \in G \subset \mathbb{R}^m_u\}$. Suppose that the following conditions hold on these sets*

1) $\lambda(x, t, u) \in C^1_{x,t}(\Pi \times I)$ and moreover $\lambda_1(x, t, \varepsilon) \leq \lambda_2(x, t, \varepsilon) \leq \cdots \leq \lambda_m(x, t, \varepsilon)$ $(\forall (x, t) \in \Pi, \varepsilon \times I)$, $0 < m_0 \leq |\lambda_i(x, t, \varepsilon)| \leq \nu_0$, $|\partial \lambda_i / \partial x| \leq \varepsilon \nu_1$, $i = 1, 2, \ldots, m$, $(\lambda, \lambda'_x, \lambda'_t) \in C_\varepsilon(\Pi \times I)$;

2) $g(x, \varepsilon) \in C^1(\mathbb{R} \times I)$, $|g'_{ix}| \leq \varepsilon \alpha$ and $g(x, \varepsilon) = u(x, 0) \in G$;

3) $F_i(x, t, u) \in C_t \times \mathrm{Lip}_{x,u}(M; \Omega)$, $i = 1, 2, \ldots, m$;

4) the conditions for the classical solution of the Cauchy problem (6.107), (6.108) to exist on the set Π [52, 90];

5) the limit (6.109) exists at every point $u \in G$ and is independent of the initial data (x_0, t_0), $|F_i^0(u)| \leq K$ and $F_i^0(u)$ being continuously differentiable with respect to u;

6) a continuous solution $v(x, t)$ of the Cauchy problem (6.110), (6.111) defiend on Π lies in the region G with a certain ρ-neighbourhood.

Then, for all $\gamma > 0$ and $L > 0$, there exists ε_0 such that for $\varepsilon < \varepsilon_0$ the inequality $|u_i(x, t) - v_i(x, t)| < \gamma\ \forall (x, t) \in \Pi_{g\varepsilon}$ holds for $i = 1, 2, \ldots, m$, where $u(x, t)$ is the classical solution of the Cauchy problem (6.107), (6.108); $\Pi_{g\varepsilon} = \{(x, t) \in \Pi : x_m\ (t; -L\varepsilon^{-1}, L\varepsilon^{-1}, \varepsilon) \leq x \leq x_1(t; -L\varepsilon^{-1}, L\varepsilon^{-1}, \varepsilon);\ 0 \leq t \leq L\varepsilon^{-1}\}$.

Proof. Observe that the functions $F_i^0(u)$ satisfy a Lipschitz condition with constant M with respect to u. Hence there exists a unique continuous solution $v(x, t)$ of the Cauchy problem (6.4), (6.5) on every bounded closed set $\Pi_{g\varepsilon}$, and this solution satisfies the integral identity (see [52, 90])

$$v_i(x,t) = g_i(x_{0i},\varepsilon) + \varepsilon \int_0^t F_i^0(v(x_i,\tau))d\tau,$$

where $x_i(\tau;x,t,\varepsilon)$, $0 \le \tau \le t$; $x_{0i} = x_i(0;x,t,\varepsilon)$. Differentiating this identity with respect to x yields

$$\frac{\partial v_i}{\partial x} = g'_{ix_{0i}}(x_{0i})'_x + \varepsilon \int_0^i \sum_{j=1}^m \frac{\partial F_i^0}{\partial v_j}\frac{\partial v_j}{\partial x_i}\frac{\partial x_i}{\partial x} d\tau.$$

Since the functions $F_i^0(v)$ are continuously differentiable with respect to v, their derivatives with respect to v are bounded by the Lipschitz constant M. Whence it follows, by the Gronwall — Bellman lemma, that

$$\left|\frac{\partial v_i}{\partial x}\right| \le \varepsilon C_1 \quad \forall (x,t) \in \Pi_{g\varepsilon}, \qquad i = 1,2,\ldots,m,$$

where $C_1 = \alpha K_0 \exp\{mMK_0L\}$, $K_0 = \exp\{\nu_1 L\}$.

From system (6.107) we derive that $|\partial v_i/\partial t| \le \varepsilon C_2 \; \forall (x,t) \in \Pi_{g\varepsilon}$ for all $i = 1,2,\ldots,m$, where $C_2 = \nu_0 C_1 + K$.

Therefore the solution $v(x,t)$ of the Cauchy probelm (6.4), (6.5) satisfies a Lipschitz condition with respect to x and t with a constant proportional to ε.

Passing from systems (6.107) and (6.110) to systems of integral equations (see [52, 90]) gives

$$|u_i(x,t) - v_i(x,t)| \le \varepsilon \int_0^t |F_i(x_i,\tau,u(x_i,\tau)) - F_i(x_i,\tau,v(x_i,\tau))|d\tau +$$

$$+\left|\varepsilon \int_0^t [F(x_i,\tau,v(x_i,\tau)) - F_i^0(v(x_i,\tau))]d\tau\right| = \mathcal{I}_{1i} + \mathcal{I}_{2i}. \tag{6.112}$$

By using the condition 3 in Theorem 6.6, we obtain

$$\mathcal{I}_{1i} \le \varepsilon M \int_0^t \sum_{j=1}^m |u_j(x_i,\tau) - v_i(x_i,\tau)|d\tau. \tag{6.113}$$

Dividing the interval $[0, L\varepsilon^{-1}]$ into p equal subintervals and using the method of [118], we obtain

$$\mathcal{I}_{2i} \le \frac{mM(2C_1\nu_0 + K)}{p} + L\left[\sum_{k=0}^{p-1} \Psi_i\left(\frac{(k+1)L}{\varepsilon p};\ v^k\right) + \right.$$

$$\left. + \sum_{k=1}^{p-1} \Psi_i\left(\frac{kL}{\varepsilon p};\ v^k\right)\right] + \max_{0\le s\le p-1} \Psi_{0i}(\varepsilon, v^s) \equiv b_i(\varepsilon,p), \tag{6.114}$$

where

$$\Psi_i(t,v) = \frac{1}{t}\int_0^t [F_i(x_i,\tau,v) - F_i^0(v)]d\tau, \qquad i = 1,2,\ldots,m;$$

$$\Psi_{0i}(\varepsilon, v) = \sup_{0\le\tau\le L} \Psi_i\left(\frac{\tau}{\varepsilon}, v\right), \qquad i = 1, 2, \dots, m;$$

$$v^k = v(x_i(t_k; x, t, \varepsilon), t_k), \quad t_k = kL(\varepsilon p)^{-1}, \quad k = 0, 1, 2, \dots, p-1.$$

Keeping t fixed, we intriduce the function $U(t) = \sup_{j;x;\tau\le t} |u_j(x,\tau) - v_j(x,\tau)|$.

Substituting (6.113) and (6.114) into (6.112) and using the Gronwall-Bellman lemma, we see that

$$U(t) \le \sup_{j;x;\tau\le t} \{b_j(\varepsilon, p)\} \exp\{mML\} \tag{6.115}$$

holds on the set Π_{gs}. Whence, assuming $\sup_{j;x;\tau\le t} \{b_j(\varepsilon,p)\} < \exp\{-mML\} \min(\rho, \eta)$, we arrive at the statement of Theorem 6.6.

2. Estimates. While proving Theorem 6.6 we obtained the estimate (6.115) of the difference between solutions of the initial and averaged systems. This estimate can be simplified if we assume that passage to the limit in (6.109) is done uniformly with respect to $(x_)), t_0) \in \Pi$, $k \in G$. Then, indeed, there exists a function $\Psi(t) = \sup_{t,v\in G} \Psi_i(t,v)$ ($\Psi(t) \to 0$ as $t \to \infty$) such that

$$\left|\varepsilon \int_0^t [F_i(x_i, \tau, v) - F_i^0(v)]d\tau\right| \le \varepsilon t \Psi(t) \le \sup_{0\le\tau\le L} \tau\Psi\left(\frac{\tau}{\varepsilon}\right) = \varphi(\varepsilon),$$

where $\varphi(\varepsilon) \to 0$ as $\varepsilon \to 0$.

By (6.114), we therefore have

$$b_i(\varepsilon, p) \le \frac{mM(2C_1\nu_0 + K)L^2}{p} + 2p\varphi(\varepsilon) \equiv b(\varepsilon, p). \tag{6.116}$$

The last estimate can be improved if p is suitably chosen. Let p be chosen so that the function $b(\varepsilon, p)$ assumes its least value. Obviously, $b(\varepsilon, p)$ attains its minimum at

$$p = \sqrt{mM(2C_1\nu_0 + K)L^2(2\varphi(\varepsilon))^{-1}}.$$

For such p we have

$$|u_i(x,t) - v_i(x,t)| \le \sqrt{2}\, Le^{mML}\sqrt{mM(2C_1\nu_0 + K)\varphi(\varepsilon)}. \tag{6.117}$$

The problem of obtaining more precise estimates amounts to the investigation into the rate of decrease of the functions

$$\Psi_i(t, v) = \left|\frac{1}{t}\int_0^t [F_i(x_i, \tau, v) - F_i^0(v)]d\tau\right|, \qquad i = 1, 2, \dots, m,$$

when $t \to \infty$. Assuming $\sup_{i;v} \Psi_i(t,v) \le \frac{a_1}{t}$, one can obtain the following estimate

$$|u_i(x,t) - v_i(x,t)| < \varepsilon P \quad \forall (x,t) \in \Pi_{g\varepsilon}, \quad P = \text{const}, \quad i = 1, 2, \dots, m. \tag{6.118}$$

Remark 6.6. If the conditions of the existence of classical solution hold for the Cauchy problem (6.110), (6.111), i. e. if the averaged function $F_i^0(u)$ satisfies the conditions of the existence of classical solution of the Cauchy problem (6.110), (6.111) [52, 90], then the estimates (6.115)–(6.118) are valid for the classical solution $v(x,t)$ of this problem.

Remark 6.7. Theorem 6.6 shows that for the method of averaging along characteristics to be applied it is not needed to require passage to the limit in (6.109) to be uniform with respect to $u \in G$. As known, if passage to the limit in (6.109) is not uniform, then the functions $F_i^0(u)$ may become discontinuous although $F(x,t,u)$ are continuous and differentiable. One may then speak about the proximity of solutions of the initial and averaged systems. We state a theorem on averaging along characteristics for the hyperbolic system (6.107) in the class of continuous solutions.

Theorem 6.7 [72]. *Suppose that the conditions 1), 3) and 6) of Theorem 6.6 and the conditions given below are fulfilled on the sets $\Pi \times I$ and Ω.*

1) $g(x,t) \in \mathrm{Lip}_x(\varepsilon\alpha; \mathbb{R} \times I) \times C_\varepsilon(I)$ and $g(x,\varepsilon) = u(x,0) \in G$;

2) The limit (6.109) exists at any point $u \in G$, is independent of the initial data (x_0, t_0) and $|F_i^0(u)| \le K$.

Then, for all $\gamma > 0$ and $L > 0$, there exists $\varepsilon_0 > 0$ such that for $\varepsilon < \varepsilon_0$ the inequality

$$|u_i(x,t) - v_i(x,t)| < \gamma \qquad \forall (x,t) \in \Pi_{g\varepsilon}, \quad i = 1,2,\dots,m$$

holds for the continuous solutions $u(x,t)$ and $v(x,t)$ of the Cauchy problems (6.107), (6.108) and (6.109), (6.111), respectively.

Proof. Observe that since the averaged function $F_i^0(u)$ satisfies a Lipschitz condition with constant M, the continuous solutions $u(x,t)$ and $v(x,t)$ of the Cauchy problems (6.107), (6.108) and (6.110), (6.111) exist on every bounded closed set $\Pi_{g\varepsilon}$ if only the conditions of Theorem 6.7 are satisfied. Hence to prove Theorem 6.7 it is necessary to prove the estimate (6.8), obtained while proving Theorem 6.6. To do this, it is necessary to prove that continuous solutions $v(x,t)$ of the Cauchy problem (6.110), (6.111) satisfy a Lipschitz condition with respect to x and t with a constant proportional to ε.

We have

$$|v_i(x,t) - v_i(x_0,t_0)| \le |g_i(x_i(0;x,t,\varepsilon),\varepsilon) - g_i(x_i(0;x_0,t_0,\varepsilon),\varepsilon)| +$$

$$+\varepsilon\left|\int_0^t F_i^0(v_i(x_i,\tau))d\tau - \int_0^{t_0} F_i^0(v(x_i^0,\tau))d\tau\right| = N_{1i} + N_{2i},$$

where $x_i = x_i(\tau; x,t,\varepsilon)$ and $x_i^0 = x_i(\tau; x_0, t_0, \varepsilon)$.

By the condition 1) of Theorem 6.7, we obtain

$$N_{1i} \le \alpha\varepsilon |x_i(0;x,t,\varepsilon) - x_i(0;x_0,t_0,\varepsilon)|. \tag{6.119}$$

Suppose that $t > t_0$ (other cases can be treated similarly). Then we have

$$N_{2i} \le \varepsilon \int_{t_0}^t |F_i^0(v(x_i,\tau))|d\tau + \varepsilon \int_0^t |F_i^0(v(x_i,\tau)) - F_i^0(v(x_i^0,\tau))|d\tau \le$$

$$\le \varepsilon K|t - t_0| + \varepsilon M \int_0^t \sum_{j=1}^m |v_j(x_i,\tau) - v_j(x_i^0,\tau)|d\tau. \tag{6.120}$$

By using the inequalities

$$\frac{\partial x_i}{\partial x} = \exp\left\{\int_t^\tau \frac{\partial \lambda_i}{\partial x} d\tau\right\}, \quad \frac{\partial x_i}{\partial t} = -\lambda_i \exp\left\{\int_t^\tau \frac{\partial \lambda_i}{\partial x} d\tau\right\},$$

we get [91]

$$\left|\frac{\partial x_i}{\partial x}\right| \le \exp\{\nu_1 L\}, \quad \left|\frac{\partial x_i}{\partial t}\right| \le \nu_0 \exp\{\nu_1 L\}, \quad 0 \le t \le L\varepsilon^{-1}.$$

Taking account of (6.119) and (6.120) we have

$$|v_i(x,t) - v_i(x_0,t_0)| \le \varepsilon\tilde{C}_1|x - x_0| + \varepsilon\tilde{C}_2|t - t_0| +$$

$$+\varepsilon M \int_0^t \sum_{j=1}^m |v_j(x_i,\tau) - v_j(x_i^0,\tau)|d\tau, \tag{6.121}$$

where $\tilde{C}_1 = \alpha K_2$; $\tilde{C}_2 = \tilde{C}_1 + K$ and $K_2 = \max_{j;x;\tau\le t}\{e^{\nu_1 L}, \nu_0 e^{\nu_1 L}\}$.

We introduce the notation $V(t) = \max_{j;x;\tau\le t}|v_j(x,\tau) - v_j(x_0,t_0)|$. By the Gronwall-Bellman lemma, from (6.121) it follows that

$$V(t) \le \varepsilon(C_1|x - x_0| + C_2|t - t_0|), \qquad \text{где } C_1 = \tilde{C}_1 e^{2mML},\ C_2 = \tilde{C}_2 e^{2mML}.$$

Hence the solution $v(x,t)$ of the Cauchy problem (6.110), (6.111) satisfies a Lipschitz condition with respect to x and t with a constant proportional to ε, the constants C_1 and C_2 occuring in (6.114)–(6.117) being determined. Now, repeating the proof of Theorem 6.6 we see that Theorem 6.7 is valid.

3. Averaging with respect to explicitly occuring time. Suppose that the limit

$$\lim_{T\to\infty} \frac{1}{T}\int_0^T F(x,t,u)dt = F_1(x,u) \tag{6.122}$$

exists. Then to system (6.107) we can set into correspondence the averaged system

$$Dv = F_1(x,v) \tag{6.123}$$

with the initial condition (6.111).

Theorem 6.8 [72]. *Suppose that the conditions 1), 2) and 4) of Theorem 6.6 and the conditions below hold on the sets $\Pi \times I$ and Ω.*

1) $F(x,t,u) \in C_t \times \mathrm{Lip}_{x,u}(M;\Omega)$, $M = \mathrm{const}$.

2) The limit (6.122) exists uniformly with respect to $x \in \mathbb{R}_x$, $u \in G$ and the functions $|F_{1i}(x,u)| \le K$, $F_{1i}(x,u)$ are continuously differentiable with respect to x and u.

3) A continuous solution $v(x,t)$ of the Cauchy problem (6.123), (6.111) defined on the set Π lies in the region G with a certain ρ-neighbourhood.

Then for all $\gamma > 0$ and $L > 0$, there exists $\varepsilon_1 > 0$ such that for $\varepsilon < \varepsilon_1$ the inequality

$$|u_i(x,t) - v_i(x,t)| < \gamma \qquad \forall (x,t) \in \Pi_{g\sqrt{\varepsilon}}, \quad i = 1,2,\ldots,m$$

holds for the classical solution $u(x,t)$ of the Cauchy problem (6.107), (6.102) with $\Pi_{g\sqrt{\varepsilon}} = \{(x,t) \in \Pi : x_m\ (t; -L\varepsilon^{-\frac{1}{2}}, L\varepsilon^{-\frac{1}{2}}, \varepsilon) \le x \le x_1(t; -L\varepsilon^{-\frac{1}{2}}, L\varepsilon^{-\frac{1}{2}}, \varepsilon),\ 0 \le t \le L\varepsilon^{-\frac{1}{2}}$.

The proof of this theorem can be found in [72].

4. Averaging with respect to fast variables. Suppose that the limit

$$\lim_{\substack{X\to\infty\\T\to\infty}} \frac{1}{XT}\int_0^X\int_0^T F(x,t,u)dxdt = F_3(u) \tag{6.124}$$

exists. Then to system (6.107) we can set into correspondence the averaged system

$$Dv = \varepsilon F_3(v). \tag{6.125}$$

Theorem 6.9 [73]. *Suppose that the conditions 1) and 2) of Theorem 6.6 and the conditions below hold on the sets $\Pi \times I$ and Ω defined in Theorem 6.6.*

1) $|\partial\lambda_i/\partial t| \le \varepsilon\nu_2$, $i = 1, 2, \dots, m$;

2) $F_i(x,t,u) \in C_t \times \mathrm{Lip}_{x,u}(M;\Omega)$, $i = 1, 2, \dots, m$; $M = \mathrm{const}$;

3) the limit (6.124) and the following limit

$$\lim_{\substack{X\to\infty \\ T\to\infty}} \frac{1}{XT} \int_X^0 \int_0^T F(x,t,u)\, dx dt = F_3(u),$$

exist uniformly with respect to $u \in G$, $|F_{3i}(u)| \le K$ and the functions $F_{3i}(u)$ are continuously differentiable with respect to u;

4) the limit

$$\lim_{T\to\infty} \frac{1}{T} \int_T^0 F(x,t,u)dt = F_1(x,u), \tag{6.126}$$

exists uniformly with respect to $x \in \mathbb{R}$, $u \in G$ and the functions $|F_1(x,u)| \le K$ $F_{1i}(x,u)$ are continuously differentable with respect to x and u.

5) The continuous solution $v(x,t)$ of the Cauchy problem (6.125), (6.111) and the continuous solution $w(x,t)$ of the following Cauchy problem

$$Dw = \varepsilon F_1(x,w),$$

$$w(x,0) = g(x,\varepsilon), \qquad x \in \mathbb{R}, \quad \varepsilon \in I, \tag{6.127}$$

which are defined on the set Π lie in the region G with a certain ρ-neighbourhood.

Then for all $\gamma > 0$ and $L > 0$ there exists $\varepsilon_3 > 0$ such that for $\varepsilon < \varepsilon_3$ the inequality $|u_i(x,t) - v_i(x,t)| < \gamma \forall (x,t) \in \Pi_{g\sqrt{\varepsilon}}$; $i = 1, 2, \dots, m$ holds for the continuous solution $u(x,t)$ of the Cauchy problem (6.107), (6.108).

The proof can be found in [73].

5. Averaging with respect to a fast coordinate. The averaging method with respect to a fast coordinate is an immediate corollary of Theorem 6.9.

Suppose that the limit

$$\lim_{X\to\infty} \frac{1}{X} \int_0^X F(x,t,u)dx = F_4(t,u)$$

exists. To system (6.107) one can set into correspondence the averaged system

$$Du = \varepsilon F_4(t,u).$$

Remark 6.8. Consider a particular case of system (6.107), namely the system

$$Du = \varepsilon F_4(t,u). \tag{6.128}$$

Suppose that the limit

$$\lim_{T\to\infty} \frac{1}{T} \int_0^T F(t,u)dt = F_2(u)$$

exists and consider the averaged system

$$Dv = \varepsilon F_2(v). \tag{6.129}$$

If the coefficients and functions occuring in the systems (6.128) and (6.129) satisfy the conditions of Theorem 6.8, then for all $\gamma > 0$ and $L > 0$ there exists $\varepsilon_2 > 0$ such that for $\varepsilon < \varepsilon_2$ the inequalities

$$|u_i(x,t) - v_i(x,t)| < \gamma \qquad \forall (x,t) \in \Pi_{g\varepsilon}, \quad i = 1, 2, \ldots, m$$

hold for solutions $u(x,t)$ and $v(x,t)$ of the Cauchy problems (6.128), (6.108) and (6.129), (6.111), respectively.

6. Partial averaging. Various schemes of partial averaging can be applied to the hyperbolic system (6.107) as it is done for ordinary differential equations [118] and for the hyperbolic systems of standard form [128, 131].

6.7. Reduction of quasilinear equations to a countable system

Suppose one needs to find a wave solution of the quasilinear equation

$$u_{tt} - a^2 u_{xx} = cu + \varepsilon f(x, t, \eta, \tau, u, p, q). \tag{6.130}$$

We assume that the nonperturbed equation $u_{0tt} - a^2 u_{0xx} = cu_0$ has a solution that can be represented as the series

$$u_0 = \sum_{m=1}^{\infty} a_m \cos(\omega_m t + \theta_m) \sin k_m x, \tag{6.131}$$

where a_m and θ_m are arbitrary constants, and the frequency ω and wave number $k = 2\pi/\lambda$ (λ a wave length) are related by the dispersion equation

$$a^2 k_m^2 - \omega^2 - c = 0, \quad a^2 k_m^2 - c > 0, \quad m = 1, 2, \ldots; \tag{6.132}$$

m being an index of normal wave (a branch of the dispersion equation).

Solution (6.131) of nonperturbed equation can be represented as

$$u_0 = \sum_{m=1}^{\infty} \sum_{s=1}^{2} a_{sm} (b e^{i\tilde{\beta}_{sm}} + \bar{b} e^{-i\tilde{\beta}_{sm}}), \tag{6.133}$$

where $a_{1m} = a_{2m} = a_m$; $b = i/4$, $\bar{b} = -i/4$ are complex conjugate (c. c.) expressions, and $\tilde{\beta}_{sm} = \tilde{\omega}_s t - k_m x + \tilde{h}_{sm}$, $s = 1, 2$, $\tilde{\omega}_1 = \omega_m$, $\tilde{\omega}_2 = -\omega_m$, $\tilde{h}_{1m} = \theta_m$, $\tilde{h}_{2m} = -\theta_m$, $m = 1, 2, \ldots$; $i = \sqrt{-1}$.

For $\varepsilon \neq 0$, the wave solution of the nonperturbed equation (6.130) is sought in the form

$$u(x,t,\varepsilon) = \sum_{m=1}^{\infty} \sum_{s=1}^{2} A_{sm}(\xi_{sm}) \left[b e^{i\beta_{sm}} + \bar{b} e^{-i\beta_{sm}} \right], \tag{6.134}$$

where $\beta_{sm} = \xi_{sm} + h_{sm}(\xi_{sm})$; $\xi_{sm} = \tilde{\omega}_s t - k_m x$; $\tilde{\omega}_1 = \omega_m$; $\tilde{\omega}_2 = -\omega_m$.

Since new unknown functions $A_{1m}, A_{2m}, h_{1m}, h_{2m}$ have been introduced, some additional conditions are necessary to determine them uniquely. These conditions can be written as

$$p = i \sum_{m=1}^{\infty} \sum_{s=1}^{2} \tilde{\omega}_s A_{sm}(\xi_{sm}) \left[b e^{i\beta_{sm}} - \bar{b} e^{-i\beta_{sm}} \right]; \tag{6.135}$$

$$q = -i \sum_{m=1}^{\infty} \sum_{s=1}^{2} k_m A_{sm}(\xi_{sm}) \left[b e^{i\beta_{sm}} - \bar{b} e^{-i\beta_{sm}} \right]. \tag{6.136}$$

Differentiating (6.111) with respect to t and equating the resulted expression to (6.112), we obtain

$$\sum_{m=1}^{\infty} \sum_{s=1}^{2} \tilde{\omega}_s \left[\frac{dA_{sm}}{d\xi_{sm}} + i \frac{dh_{sm}}{d\xi_{sm}} A_{sm} \right] b e^{i\beta_{sm}} + \text{c.c.} = 0. \tag{6.137}$$

Differentiating one more time to get the second derivatives $u_{tt} = p_t$ and $u_{xx} = q_x$, substituting them into (6.107) and taking account of (6.134)–(6.136), we obtain

$$i \sum_{m=1}^{\infty} \sum_{s=1}^{2} (\omega_s^2 - a^2 k_m^2) \left[\frac{dA_{sm}}{d\xi_{sm}} + i \frac{dh_{sm}}{d\xi_{sm}} A_{sm} \right] b e^{i\beta_{sm}} + \text{c.c.} =$$

$$= \varepsilon F(x, t, \eta, \tau, A_1, A_2, \beta_1, \beta_2), \tag{6.138}$$

Here

$$F(x, t, \eta, \tau, A_1, A_2, \beta_1, \beta_2) = f \left(x, t, \eta, \tau, \sum_{m=1}^{\infty} \sum_{s=1}^{2} A_{sm} b e^{i\beta_{sm}} + \right.$$

$$\left. + \text{c.c.}, i \sum_{m=1}^{\infty} \sum_{s=1}^{2} \tilde{\omega}_s A_{sm} b e^{i\beta_{sm}} + \text{c.c.}, - i \sum_{m=1}^{\infty} \sum_{s=1}^{2} k_m A_{sm} b e^{i\beta_{sm}} + \text{c.c.} \right), \tag{6.139}$$

where $A_s, \beta_s, s = 1, 2$, are infinite-dimensional vectors.

Suppose that the function F can be expanded into a Fourier series

$$F(\ldots) = \sum_{m=1}^{\infty} F_m(t, \eta, \tau, A_1, A_2, \psi_1, \psi_2) e^{-ik_m x} + \text{c.c.}, \tag{6.140}$$

where $\psi_1 = (\omega_1 t + h_{11}, \omega_2 t + h_{12}, \ldots)$; $\psi_2 = (-\omega_1 t + h_{21}, -\omega_2 t + h_{22}, \ldots)$;

$$F_m = \frac{1}{2X} \int_0^X F(x, t, \eta, \tau, A_1, A_2, \beta_1, \beta_2) e^{ik_m x} dx. \tag{6.141}$$

In view of (6.11) and the notation $\tilde{\omega}_1 = \omega_m$ and $\tilde{\omega}_2 = -\omega_m$, (6.8) and (6.9) can be written in an expanded form as

$$\sum_{m=1}^{\infty} \left(\omega_m \left[\frac{dA_{1m}}{d\xi_{1m}} + i \frac{dh_{1m}}{d\xi_{1m}} A_{1m} \right] b e^{i(\omega_m t + h_{1m})} - \omega_m \left[\frac{dA_{2m}}{d\xi_{2m}} + \right. \right.$$

$$\left. \left. + i \frac{dh_{2m}}{d\xi_{2m}} A_{2m} \right] b e^{i(-\omega_m t + h_{2m})} \right) e^{-ik_m x} + \text{c.c.} = 0,$$

$$i \sum_{m=1}^{\infty} (\omega_m^2 - a^2 k_m^2) \left\{ \left[\frac{dA_{1m}}{d\xi_{1m}} + i \frac{dh_{1m}}{d\xi_{1m}} \right] b e^{i(\omega_m t + h_{1m})} + \left[\frac{dA_{2m}}{d\xi_{2m}} + \right. \right.$$

$$\left. \left. + i \frac{dh_{2m}}{d\xi_{2m}} A_{2m} \right] b e^{i(-\omega_m t + h_{2m})} \right\} e^{-ik_m x} + \text{c.c.} =$$

$$= \varepsilon \sum_{m=1}^{\infty} (F_m e^{-ik_m x} + \overline{F}_m e^{ik_m x}).$$

Whence equating the coefficients at the equal exponents it follows that

$$\gamma_{1m}e^{i(\omega_m t+h_{1m})} - \gamma_{2m}e^{i(-\omega_m t+h_{2m})} = 0,$$

$$\gamma_{1m}e^{i(\omega_m t+h_{1m})} + \gamma_{2m}e^{i(-\omega_m t+h_{2m})} = -\frac{4\varepsilon F_m}{\omega_m^2 - a^2k_m^2}, \tag{6.142}$$

where

$$\gamma_{sm} = \frac{dA_{sm}}{d\xi_{sm}} + i\frac{dh_{sm}}{d\xi_{sm}}A_{sm}, \quad s = 1,2; \quad m = 1,2,\ldots \tag{6.143}$$

Solving system (6.142) we get

$$\gamma_{1m} = -\frac{2\varepsilon F_m \exp\{i(-\omega_m t - h_{1m})\}}{\omega_m^2 - a^2k_m^2}, \quad \gamma_{2m} = -\frac{2\varepsilon F_m \exp\{i(\omega_m t - h_{2m})\}}{\omega_m^2 - a^2k_m^2}$$

or, in view of (6.143),

$$\frac{dA_{sm}}{d\xi_{sm}} = -\varepsilon Re\frac{2F_m e^{-i\psi_{sm}}}{\omega_m^2 - a^2k_m^2}, \quad A_{sm}\frac{dh_{sm}}{d\xi_{sm}} = -\varepsilon \operatorname{Im}\frac{2F_m e^{-i\psi_{sm}}}{\omega_m^2 - a^2k_m^2}, \tag{6.144}$$

where $\psi_{sm} = \tilde{\omega}_s t + h_{sm}$, $s = 1,2$; $\tilde{\omega}_1 = \omega_m$, $\tilde{\omega}_2 = -\omega_m$, $m = 1,2,\ldots$

Using the relations $A_{sm}(\xi_{sm})$, $h_{sm}(\xi_{sm})$, $\xi_{sm} = \tilde{\omega}_s t - k_m x$ and equality

$$\tilde{\omega}_s\frac{\partial A_{sm}}{\partial t} + a^2k_m\frac{\partial A_{sm}}{\partial x} = (\omega_m^2 - a^2k_m^2)\frac{dA_{sm}}{d\xi_{sm}}, \quad s = 1,2; \ m = 1,2,\ldots,$$

system (6.144) can be written down as

$$\frac{\partial A_{sm}}{\partial t} + \frac{a^2k_m}{\tilde{\omega}_s}\frac{\partial A_{sm}}{\partial x} = -\frac{\varepsilon}{\tilde{\omega}_s}\operatorname{Re}[2F_m e^{-i\psi_{sm}}],$$

$$\frac{\partial h_{sm}}{\partial t} + \frac{a^2k_m}{\tilde{\omega}_s}\frac{\partial h_{sm}}{\partial x} = -\frac{\varepsilon}{A_{sm}\tilde{\omega}_s}\operatorname{Im}[2F_m e^{-i\psi_{sm}}]. \tag{6.145}$$

Thus system (6.145) is a real countable system of first order partial differential equations of hyperbolic type $(a^2k_m/\tilde{\omega}_s \in \mathbb{R})$.

6.8. Truncation of a countable system of partial differential equations. Problems of mathematical justification

Suppose we are given the countable system

$$\frac{\partial z_s}{\partial t} + \lambda_s\frac{\partial z_s}{\partial x} = f_s(t,\eta,\tau,z_1,z_2,\ldots), \quad s = 1,2,\ldots, \tag{6.146}$$

on the set $\Pi = \{x \in \mathbb{R}, t \geq 0\} \times I$ with the initial condition

$$z_s(x,0) = g_s(x), \quad s = 1,2,\ldots; \ x \in \mathbb{R}, \tag{6.147}$$

where λ_s, $s = 1,2,\ldots$, are known real numbers.

Let $\mathbb{R}^2$ be a closed region in the plane Π_g such that all the characteristics L_s of system (6.146) drawn from the point P into the region Π_g in the direction $t = 0$ meet the given interval $[a,b]$ of the axis x at the points P_s with the coordinates $(x_s^0, 0)$, where $x_s^0 = x - \lambda_s t$ (see §6.6 for the definition of Π_g).

Besides the system (6.146) we shall consider the following system

$$\frac{\partial \omega_k}{\partial t} + \lambda_k \frac{\partial \omega_k}{\partial x} = f_k(t, \eta, \tau, w_1, w_2, \ldots, w_n, g_{n+1}, g_{n+2}, \ldots), \tag{6.148}$$

$$w_{n+j}(x,t) = z^0_{n+j}(x,t), \; k = 1, 2, \ldots, n; \; j = 1, 2, \ldots,$$

with the initial conditions

$$w_k(x, 0) = g_k(x), \; k = 1, 2, \ldots, n, \tag{6.149}$$

where $z^0_{n+j} = g_{n+j}(x^0_{n+j})$; $x^0_{n+j} = x - \lambda_{n+j} t$, $j = 1, 2, \ldots$

System (6.148) is said to be a truncated system of differential equations for system (6.146). It is obtained from system (6.148) by setting the unknown functions beginning with $(n+1)$th to equal the initial conditions and by taking the solution of the equation of characteristics $dx/dt = \lambda_s$ at $t = 0$ as x.

Consider system (6.146) in the space M^∞ a point of which is a countable collection of continuous functions uniformly bounded by a constant. The vector-functions $f = (f_1, \ldots, f_n, \ldots)$, $z = (z_1, \ldots, z_n, \ldots)$ in system (6.146) are thus points of the space M^∞. We introduce the norm in M^∞ as follows

$$\|z_k(x,t)\| = \sup_k \max_{x,t} |z_k(x,t)|$$

Continuity of functions is thought of as continuity with respect to the norm.

Theorem 6.10 [126]. *Suppose the following conditions are satisfied:*

1) the functions $f_s(t, \eta, \tau, z_1, z_2, \ldots)$, $s = 1, 2, \ldots$, are defined on the set $\Omega = \Pi_g \times I \times D$, where D is a bounded region of the space M^∞, and satisfy the conditions

a) the functions $f_s(t, \eta, \tau, z_1, z_2, \ldots)$, $s = 1, 2, \ldots$, are continuous with respect to the collection of variables $t, \eta, \tau, z_1, z_2, \ldots$ in the region Ω and satisfy a Lipschitz condition there with respect to $z_1, z_2, \ldots$, i. e.

$$|f_s(t, \eta, \tau, z_1'', z_2'', \ldots) - f_s(t, \eta, \tau, z_1', z_2', \ldots)| \le K \Delta z$$

the constant K being independent of t, η, τ. Here

$$\Delta z = \sup\,[|z_1'' - z_1'|, |z_2'' - z_2'|, \ldots] \ldots;$$

b) the functions $f_s(t, \eta, \tau, z_1, z_2, \ldots)$, $s = 1, 2, \ldots$, satisfy the condition $|f_s(t, \eta, \tau, z_1, z_2, \ldots)| \le \alpha_s$, where $\alpha_s = \text{const}$, and $\alpha_s \to 0$ as $s \to \infty$;

2) the functions $g_s(x)$ are continuous and $\lim_{s \to \infty} \lambda_s = a < \infty$.

Then the solution $z_s(x,t)$ of the exact system (6.146) and solution $w_s(x,t)$ of the truncated system (6.148) satisfying the same initial conditions $z_s(x,0) = w_s(x,0) = g_s(x)$, $s = 1, 2, \ldots$, come arbitrarily close to each other in Π_g as n becomes sufficiently large.

The reader can find the proof in [126].

We note that Theorem 6.10 establishes only an estimate of the difference between the solutions of countable (exact) and truncated systems of first order hyperbolic equations. It does not solve the problem on esimating the difference between a solution of the perturbed partial differential equation of second order (6.130) and its approximate solution under the assumption that this equation can be reduced to a countable system and this system admits truncation. This is a so-called problem on summing trigonometric series whose coefficients are given approximately (see §6.2 in the present chapter). It is therefore more natural to obtain results that provide some estimates for solutions of equation (6.130) and their approximations.

So, besides the countable system (6.145), we shall consider the truncated system

$$\frac{\partial A_{smn}}{\partial t} + \frac{a^2 k_m}{\tilde{\omega}_s}\frac{\partial A_{smn}}{\partial x} = -\frac{\varepsilon}{\tilde{\omega}_s}\operatorname{Re}\,[2F_{mn}e^{-i\psi_{smn}}],$$

$$\frac{\partial h_{smn}}{\partial t} + \frac{a^2 k_m}{\tilde{\omega}_s}\frac{\partial h_{smn}}{\partial x} = -\frac{\varepsilon}{A_{smn}\tilde{\omega}_s}\operatorname{Im}\,[2F_{mn}e^{-i\psi_{smn}}], \tag{6.150}$$

$$s = 1,2; \quad m = 1,2,\ldots,n,$$

obtained from (6.16) by setting the functions sought for beginning with $(n+1)$th to be zero and by discarding the equations beginning with $(n+1)$th.

Consider the function

$$u_n(x,t,\varepsilon) = \sum_{m=1}^{n}\sum_{s=1}^{2} A_{smn}(\xi_{sm})[be^{i\beta_{smn}} + \bar{b}e^{-i\beta_{smn}}], \tag{6.151}$$

where $\beta_{smn} = \xi_{sm} + h_{smn}(\xi_{sm})$; $\xi_{sm} = \tilde{\omega}_s t - k_m x$, $s = 1,2$; $m = 1,2,\ldots,n$; $\tilde{\omega}_1 = \omega_m$, $\tilde{\omega}_2 = -\omega_m$; A_{smn}, h_{smn}, $s = 1,2$ and $m = 1,\ldots,n$ is a solution of the truncated system (6.150). This function is said to be an approximate solution of equation (6.130).

We are now in a position to state the problem: find a number n beginning with which the difference between the exact solution (6.134) of equation (6.130) and its approximate solution (6.151) is becoming arbitrarily small over a finite time interval.

One can readily prove the following theorem.

Theorem 6.11 [134]. *Suppose that the following conditions are fulfilled on the set* $\Omega = \{x \in \mathbb{R}_x, 0 \le t \le T,\ (u,p,q) \in G \subset \mathbb{R}^3\} \times \{\varepsilon : 0 \le \varepsilon < \varepsilon_0\}$:

1) the function $f(x,t,\eta,\tau,u,p,q)$ *(the right-hand side in (6.130)) is continuous with respect to* x,t,η,τ *and satisfies a Lipschitz condition with respect to* u,p,q *with a constant independent of* x,t,η,τ, *i. e.* $f(x,t,\eta,\tau,u,p,q) \in \mathrm{Lip}_{u,p,q}(K;\Omega)$, $K = \mathrm{const}$;

2) equation (6.130) has a solution $u(x,t,\varepsilon)$ *representable as the series (6.5), moreover,* $A_{sm}|_{t=0} = a_m$, $h_{sm}|_{t=0} = (-1)^{s-1}\theta_m$, $s = 1,2$; $m = 1,2,\ldots$, *the series* $\sum_{m=1}^{\infty} \nu a_m$, $\nu = 1, \omega_m, k_m$ *is convergent and the series* $\sum_{m=1}^{\infty}\sum_{s=1}^{2} \nu A_{sm}$, $\nu = 1, k_m, \omega_m$ *is uniformly convergent;*

3) the function $f(x,t,\eta,\tau,u(x,t,\varepsilon),\ u_t(x,t,\varepsilon),\ u_x(x,t,\varepsilon))$ *can be expanded into the Fourier series (6.140) where* $u(x,t,\varepsilon)$ *is the solution (6.134),* u_t *and* u_x *are partial derivatives of* $u(x,t,\varepsilon)$;

4) the Fourier coefficients (6.12) satisfy the condition $|\, F_m\,(t,\eta,\tau,A_1,A_2,\psi_1,\psi_2)\,| \le \alpha_m$, α_m, $m = 1,2,\ldots$, *the series* $\sum_{m=1}^{\infty}\alpha_m$ *being convergent;*

5) the series $\sum_{m=1}^{\infty}\frac{1+\omega_m+k_m}{\omega_m}\alpha_m$ *is convergent.*

Then, for any T, $0 < T \le T_1 < \infty$ *there exists a number* n *such that*

$$|u_i(x,t,\varepsilon) - u_{ni}(x,t,\varepsilon)| < \varepsilon P_n(T,\varepsilon), \quad i = 0,t,x,$$

for all $(x,t) \in \Pi_g$, *where* $u_0(x,t,\varepsilon) \equiv u(x,t,\varepsilon)$ *is a solution of equation (6.130) and the function* $u_{n0}(x,t,\varepsilon) \equiv u_n(x,t,\varepsilon)$ *is given by (6.151),*

$$P_n(T,\varepsilon) = \exp\{\varepsilon S_n T\}\left(1 + \frac{1}{S_n}\right) - \frac{1}{S_n}, \quad S_n = 4K\sum_{m=1}^{n}\frac{1+\omega_m+k_m}{\omega_m},$$

with $T_1 (T \le T_1)$ *being the largest value of the coordinate in* Π_g (see the analogous theorem in Chapter 7).

6.9. Investigation into the multifrequency oscillation modes of the quasiwave equation

As established in §6.1 of this chapter, the phase-amplitude parameters of the quasilinear hyperbolic equation of second order [134]

$$u_{tt} - a^2 u_{xx} = cu + \varepsilon f(x, t, \eta, \tau, u, p, q, q_x, q_t, \varepsilon) \tag{6.152}$$

depend on slow time and a spatial coordinate; this dependence has a specific character: in the first approximation, the amplitude and phase shift of the normal wave depend on the phase of nonperturbed wave.

In this section we present an algorithm for constructing multifrequency oscillation modes for equation (6.152). For $\varepsilon \neq 0$, the results in §§6.1 and 6.7 of this chapter allows us to write down a multifrequency solution of equation (6.152) as follows

$$u(x,t,\varepsilon) = \sum_{m=1}^{r} \sum_{s=1}^{2} A_{sm}(x,t,\varepsilon) \left[b e^{i\theta_{sm}(x,t,\varepsilon)} + \bar{b} e^{-i\theta_{sm}(x,t,\varepsilon)} \right] +$$

$$+\varepsilon w_1(x,t,\eta,\tau) + \varepsilon^2 w_2(x,t,\eta,\tau) + \ldots, \tag{6.153}$$

Here

$$A_{sm} = \alpha_{sm} + \varepsilon u^s_{1m}(\eta,\tau,\alpha,h,\xi_{sm}) + \varepsilon^2 u^s_{2m}(\eta,\tau,\alpha,h,\xi_{sm}) + \ldots,$$

$$\theta_{sm} = \xi_{sm} + h_{sm} + \varepsilon v^s_{1m}(\eta,\tau,\alpha,h\xi_{sm}) + \varepsilon^2 v^s_{2m}(\eta,\tau,\alpha,h\xi_{sm}) + \ldots,$$

$$\xi_{sm} = \tilde{\omega}_s t - k_m x, \; s = 1,2; \; \tilde{\omega}_1 = \omega_m, \; \tilde{\omega}_2 = -\omega_m, \; m = 1,2,\ldots,r; \tag{6.154}$$

α_{sm}, $\beta = \xi_{sm} + h_{sm}$ are averaged amplitudes and phases describing the regular motion (changing), u^s_{jm} and v^s_{jm}, $s = 1,2; m = 1,\ldots,r, j = 1,2,\ldots$, are bounded smooth functions corresponding to small oscillations A_{sm} and θ_{sm} around their first approximations α_{sm} and β_{sm}; $b = i/4$, $i = \sqrt{-1}$.

We set

$$p = i \sum_{m=1}^{r} \sum_{s=1}^{2} \tilde{\omega}_s A_{sm}(x,t,\varepsilon) \left[b e^{i\theta_{sm}} - \bar{b} e^{-i\theta_{sm}} \right] + \varepsilon w_{1t} + \varepsilon^2 w_{2t} + \ldots,$$

$$q = -i \sum_{m=1}^{r} \sum_{s=1}^{2} k_m A_{sm}(x,t,\varepsilon) \left[b e^{i\theta_{sm}} - \bar{b} e^{-i\theta_{sm}} \right] + \varepsilon w_{1x} + \varepsilon^2 w_{2x} + \ldots \tag{6.155}$$

The functions u, p, q and q can be represented as

$$u = \sum_{m=1}^{r} \sum_{s=1}^{2} [\alpha_{sm} + \varepsilon u^s_{1m} + i\varepsilon v^s_{1m} \alpha_{sm}] b e^{i\beta_{sm}} + \text{c.c.} + \varepsilon w_1; \tag{6.156}$$

$$p = i \sum_{m=1}^{r} \sum_{s=1}^{2} \tilde{\omega}_s [\alpha_{sm} + \varepsilon u^s_{1m} + i\varepsilon v^s_{1m} \alpha_{sm}] b e^{i\beta_{sm}} + \text{c.c.} + \varepsilon w_{1t}; \tag{6.157}$$

$$q = -i \sum_{m=1}^{r} \sum_{s=1}^{2} k_m [\alpha_{sm} + \varepsilon u^s_{1m} + i\varepsilon v^s_{1m} \alpha_{sm}] b e^{i\beta_{sm}} + \text{c.c.} + \varepsilon w_{1x}. \tag{6.158}$$

Suppose that $\alpha_{sm} = \alpha_{sm}(\xi_{0sm})$, $\beta_{sm} = \xi_{sm} + h_{sm}(\xi_{0sm})$ ($\xi_{0sm} = \varepsilon \xi_{sm}; \xi_{sm} = \tilde{\omega}_s t - k_m x$; $\tilde{\omega}_1 = \omega_m, \tilde{\omega}_2 = -\omega_m; m = 1,\ldots,r$).

Differentiating (6.156) with respect to t and setting the resulted expression equal to (6.157) yield

$$\sum_{m=1}^{r}\sum_{s=1}^{2}\tilde{\omega}_s\left[\frac{d\alpha_{sm}}{d\xi_{0sm}}+i\frac{dh_{sm}}{d\xi_{0sm}}\alpha_{sm}+\frac{\partial u_{1m}^{s}}{\partial\xi_{sm}}+i\frac{\partial v_{1m}^{s}}{\partial\xi_{sm}}\alpha_{sm}\right]be^{i\beta_{sm}}+c.c.=0. \tag{6.159}$$

Finding the second derivatives u_{tt}, $u_{tx}=q_t$, $u_{xx}=q_x$ substituting them and expressions (6.156)–(6.158) into equation (6.152) and comparing the coefficients at equal powers of ε (ε^0 and ε), we obtain

$$i\sum_{m=1}^{r}\sum_{s=1}^{2}(\tilde{\omega}_s^2-k_m^2a^2)\left[\frac{d\alpha_{sm}}{d\xi_{0sm}}+i\frac{dh_{sm}}{d\xi_{0sm}}\alpha_{sm}+\frac{\partial u_{1m}^{s}}{\partial\xi_{sm}}+\right.$$

$$\left.+i\frac{\partial v_{1m}^{s}}{\partial\xi_{sm}}\alpha_{sm}\right]be^{i\beta_{sm}}+c.c.+\omega_{1tt}-a^2w_{1xx}=cw_1+f_1^*(x,t,\eta,\tau,\alpha_1,\alpha_2,\beta_1,\beta_2), \tag{6.160}$$

where

$$f_1^*(x,t,\eta,\tau,\alpha_1,\alpha_2,\beta_1,\beta_2)=f\left(x,t,\eta,\tau,\sum_{m=1}^{r}\sum_{s=1}^{2}\alpha_{sm}be^{i\beta_{sm}}+\right.$$

$$+c.c,i\sum_{m=1}^{r}\sum_{s=1}^{2}\tilde{\omega}_s\alpha_{sm}be^{i\beta_{sm}}+c.c,-\sum_{m=1}^{r}\sum_{s=1}^{2}\tilde{\omega}_sk_m\alpha_{sm}be^{i\beta_{sm}}+$$

$$\left.+c.c.,-\sum_{m=1}^{r}\sum_{s=1}^{2}k_m^2\alpha_{sm}be^{i\beta_{sm}}+c.c.,0\right);$$

$$\alpha_s=(\alpha_{s1},\alpha_{s2},\ldots,\alpha_{sr}),\ \beta_s=(\beta_{s1},\beta_{s2},\ldots,\beta_{sr}),\ s=1,2.$$

The function f_1^* is assumed to be expandable into the Fourier series of the form

$$f_1^*=\sum_{m=1}^{d_0}f_{1m}^*(t,\eta,\tau,\alpha_1,\alpha_2,\psi_1,\psi_2)e^{-ik_mx}+c.c., \tag{6.161}$$

where $\psi_1=(\omega_1t+h_{11},\ldots,\omega_rt+h_{1r})$; $\psi_2=(-\omega_1t+h_{21},\ldots,-\omega_rt+h_{2r})$;

$$f_{1m}^*=\frac{1}{2X}\int_0^X f_1^*(x,t,\eta,\tau,\alpha_1,\alpha_2,\beta_1,\beta_2)e^{ik_mx}dx,\quad d_0\le\infty. \tag{6.162}$$

Then, taking account of (6.159)–(6.162), we obtain

$$\sum_{m=1}^{\infty}\omega_m\left(\left[\frac{d\alpha_{1m}}{d\xi_{01m}}+i\frac{dh_{1m}}{d\xi_{01m}}\alpha_{1m}+\frac{\partial u_{1m}^{1}}{\partial\xi_{1m}}+i\frac{\partial v_{1m}^{1}}{\partial\xi_{1m}}\alpha_{1m}\right]be^{i(\omega_mt+h_{1m})}-\right.$$

$$\left.-\left[\frac{d\alpha_{2m}}{d\xi_{02m}}+i\frac{dh_{2m}}{d\xi_{02m}}\alpha_{2m}+\frac{\partial u_{1m}^{2}}{\partial\xi_{2m}}+i\frac{\partial v_{1m}^{2}}{\partial\xi_{2m}}\alpha_{2m}\right]be^{i(-\omega_mt+h_{2m})}\right)\times$$

$$\times e^{-ik_mx}+c.c.=0,$$

$$i\sum_{m=1}^{r}b(\omega_m^2-a^2k_m^2)\left(\left[\frac{d\alpha_{1m}}{d\xi_{01m}}+i\frac{dh_{1m}}{d\xi_{01m}}\alpha_{1m}+\frac{\partial u_{1m}^{1}}{\partial\xi_{1m}}+\right.\right.$$

$$+i\frac{\partial v_{1m}^1}{\partial \xi_{1m}}\alpha_{1m}\Bigg] e^{i(\omega_m t+h_{1m})} + \left[\frac{d\alpha_{2m}}{d\xi_{02m}} + i\frac{dh_{2m}}{d\xi_{02m}}\alpha_{2m} + \frac{\partial u_{1m}^2}{\partial \xi_{2m}} + \right.$$

$$\left. + i\frac{\partial v_{1m}^2}{d\xi_{2m}}\alpha_{2m}\right] e^{i(-\omega_m t+h_{2m})}\Bigg) e^{-ik_m x} + c.c. +$$

$$+w_{1tt} - a^2 w_{1xx} = cw_1 + \sum_{m=1}^{d_0}(f_{1m}^* e^{-ik_m x} + c.c.). \tag{6.163}$$

Let $w_1(x,t,\eta,\tau)$ be a solution of the equation

$$w_{1tt} - a^2 w_{1xx} = cw_1 + \sum_{m=1}^{d_0}(f_{1m}^* e^{-ik_m x} + c.c.). \tag{6.164}$$

Comparing coefficients at equal exponents in (6.163) and using (6.164), we can write down

$$\gamma_{1m}e^{i(\omega_m t+h_{1m})} + \gamma_{2m}e^{i(-\omega_m t+h_{2m})} = 0,$$

$$\gamma_{1m}e^{i(\omega_m t+h_{1m})} + \gamma_{2m}e^{(-\omega_m t+h_{2m})} = \frac{4f_{1m}^*(t,\eta,\tau,\alpha_1,\alpha_2,\psi_1,\psi_2)}{\omega_m^2 - a^2k_m^2}, \tag{6.165}$$

where

$$\gamma_{sm} = \frac{d\alpha_{sm}}{d\xi_{0sm}} + i\frac{dh_{sm}}{d\xi_{0sm}}\alpha_{sm} + \frac{\partial u_{1m}^s}{\partial \xi_{sm}} + i\frac{\partial v_{1m}^s}{\partial \xi_{sm}}\alpha_{sm}, \quad s = 1,2; \; m = 1,2,\ldots,r. \tag{6.166}$$

Solving system (6.165), we obtain

$$\gamma_{sm} = -\frac{2f_{1m}^*\exp\{-i\psi_{sm}\}}{\omega_m^2 - a^2k_m^2}, \quad s = 1,2; \; m = 1,2,\ldots,r,$$

where $\psi_{sm} = \tilde{\omega}_s t + h_{sm}$, $\tilde{\omega}_1 = \omega_m$, $\tilde{\omega}_2 = -\omega_m$; $s = 1,2$, $m = 1,2,\ldots,r$. By (6.166), we have

$$\frac{\partial u_{1m}^s}{\partial \xi_{sm}} = -\operatorname{Re}\{2f_{1m}^* e^{-i\psi_{sm}}/(\omega_m^2 - a^2k_m^2)\} - \frac{d\alpha_{sm}}{d\xi_{0sm}},$$

$$\frac{\partial v_{1m}^s}{\partial \xi_{sm}} = -\frac{1}{\alpha_{sm}}\operatorname{Im}\{2f_{1m}^* e^{-i\psi_{sm}}/(\omega_m^2 - a^2k_m^2)\} - \frac{dh_{sm}}{d\xi_{0sm}}, \tag{6.167}$$

$$s = 1,2; \quad m = 1,2,\ldots,r.$$

Suppose that the following average values

$$\underset{t}{M}\{f_{1m}^* e^{-i\psi_{sm}}\} = \lim_{t\to\infty}\frac{1}{T}\int_0^T f_{1m}^* e^{-i\psi_{sm}}dt, \quad s = 1,2; \; m = 1,2,\ldots,r. \tag{6.168}$$

exist. In view of (6.167) and (6.168) we set

$$(\omega_m^2 - a^2k_m^2)\frac{d\alpha_{sm}}{d\xi_{0sm}} = -2\operatorname{Re}[\underset{t}{M}\{f_{1m}^* e^{-i\psi_{sm}}\}],$$

$$\alpha_{sm}(\omega_m^2 - a^2k_m^2)\frac{dh_{sm}}{d\xi_{0sm}} = -2\operatorname{Im}[\underset{t}{M}\{f_{1m}^* e^{-i\psi_{sm}}\}]$$

or

$$\frac{\partial \alpha_{sm}}{\partial t} + \frac{a^2k_m}{\tilde{\omega}_s}\frac{\partial \alpha_{sm}}{\partial x} = -\frac{2\varepsilon}{\tilde{\omega}_s}\operatorname{Re}[\underset{t}{M}\{f_{1m}^* e^{-i\psi_{sm}}\}],$$

$$\frac{\partial h_{sm}}{\partial t} + \frac{a^2 k_m}{\tilde{\omega}_s}\frac{\partial h_{sm}}{\partial x} = -\frac{2\varepsilon}{\alpha_{sm}\tilde{\omega}_s}\,\mathrm{Im}\,[\underset{t}{M}\{f_{1m}^{*}e^{-i\psi_{sm}}\}], \tag{6.169}$$

$$s = 1, 2; \quad m = 1, 2, \ldots, r.$$

Thus the first approximation of the asymptotic solution of perturbed equation (6.152) has the form

$$u_I(x, t, \varepsilon) = \sum_{m=1}^{r} b\alpha_{sm}(\xi_{0sm})e^{i\xi_{sm}+ih_{sm}(\xi_{0sm})} + c.c.,$$

where $\alpha_{sm}(\xi_{0sm})$ and $h_{sm}(\xi_{0sm})$ are to be found from system (6.169).

Remark 6.8. The above asymptotic method for finding multifrequency oscillation modes can easily be carried over to general second order equations of the form (6.43)

$$\sum_{i,j=1}^{N} a_{ij}(\eta)\frac{\partial^2 u}{\partial x_i \partial x_j} + \lambda(\eta)u = \varepsilon f\left(x, t, \eta, u, \frac{\partial u}{\partial x_1}, \ldots, \frac{\partial u}{\partial x_N}, \varepsilon\right),$$

if its coefficients and functions satisfy the condition

$$\lambda(\eta) \neq 0, \quad s(\eta) \equiv \sum_{i,j=1}^{N} a_{ij}(\eta)\omega_j\omega_i \neq 0, \quad s(\eta)\lambda(\eta) > 0,$$

$$s(\eta) - \lambda(\eta) = 0, \quad \frac{\partial \omega_i}{\partial \eta_j} = \frac{\partial \omega_j}{\partial \eta_i}, \quad i, j = 1, 2, \ldots, N,$$

and the function f is periodic or almost periodic in fast variables x and t. Here η can be either one-dimensional or mani-dimensional slow variable (see §6.1 in this chapter). Moreover, the asymptotic method presented above for finding multifrequency oscillation modes shows the modification of asymptotic methods of nonlinear mechanics to be correct for investigating oscillations of one-dimensional systems with distributed parameters [29, 104]

$$a_k(\eta, \tau)\frac{\partial u_k}{\partial t} + \left(\sum_{l=1}^{n} a_{kl}(\eta, \tau)\frac{\partial}{\partial x} + b_{kl}(\eta, \tau)\right)u_l =$$

$$= \varepsilon f_k\left(x, t, \eta, \tau, u, \frac{\partial u}{\partial t}, \frac{\partial u}{\partial x}, \varepsilon\right), \quad k = 1, \ldots, n; \quad \tau = \varepsilon t; \quad \eta = \varepsilon x,$$

which describe fields in weakly nonlinear medium or motion of continuous medium considered in the next section.

6.10. Asymptotic solution of nonlinear systems of first order partial differential equations

As shown in [29, 96, 97, 104], a number of problems concerning propagation and interaction of waves in unbounded weakly nonlinear systems (media) is not reduced to the investigation into interaction of a small number of normal oscillations with fixed spatial structure, and it is difficult to apply to them asymptotic methods [14, 24, 63, 74] elaborated for ordinary differential equations. One has therefore to consider the amplitude-phase parameters as slow functions of not only time [62, 99, 117] but spatial coordinate too [116]. As a result, one obtains approximate partial differential equations in parameters; and according to the Gaponov-Ostrovkii-Rabinovich method [29, 104], the type of equation changes and the order of equation increases as approximations go forward.

In the present section, we set forth an asymptotic method, which treats time and spatial coordinates "equally" and can be applied to equations that depend on many spatial coordinates [135, 137, 146].

1. One-dimensional case. The equations describing fields in weakly nonlinear medium can be represented in the form

$$Au_t + Bu_x + Cu = \varepsilon f^{(1)}(x, t, \eta, \tau, u, u_x, u_t) +$$

$$+\varepsilon^2 f^{(2)}(x, t, \eta, \tau, u, u_t, u_x) + \ldots, \quad \eta = \varepsilon x, \quad \tau = \varepsilon t, \quad \varepsilon \ll 1, \tag{6.170}$$

where A, B, C are square matrices, u and f are n-dimensional vector-functions (f being polynomial in u, u_x, u_t and periodic in explicitly occuring x and t).

In a weakly nonlinear medium, a wave effectively rouses harmonics only if they are resonant to proper (normal) waves of the medium, i. e. if their frequency $m\omega$ and wave number mk satisfy the dispersion equation $D(m\omega, mk) \equiv \det|Am\omega - Bmk - iC| = 0$. This condition particularly holds if, in the medium, there is no dispersion in the interval of frequencies under consideration. Multi-wave interaction is however possible in the presence of intensive dispersion [104].

Suppose that the waves $\omega = \omega_j$ and $\vec{k}_j = \vec{k}_j(\omega)$ are propagating in a medium with dispersion and weak nonlinearity (the relation $\vec{k}(\omega)$ is determined by the dispersion equation $D(\omega_j, k) = 0$). Since the medium is nonlinear, a forced combinative wave of the frequency and wave vector

$$\omega_H = \sum_{j=1}^{m} n_j \omega_j, \quad \vec{k}_H = \sum_{j=1}^{m} n_j \vec{k}_j$$

arises. This wave remains small (of order of magnitude of nonlinearity) if $D(\omega_H, \vec{k}_H) \neq 0$ and rises if $D(\omega_H, \vec{k}_H) \approx 0$ or, in another form,

$$\omega_H = \omega_i, \quad \vec{k}_i = \vec{k}(\omega_H). \tag{6.171}$$

Regarding the wave ω_i, k_i as interacting with others, relations in (6.171) can be considered as the conditions of resonance of frequency and wave vector for the wave interaction to be effective. Relations (6.171) are often called the conditions of synchronism. Since the amplitudes and phases of quasiharmonic waves change slowly (because of weak nonlinearity) even if the wave interaction is effective, a method connected with averaging with respect to time and spatial variables is used to investigate these processes. The method enables approximate nonlinear partial differential equations to be obtained for the slowly changing amplitude and phase of interacting waves. We shall consider a general scheme of constructing this method, assuming the waves to be one-dimensional.

For $\varepsilon = 0$, the field in a medium possesses the superposition of harmonic waves of the form

$$u = \psi^\nu \exp\{i(\omega t - kx)\} + c.c., \tag{6.172}$$

c. c. standing here and further for complex conjugate. In (6.3), ψ^ν is determined by the system

$$(A\omega - Bk_\nu - iC)\psi^\nu = 0, \tag{6.173}$$

and ω and k_ν are related by the dispersion equation

$$D(\omega, k_\nu) \equiv \det|A\omega - Bk_\nu - iC| = 0,$$

where ν is the index of the normal wave (a branch of the dispersion equation).

We shall consider the interaction of a finite number of waves of the form (6.172). It is possible if the initial or boundary conditions for (6.170) are represented as a finite number of waves and the condition of synchronism (6.171) holds for a finite number of waves, too, i. e. dispersion is present in the system. For $\varepsilon \neq 0$, a multi-wave solution is sought in the form

$$u(x,t,\varepsilon) = \sum_{\nu=1}^{q} \psi^{\nu} a_{\nu}(\varepsilon_{\nu}) \exp\{i[\xi_{\nu} + \varphi(\xi_{\nu})]\} + c.c. +$$

$$+ \sum_{n=1}^{\infty} \varepsilon^{n} W^{(n)}(a, \varphi, \eta, \tau, x, t), \tag{6.174}$$

where $\xi_{\nu} = \omega_{\nu} t - k(\omega_{\nu})x$. Solution (6.171) ought to take into account all waves caused by nonlinearity which are in synchronism with the normal waves of linear system.

The equations for new unknown functions a_{ν} and φ_{ν} are sought in the form

$$\frac{da_{\nu}}{d\xi_{\nu}} = \sum_{m=1} \varepsilon^{m} F_{m}^{\nu}(a, \varphi, \eta, \tau), \quad \frac{d\varphi_{\nu}}{d\xi_{\nu}} = \sum_{m=1} \varepsilon^{m} \Phi(a, \varphi, \eta, \tau), \tag{6.175}$$

where F_m and Φ_m are unknown functions. The functions $W^{(n)}$, F_m and Φ_m being known, the desired solution (6.174) can be determined with any prescribed accuracy. To obtain an equation for $W^{(n)}$, we substitute (6.174) and (6.175) into (6.170) and compare the coefficients at equal powers of ε. As the first approximation, we have

$$A\frac{\partial W^{(1)}}{\partial t} + B\frac{\partial W^{(1)}}{\partial x} + CW^{(1)} = h^{(1)}(\eta, \tau, x, t),$$

$$h^{(1)} = -\sum_{\nu=1}^{q} e^{\{i(\xi_{\nu}+\varphi_{\nu})\}}(A\psi^{\nu}\omega_{\nu} - B\psi^{\nu}k(\omega_{\nu}))[F_1^{\nu} + ia_{\nu}\Phi_1^{\nu}] +$$

$$+ c.c. + f^{(1)}(x, t, \eta, \tau, u^0, u_x^0, u_t^0), \tag{6.176}$$

where u^0 is a solution corresponding to $\varepsilon = 0$. Making use of periodicity in x and t of the right-hand side of this equation, we can represent it as follows

$$h^{(1)} = \sum_{\nu=1}^{q} H^{\nu}(\eta, \tau)e^{i\xi_{\nu}} + \sum_{d=1}^{d_0} H^{d}(\eta, \tau)e^{i\xi_{d}} + c.c. \tag{6.177}$$

The first group of summands includes those corresponding to proper waves of the system, which have been taken into account in the principal term of solution (6.174) and for which

$$D\left(\omega^{\nu}, k(\omega^{\nu})\right) = 0; \tag{6.178}$$

the second group includes the summands that appear because of nonlinearity of the waves which are not in synchronism with the former and for which

$$D\left(\omega^{d}, k(\omega^{d})\right) \neq 0. \tag{6.179}$$

The functions $H^{\nu,d}$ are the Fourier coefficients

$$H^{\nu,d} = \frac{1}{TX} \int_{t}^{t+T} \int_{x}^{x+X} h^{(1)} e^{-i(\omega^{\nu,d} t - k^{\nu,d} x)} dx dt.$$

The periodic functions $W^{(1)}$ are represented as in (6.177)

$$W^{(1)} = \sum_{\nu=1}^{q} W^{\nu}(\eta,\tau)e^{i\xi_\nu} + \sum_{d=1}^{d_0} W^d e^{i\xi_d} + \text{c.c.} \tag{6.180}$$

Substituting (6.177) and (6.180) into (6.176) and comparing coefficients at equal exponents, we obtain a nonhomogeneous system of algebraic equations to determine $W^{\nu,d}$. According to (6.179), the amplitudes W^d are bounded

$$W_k^d = \sum_{j=1}^{n} \frac{D_{jk}H_j^d}{D\left(\omega_d, k(\omega_d)\right)}$$

(D_{jk} is a minor of D). In view of (6.178), bounded solutions for W^ν may exist if only the conditions of orthogonality $\sum_{s=1}^{n} \zeta_s^\nu H_s^\nu = 0$ holds where ζ are eigenfunctions of the system conjugate to (6.173). Using (6.176) and (6.177), we derive from these conditions that

$$\left[(\zeta^*, A\psi^\nu)\omega_\nu - (\zeta^{*\nu}, B\psi^\nu)k(\omega_\nu)\right]\{F_1^\nu + ia\nu_\nu\Phi_1^\nu\} =$$

$$= \sum_{k=1}^{n} \zeta_k^\nu \langle f_k^{(1)} \exp\{-i[\xi_\nu + \varphi_\nu]\}\rangle \equiv f^{(1)\nu}, \tag{6.181}$$

where

$$\langle f_k^1 e^{-i(\xi^\nu+\varphi^\nu)}\rangle = \frac{1}{TX}\int_t^{t+T}\int_x^{x+X} f_k^{(1)}(x,t,\eta,\tau,u^0,u_x^0,u_t^0)e^{-i(\xi^\nu+\varphi^\nu)}dx\,dt;$$

ζ^* is a vector conjugate to the vector ζ, (a,b) is the scalar product of vectors a and b.

By using the relations $a_\nu = a_\nu(\xi_\nu)$, $\varphi_\nu = \varphi_\nu(\xi_\nu)$, $\xi_\nu = \omega_\nu t - k(\omega_\nu)x$, we obtain the equality

$$\left[(\zeta^{*\nu}, A\psi^\nu)\omega_\nu - (\zeta^{*\nu}, B\psi^\nu)k(\omega_\nu)\right]\left(\frac{da_\nu}{d\xi_\nu} + ia_\nu\frac{d\varphi}{d\xi_\nu}\right) =$$

$$= (\zeta^{*\nu}, A\psi^\nu)\left[\frac{\partial a_\nu}{\partial t} + ia_\nu\frac{\partial\varphi_\nu}{\partial t}\right] + (\zeta^*, B\psi^\nu)\left[\frac{\partial a_\nu}{\partial x} + ia_\nu\frac{\partial\varphi_\nu}{\partial x}\right]. \tag{6.182}$$

By (6.175), (6.181) and (6.182), the first approximation equations have the form

$$\frac{\partial a_\nu}{\partial t} + v_\nu\frac{da_\nu}{\partial x} = \frac{\varepsilon\,\mathrm{Re}\,f^{(1)\nu}}{(\zeta^{*\nu}, A\psi^\nu)},\quad \frac{\partial\varphi_\nu}{\partial t} + v_\nu\frac{\partial\varphi_\nu}{\partial x} = \frac{\varepsilon\,\mathrm{Im}\,f^{(1)\nu}}{a_\nu(\zeta^{*\nu}, A\psi^\nu)}, \tag{6.183}$$

where $v_\nu = \dfrac{d\omega_k}{dk_\nu}$ is the group velocity. Using the relations

$$\frac{d\omega_k}{dk_\nu} = \frac{(\zeta^{*\nu}, B\psi^\nu)}{(\zeta^{*\nu}, A\psi^\nu)},\quad \frac{\zeta^\nu}{(\zeta^{*\nu}, A\psi^\nu)} = \frac{D_{kl}^\nu}{D'_{\omega_\nu}\psi_l^\nu}$$

one can readily verify that equations (6.183) coincides with the corresponding first approximation equations obtained in [29, 104].

2. Mani-dimensional case. Consider the system

$$Au_t + \sum_{j=1}^{n} B^j u_{x_j} + Cu = \varepsilon f(x,t,\eta,\tau,u,u_{x_1},\ldots,u_{x_n},u_t) + \varepsilon^2\ldots, \tag{6.184}$$

where A, B^j, C are square matrices, u and f are n-dimensional vectors, $x = (x_1, \ldots, x_n)$ an n-dimensional vector, $\eta = \varepsilon x$; $\tau = \varepsilon t$; $\varepsilon \ll 1$.

Suppose that for $\varepsilon = 0$, the field in the medium can be represented as the superposition of harmonic waves

$$u = \psi^s e^{i(\omega_s t - (k^s, x))} + c.c.,$$

where ψ^s is obtained from the system

$$\left(A\omega_s - \sum_{j=1}^{n} B^j k_j(\omega_s) - iC \right) \psi^s = 0,$$

$(k^s, x) = k_1^s x_1 + \ldots + k_n^s x_n$, ω_s and $\vec{k}^s = \vec{k}(\omega_s)$ are related by the dispersion equation

$$D\left(\omega_s, \vec{k}^s\right) \equiv \det \left| A\omega_s - \sum_{j=1}^{n} B^j k_j(\omega_s) - iC \right| = 0.$$

As in item 1 of the present section, the multi-wave solution of system (6.184) for $\varepsilon \neq 0$ is sought in the form

$$u = \sum_{s=1}^{q} \psi^s a_s(\xi_s) e^{i(\xi_s + \varphi_s(\xi_s))} + c.c. + \sum_{r=1}^{\infty} \varepsilon^r W^{(r)}(a, \varphi, \eta, \tau, x, t),$$

where $\xi_s = \omega_s t - (k^s, x)$.

The equations for new unknown a_s and φ_s can be obtained from the system

$$\frac{da_s}{d\xi_s} = \sum_{m=1} \varepsilon^m F_m^s(a, \varphi, \eta, \tau), \quad \frac{d\varphi_s}{d\xi_s} = \sum_{m=1} \varepsilon^m \Phi_m^s(a, \varphi, \eta, \tau),$$

where F_m and Φ_m are unknown functions.

3. Proper single-frequency oscillations in distributed parameter systems. Distributed parameter systems appear in various problems in radiophysics and modern technology [66]. As known, the application of the asymptotic methods of nonlinear mechanics, the basic idea of which has been set forth above (see also [14]), to these systems requires solving many averaged differential equations in large number of unknowns even if oscillations in these systems are described by almost linear partial differential equations. This circumstance leads to substantial difficulties in practical usage of these methods.

In oscillating distributed parameter systems, the presence of unavoidable internal friction as well as external perturbing forces usually result in some wave packets rapidly disappearing, i. e. the main tone of oscillations (or oscillations with a certain single frequency ω_s and wave number $k(\omega_s)$) being set. Then, if the right-hand side in system (6.170) is independent of x and t, the asymptotic method of constructing single wave solutions, presented in items 1 and 2, is similar to the known Bogolyubov-Mitropol'skii asymptotic method [14, 63]. Indeed, the solution of perturbed system (6.170) is sought in the form

$$u = a\psi^\nu e^{i\theta_s} + c.c. + \sum_{m=1} \varepsilon^m W(a, \theta_s, \eta, \tau),$$

where $a = a(\xi_s)$, $\theta_s = \theta(\xi_s)$, $\xi_s = \omega_s t - k^s x$ are determined by the differential equations

$$\frac{da}{d\xi_s} = \sum_{m=1} \varepsilon^m F_m(a, \eta, \tau), \quad \frac{d\theta_s}{d\xi_s} = 1 + \sum_{m=1} \varepsilon^m \Phi_m(a, \eta, \tau).$$

It is thus necessary to determine the functions $W^{(1)}(a, \theta_s, \eta, \tau)$, $W\,(a, \theta_s, \eta, \tau)$ $\ldots$, 2π-periodic, which are with respect to θ_s, and the functions $F_1(a, \eta, \tau)$, $F_2(a, \eta, \tau)$, $\ldots$, $\Phi_1(a, \eta, \tau)$, $\Phi_2(a, \eta, \tau)$, $\ldots$. The method of constructing these functions, i. e. single frequency oscillations, fully coincides with the common method for systems of ordinary differential equations [14, 74].

CHAPTER 7

ASYMPTOTIC METHODS FOR THE SECOND ORDER PARTIAL DIFFERENTIAL EQUATIONS OF HYPERBOLIC TYPE

This chapter is devoted to the construction and justification of asymptotic expansions of solutions of hyperbolic equations, with some additional conditions (initial or initial and boundary) imposed on them. The systematic application of the asymptotic methods to the investigation of nonstationary processes in nonlinear distributed parameter systems has begun in [62, 66].

7.1. The reduction of quasilinear equations of hyperbolic type to a countable system of ordinary differential equations in standard form

We are searching for a solution of the equation

$$u_{tt} - a^2 u_{xx} = cu + \varepsilon f(x, t, u, u_t, u_x) \tag{7.1}$$

satisfying the following initial and boundary conditions

$$u(x,0) = \varphi(x), \quad u_t(x,0) = \chi(x); \tag{7.2}$$

$$u(0,t) = 0, \quad u(l,t) = 0, \tag{7.3}$$

where $\varphi(x)$ and $\chi(x)$ are sufficiently smooth functions defined on interval $[0, l]$.

Let us first consider the nonperturbed equation obtained from (7.1) by setting $\varepsilon = 0$

$$u_{tt} - a^2 u_{xx} = cu, \tag{7.4}$$

with the initial and boundary conditions (7.2) and (7.3).

If we assume that $(\lambda_k a)^2 - c > 0$, $\lambda_k = k\pi/l$, $k = 1, 2, \ldots$, the Fourier method gives the solution of equation (7.4) as the series

$$u(x,t) = \sum_{k=1}^{\infty} (A_k \cos\omega_k t + B_k \sin\omega_k t) \sin\lambda_k x, \tag{7.5}$$

where $\omega_k = \sqrt{(\lambda_k a)^2 - c}$ are the frequences of normal oscillations, A_k and B_k certain constants determined by the initial condition (7.2).

We may assume that the forms of normal oscillations in the presence of perturbations are determined with sufficient accuracy by the same functions $\sin\lambda_k x$, $k = 1, 2, \ldots$, since the parameter ε is small. Now that we have the solution (7.5) of the nonperturbed equation (7.4), we seek the solution of the perturbed equation (7.1) in the form of the series

$$u(x,t,\varepsilon) = \sum_{k=1}^{\infty} \Big(z_k(t,\varepsilon) \cos\omega_k t + w_k(t,\varepsilon) \sin\omega_k t \Big) \sin\lambda_k x, \tag{7.6}$$

where $z_k(t,\varepsilon)$ and $w_k(t,\varepsilon)$, $k = 1, 2, \ldots$, are unknwon functions to be determined.

Suppose that the series (7.6) admits double differentiation with respect to x and t. As the new functions z_k and w_k, $k = 1, 2, \ldots$, have been introduced, some additional conditions are necessary to determine them uniquely. These conditions are introduced similarly to the known change of variables used in the averaging method

$$u_t = \sum_{k=1}^{\infty} \omega_k(-z_k \sin \omega_k t + w_k \cos \omega_k t) \sin \lambda_k x. \tag{7.7}$$

Differentiating (7.6) twice with respect to t and equating the resulted expression to (7.7) yield

$$\sum_{k=1}^{\infty} (\dot{z}_k \cos \omega_k t + \dot{w}_k \sin \omega_k t) \sin \lambda_k x = 0. \tag{7.8}$$

Calculating the second derivatives u_{tt}, u_x, u_{xx} and taking account of (7.1), (7.6) and (7.7), we obtain

$$\sum_{k=1}^{\infty}(-\dot{z}_k \sin \omega_k t + \dot{w}_k \cos \omega_k t) \sin \lambda_k x = \varepsilon f_0(x, t, z, w), \tag{7.9}$$

where

$$f_0(x,t,z,w) = f\left(x, t, \sum_{k=1}^{\infty}(z_k \cos \omega_k t + w_k \sin \omega_k t) \sin \lambda_k x, \right.$$
$$\sum_{k=1}^{\infty} \omega_k(-z_k \sin \omega_k t + w_k \cos \omega_k t) \sin \lambda_k x,$$
$$\left. \sum_{k=1}^{\infty} \lambda_k(z_k \cos \omega_k t + w_k \sin \omega_k t) \cos \lambda_k x \right). \tag{7.10}$$

Suppose that $f_0(x,t,z,w)$ can be expanded into the Fourier series by the eigenfunctions $\sin \lambda_k x$, $k = 1, 2, \ldots$, i. e.

$$f_0(x,t,z,w) = \sum_{k=1}^{\infty} f_{0k}(t,z,w) \sin \lambda_k x, \tag{7.11}$$

where

$$f_{0k} = \frac{2}{l} \int_0^l f_0(x,t,z,w) \sin \lambda_k x dx, \quad k = 1, 2, \ldots \tag{7.12}$$

Now combining (7.11) with (7.8) and (7.9), we arrive at the following system of equations

$$\dot{z}_k = -\frac{\varepsilon}{\omega_k} f_{0k}(t,z,w) \sin \omega_k t,$$
$$\dot{w}_k = \frac{\varepsilon}{\omega_k} f_{0k}(t,z,w) \cos \omega_k t, \quad k = 1, 2, \ldots \tag{7.13}$$

So in order to find the solution of the perturbed equation (7.1), one should solve the countable system of differential equations (7.13) in standard form.

Suppose that the functions $\varphi(x)$ and $\chi(x)$, occuring in the initial condition (7.2), can be expanded into the Fourier series by the eigenfunctions $\sin \lambda_k x$, $k = 1, 2, \ldots$, i. e.

$$\varphi(x) = \sum_{k=1}^{\infty} \varphi_k \sin \lambda_k x, \quad \chi(x) = \sum_{k=1}^{\infty} \chi_k \sin \lambda_k x, \tag{7.14}$$

where

$$\varphi_k = \frac{2}{l}\int_0^l \varphi(x)\sin\lambda_k x dx, \quad \chi_k = \frac{2}{l}\int_0^l \chi(x)\sin\lambda_k x dx, \quad k = 1,2,\ldots \tag{7.15}$$

For the function (7.6) to be a solution of the perturbed mixed boundary value problem (7.1)–(7.3) one should solve the countable system of equations (7.13) with the following initial condition

$$z_k(0,\varepsilon) = \varphi_k, \quad w_k(0,\varepsilon) = \frac{\chi_k}{\omega_k}, \quad k = 1,2,\ldots \tag{7.16}$$

The change (7.6) and (7.7), used in obtaining the countable system (7.13), might seem to violate certain properties of finctions of several variables, for example, the equality of mixed variables. It is easy to show that the equality of mixed variables holds true provided (7.8) is satisfied [136].

7.2. The reduction method in application to a countable system of differential equations

The method of separation of variables can be applied to the oscillation equation for both homogeneous and nonhomogeneous string.

Consider the following problem: find the solution of the equation

$$L[u] \equiv \frac{\partial}{\partial x}\left[k(x)\frac{\partial u}{\partial x}\right] - q(x)u = \rho(x)\frac{\partial^2 u}{\partial t^2}, \tag{7.17}$$

satisfying the conditions

$$u(0,t) = 0, \quad u(l,t) = 0, \quad t \geq 0, \tag{7.18}$$

$$u(x,0) = \varphi(x), \quad u_t(x,0) = \chi(x), \quad 0 \leq x \leq l. \tag{7.19}$$

Here k, q and p are positive functions ($k > 0$, $\rho > 0$, $q\epsilon 0$) continuous on interval $0 \leq x \leq l$. (The case that $k(x)$ vanishes at some points is considered in [115]).

We shall solve the above problem by using the method of separation of variables, i. e. we shall seek the solution in the form $u(x,t) = X(x)\,T(t)$. Substituting the form of the solution into equation (7.17) and using the boundary condition, we obtain

$$\frac{d}{dx}\left[k(x)\frac{dX}{dx}\right] - qX + \lambda\rho X = 0, \quad X(0) = X(l) = 0, \quad T'' + \lambda T = 0.$$

after separating variables.

To determine the function $X(x)$, we thus have the following boundary value problem: find those values of the parameter λ, for which nontrivial solutions of the problem

$$L(x) + \lambda\rho X = 0; \tag{7.20}$$

$$X(0) = 0, \quad X(l) = 0, \tag{7.21}$$

exist and determine these solutions. These values of the parameter λ are called eigenvalues, and nontrivial solutions corresponding to them are called eigenfunctions of the boundary value problem (7.20), (7.21).

We first state basic properties of the eigenfunctions and eigenvalues of the boundary value problem (7.20), (7.21), which we shall need later.

1. There is a countable set of eigenvalues $\lambda_1 < \lambda_2 < \ldots < \lambda_n < \ldots$, to which correspond the nontrivial solutions, namely the eigenfunctions $X_1(x), X_2(x), \ldots, X_n(x), \ldots$

2. For $q \geq 0$, all the eigenvalues λ_n are positive.

3. For $m \neq 0$, the eigenfunctions X_m and X_n are orthogonal with weight $\rho(x)$ on interval $0 \leq x \leq l$

$$\int_0^l X_m(x)X_n(x)\rho(x)dx = 0, \quad m \neq n.$$

Steklov's expansion theorem. An arbitrary function $F(x)$ that is double continuously differentiable and satisfies the boundary condition $F(0) = F(l) = 0$ can be expanded into a uniformly and absolutely converging series by the eigenfunctions X_n

$$F(x) = \sum_{n=1}^{\infty} F_n X_n(x), \quad F_n = \left(\int_0^l X_n^2(x)\rho(x)dx \right)^{-1} \int_0^l F(x)X_n(x)\rho(x)dx.$$

The solution of the equation $T'' + \lambda T = 0$ is of the form

$$T_n(t) = A_n \cos\sqrt{\lambda_n}t + B_n \sin\sqrt{\lambda_n}t,$$

because λ_n is positive. The general solution of (7.17) can therefore be sough in the form

$$u(x,t) = \sum_{k=1}^{\infty}(A_k \cos\sqrt{\lambda_k}t + B_k \sin\sqrt{\lambda_k}t)X_k(x). \tag{7.22}$$

To satisfy the initial condition, we may use the expansion theorem and proceed similarly to the case of homogeneous string. From the equalities

$$u(x,0) \equiv \varphi(x) = \sum_{k=1}^{\infty} A_k X_k(x), \quad u_t(x,0) \equiv \chi(x) = \sum_{k=1}^{\infty} B_k \sqrt{\lambda_k} X_k(x)$$

it follows that

$$A_k = \varphi_k, \quad B_k = \chi_k/\sqrt{\lambda_k},$$

where φ_k and χ_k are the Fourier coefficients of the functions $\varphi(x)$ and $\chi(x)$ expanded by the system of functions $\{X_k(x)\}$, which are orthogonal with weight $\rho(x)$.

We confine ourselves to a general scheme of the method of separation of variables and do not give any condition on both equation coefficients and initial functions ensuring this method to be applicable to various equations. The justification of this method was given by Steklov in his fundamental works [110].

A method for reducing hyperbolic equations to countable systems. If we analyse how equation (7.17) reduces to a countable system, we shall see that the key point in this method is the assertion that the functions of a certain class can be expanded into Fourier series by eigenfunctions of the boundary value problem. Therefore this method can be applied to the following nonhomogeneous mixed problem: find the solution of the equation

$$\frac{\partial}{\partial x}\left[k(x)\frac{\partial u}{\partial x}\right] - q(x)u - \rho(x)\frac{\partial^2 u}{\partial t^2} = \varepsilon f(x,t,u,u_t,u_x), \tag{7.23}$$

satisfying the conditions

$$u(x,0) = \varphi(x), \quad u_t(x,0) = \chi(x), \quad 0 \leq x \leq l; \tag{7.24}$$

$$u(0,t) = 0, \quad u(l,t) = 0, \quad t \geq 0. \tag{7.25}$$

Here k, q, ρ are positive functions $k > 0$, $\rho > 0$, $q\epsilon 0$ continuous on interval $0 \le x \le l$. Since the solution of the nonperturbed mixed problem (7.17)–(7.19) can be represented as the series (7.22), the solution of the pertûrbed equation (7.23) is sought, too, as the series

$$u(x,t,\varepsilon) = \sum_{k=1}^{\infty}[z_k(t,\varepsilon)\cos\sqrt{\lambda_k}t + w_k(t,\varepsilon)\sin\sqrt{\lambda_k}t]X_k(x), \tag{7.26}$$

where $z_k(t,\varepsilon)$, $w_k(t,\varepsilon)$, $k = 1,2,\ldots$, are unknown functions to be determined, $X_k(x)$, $k = 1,2,\ldots$, are the eigenfunctions of the boundary value problem (7.20), (7.21).

Assuming that

$$u_t = \sum_{k=1}^{\infty}\sqrt{\lambda_k}[-z_k\sin\sqrt{\lambda_k}t + w_k\cos\sqrt{\lambda_k}t]X_k(x)$$

and the function

$$f_0(x,t,z,w) = f\Bigg(x, t, \sum_{k=1}^{\infty}(z_k\cos\sqrt{\lambda_k}t + w_k\sin\sqrt{\lambda_k}t)X_k(x),$$

$$\sum_{k=1}^{\infty}\sqrt{\lambda_k}X_k(x)[-z_k\sin\sqrt{\lambda_k}t + w_k\cos\sqrt{\lambda_k}t],$$

$$\sum_{k=1}^{\infty}X_k'(x)(z_k\cos\sqrt{\lambda_k}t + w_k\sin\sqrt{\lambda_k}t)\Bigg) \tag{7.27}$$

can be expanded into Fourier series by the eigenfunctions

$$f_0(x,t,z,w) = \sum_{k=1}^{\infty} f_{0k}(t,z,w)X_k(x),$$

where

$$f_{0k} = \frac{1}{m_k}\int_0^l f_0(x,t,z,w)\rho(x)X_k(x)dx, \quad k = 1,2,\ldots,$$

we obtain the following countable system of equations

$$z_k = -\frac{\varepsilon}{\sqrt{\lambda_k}}f_{0k}(t,z,w)\sin\sqrt{\lambda_k}t,$$

$$\dot{w}_k = \frac{\varepsilon}{\sqrt{\lambda_k}}f_{0k}(t,z,w)\cos\sqrt{\lambda_k}t, \quad k = 1,2,\ldots \tag{7.28}$$

According to Steklov's theorem, one of the conditions necessary for the hyperbolic equation to be reducible to a countable system is that the following equality holds

$$f_0(0,t,z,w) = f_0(l,t,z,w), \tag{7.29}$$

i. e. the function $f_0(x,t,z,w)$ given by (7.27) must satify the boundary condition. Hence the function $f(x,t,u,u_t,u_x)$ in the right-hand side of the perturbed equation (7.23) must not contain a free term depending only on t.

If the right-hand side of equation (7.23) includes a free term that explicitly depends only on t, this equation can be reduced, by changing the functions, to a new equation that does not contain this free term.

Examples. We shall consider some classes of the previously studied equations.

1^0. The equation describing propagation of electromagnetic oscillations in cable [63]

$$u_{tt} - a^2 u_{xx} = cu + \varepsilon \sin \frac{\pi}{l} x \cos pt (uu_x) \tag{7.30}$$

or

$$u_{tt} - a^2 u_{xx} = cu + \varepsilon \cos \frac{\pi}{l} x \cos pt (uu_x) \tag{7.31}$$

with the boundary condition

$$u(0,t) = u(l,t) = 0. \tag{7.32}$$

The function $f_0(x,t,z,w)$ given by (7.27) for equation (7.30) or (7.31) satisfies the condition (7.29) that is necessary for the function to be expandable into Fourier series, i. e. the boundary condition

$$f_0(0,t,z,w) = f_0(l,t,z,w),$$

the Fourier series for equation (7.31) being trigonometric.

2^0. The equation describing oscillations of ocean waves [63]

$$\frac{\partial}{\partial x}\left[A(x)\frac{\partial u}{\partial x}\right] - B(x)u - C(x)\frac{\partial^2 u}{\partial t^2} = \varepsilon u^3 \tag{7.33}$$

with the boundary condition (7.32). In this case, the function $f_0(x,t,z,w)$ given by (7.27) for equation (7.33) satisfies the condition (7.29) that is necessary for the function to be expandable into Fourier series.

7.3. Summation of trigonometric Fourier series with coefficients given approximately

The reason for the theory of Fourier series being used in the reduction of partial differential equations to countable systems is that these systems are subsequently replaced by so-called truncated systems comprising n differential equations in the first n unknown functions.

Indeed, if φ_k χ_k and $f_{0k}(t,z,w)$ are the coefficients of the uniformly converging Fourier series

$$\sum_{k=1}^{\infty} \varphi_k X_k(x), \quad \sum_{k=1}^{\infty} \chi_k X_k(x), \quad \sum_{k=1}^{\infty} f_{0k}(t,z,w) X_k(x),$$

then we have $\varphi_k \to 0$ $(k \to \infty)$, $\chi_k \to 0$ $(k \to \infty)$, $f_{0k}(t,z,w) \to 0$ $(k \to \infty)$ for fixed t,z,w. Therefore the solution of the countable system (7.28) can be written as

$$z_k(t,\varepsilon) = \varphi_k - \frac{\varepsilon}{\sqrt{\lambda_k}} \int_0^t f_{0k}(t,z,w) \sin \sqrt{\lambda_k} t dt \to 0 \quad (k \to \infty),$$

$$w_k(t,\varepsilon) = \chi_k - \frac{\varepsilon}{\sqrt{\lambda_k}} \int_0^t f_{0k}(t,z,w) \cos \sqrt{\lambda_k} t dt \to 0 \quad (k \to \infty).$$

Hence, for a fixed value of t, $0 < t \leq T$ and sufficiently large n, we may assume that $z_k(t,\varepsilon) \sim 0$, $w_k(t,\varepsilon) \sim 0$ for all $k\varepsilon 0$. However, before speaking about the replacement of the countable system by a finite one, we shall pay some attention to the summation of trigonometric series with coefficients given approximately.

We first note that there are two basic methods, based on the Bogolyubov method, for solving countable systems of ordinary differential equations in standard form.

The first method employs the generalised Bogolyubov first and second main theorems for countable systems, the second one "shortens" these systems. The generalisation of the Bogolyubov first main theorem to countable systems was obtained in [56], and [106] generalised the second main theorem. The theorem on shortening countable systems of ordinary differential equations was proved in [40].

Note that the results obtained so far apply only to the estimation of the difference between solutions of the exact and averaged countable or truncated systems of ordinary differential equations. They do not provide estimates of the difference between the solution of the perturbed mixed problem (7.1)–(7.3) and its approximate solution. To see this, one can consider the problem on summation of trigonometric Fourier series, the coefficients of which are given approximately [44].

Suppose that the function $f(x)$ satisfies the condition ensuring the uniform convergence of the trigonometric Fourier series

$$\frac{a_0}{2}+\sum_{k=1}^{\infty}(a_k\cos kx+b_k\sin kx) \tag{7.34}$$

on interval $[-\pi,\pi]$. Next, suppose that we are given approximate values a_k and b_k of the Fourier coefficients of this function in place of the exact values $\tilde{a}_k$ and $\tilde{b}_k$. It is just the case that most often occurs in applications.

We shall assume that errors in the approximate Fourier coefficients are small in the sense of the norm in l_2. This amounts to the following inequality being satisfied

$$\frac{(a_0-\tilde{a}_0)^2}{2}+\sum_{k=1}^{\infty}\left((a_k-\tilde{a}_k)^2+(b_k-\tilde{b}_k)^2\right)\leq\delta^2, \tag{7.35}$$

where δ is a sufficiently small number called the error in Fourier coefficients.

The following problem significant for applications naturally arises: given approximate values of the Fourier coefficients $\tilde{a}_k$ and $\tilde{b}_k$ find the value of the function $f(x)$ at a given fixed point x with error $\varepsilon(\delta)$ tending to zero as $\delta\to 0$.

We shall show that, in general, it is impossible to find the value $f(x)$ at the point x with arbitrary accuracy by straightforward summation of the Fourier series with approximate coefficients

$$\frac{\tilde{a}_0}{2}+\sum_{k=1}^{\infty}(\tilde{a}_k\cos kx+\tilde{b}_k\sin kx), \tag{7.36}$$

Let us keep the error $\delta>0$ arbitrarily fixed and take $C=\sqrt{\sum\limits_{k=1}^{\infty}\frac{1}{k^2}}$. Suppose that the errors in the Fourier coefficients are as follows

$$\tilde{a}_0-a_0=0,\quad a_k-\tilde{a}_k=b_k-\tilde{b}_k=\frac{\delta}{kC\sqrt{2}},\quad k=1,2,\ldots$$

The errors satisfy relation (7.35) and moreover, turn it into strict equality. Replacing the exact Fourier series (7.34) by the one with approximate coefficients (7.36) results in an error, which is equal to the sum of the series

$$\sum_{k=1}^{\infty}\left((\tilde{a}_k-a_k)\cos kx+(\tilde{b}_k-b_k)\sin kx\right).$$

At the point $x=0$, this error is equal to the sum of the series

$$\sum_{k=1}^{\infty}(\tilde{a}_k-a_k)=\frac{\delta}{C\sqrt{2}}\sum_{k=1}^{\infty}\frac{1}{k}=\infty$$

(however small $\delta > 0$ may be).

So, however rapidly the exact trigonometric Fourier series (7.34) converges to the function $f(x)$ and however small the error δ is in relation (7.35), which determines the deviation of the approximate coefficients from the exact ones, it is impossible to restore the function $f(x)$ at any given point of interval $[-\pi, \pi]$ with any accuracy by using the straightforward summation of the Fourier series with approximately given coefficients (7.36).

In fact, we have proved that however small $\delta > 0$, which characterises the difference between the two sets of the Fourier coefficients $\{a_k, b_k\}$ and $\{\tilde{a}_k, \tilde{b}_k\}$ in the sense of (7.35), is, the straightforward sums of the corresponding Fourier series (7.34) and (7.36) may differ arbitrarily. This kind of problem when arbitrarily small error in initial data may result in arbitrarily large deviation in the solutions corresponding to these data is often encountered in mathematics and its applications and called the ill-posed problem. In other words, the above problem on straightforward summation of trigonometric Fourier series is ill-posed.

A general method for solving a wide class of ill-posed problems was developed by Tikhonov and is called the method of regularisation [114].

7.4. Shortening countable systems

Following Persidskii [89] and Zhautykov [40], along with the countable system in standard form (7.13) we shall consider the so-called truncated system

$$\dot{z}_{kn} = -\frac{\varepsilon}{\omega_k} f_{0k}^{n}(t, z_n, w_n) \sin \omega_k t, \quad z_{kn}(0, \varepsilon) = \varphi_k,$$

$$\dot{w}_{kn} = \frac{\varepsilon}{\omega_k} f_{0k}^{n}(t, z_n, w_n) \cos \omega_k t, \quad w_{kn}(0, \varepsilon) = \frac{\chi_k}{\omega_k}, \quad k = 1, 2, \ldots, n \tag{7.37}$$

which is obtained from the initial system (7.13) by setting the unknown functions beginning with $(n+1)$th equal to zero and dropping the equations beginning with $(n+1)$th. In system (7.37)

$$f_{0k}^{n}(t, z_n, w_n) = f_{0k}(t, z_{1n}, \ldots, z_{nn}, 0, 0, \ldots, w_{1n}, \ldots, w_{nn}, 0, 0, \ldots),$$
$$k = 1, 2, \ldots, n.$$

Let us introduce the function

$$u_n(x, t, \varepsilon) = \sum_{k=1}^{n} (z_{kn} \cos \omega_k t + w_{kn} \sin \omega_k t) \sin \lambda_k x, \tag{7.38}$$

where z_{kn} and w_{kn}, $k = 1, 2, \ldots, n$ constitute the solution of the truncated system (7.37), and call it the approximate solution of the mixed problem (7.1)–(7.3).

We are interested in the following problem: find a number n such that, for numbers greater than n, the difference $|u(x,t,\varepsilon) - u_n(x,t,\varepsilon)|$ between the exact (7.6) and approximate (7.38) solutions of the mixed problem (7.1)–(7.3) is smaller, on a finite time interval, than any given number.

Theorem 7.1. *Suppose that on the set* $\Omega = \{(x, t, u, u_t, u_x) \subset \mathbb{R}^5 : 0 \le x \le l,\ 0 \le t \le T,\ (u, u_t, u_x) \in G \subset \mathbb{R}^3\} \times \{\varepsilon : 0 \le \varepsilon < \varepsilon_0\}$ *the following conditions hold*

1) the function $f(x, t, u, u_t, u_x)$ *is continuous with respect to* x *and* t *and satisfies a Lipschitz condition with respect to* u, u_t, u_x *with a constant independent of* x *and* t, *i. e.*

$$|f(x, t, \tilde{u}, \tilde{u}_t, \tilde{u}_x) - f(x, t, u, u_t, u_x)| \le$$
$$\le K\{|\tilde{u} - u| + |\tilde{u}_t - u_t| + |\tilde{u}_x - u_x|\};$$

2) the mixed problem (7.1)–(7.3) has a classical solution $u(x,t,\varepsilon)$ representable as series (7.6), with $z_k(0,\varepsilon) = \varphi_k$, $w_k(0,\varepsilon) = \dfrac{\chi_k}{\omega_k}$, $k = 1,2,\ldots$, the series $\sum_{k=1}^{\infty} k^{\alpha}\varphi_k$, $\alpha = 0,1,2$, and $\sum_{k=1}^{\infty} k^{\alpha}\chi_k$, $\alpha = -1,0,1$, being convergent, and the series $\sum_{k=1}^{\infty} k^{\nu}(|z_k(t,\varepsilon)| + |w_k(t,\varepsilon)|)$, $\nu = 0,1,2$, being uniformly convergent;

3) the function $f(x,t,u(x,t),u_t(x,t),u_x(x,t))$ can be expanded into the Fourier series (7.11), where $u(x,t)$ is the solution of the mixed problem (7.1)–(7.3), u_t, u_x are the partial derivatives of the function $u(x,t,\varepsilon)$;

4) the Fourier coefficients (7.12) satisfy the condition $|f_{0k}(t,z,w)| \le \alpha_k$, the series $\sum_{k=1}^{\infty} \alpha_k$ formed of the coefficients α_k, $k = 1,2,\ldots$, being convergent.

Then, for any T $(0 < T < \infty)$, there exists a number n such that $|u(x,t,\varepsilon) - u_n(x,t,\varepsilon)| < \varepsilon P_n(T,\varepsilon)$, $|u_t(x,t,\varepsilon) - u_{nt}(x,t,\varepsilon)| < \varepsilon P_n(T,\varepsilon)$, $|u_x(x,t,\varepsilon) - u_{nx}(x,t,\varepsilon)| < \varepsilon P_n(T,\varepsilon)$, $0 \le x \le l$, $0 \le t \le T$ hold, where $u(x,t,\varepsilon)$ is the solution of the mixed problem (7.1)–(7.3) and $u_n(x,t,\varepsilon)$ is given by (7.38).

Proof. We observe that the Fourier coefficients (7.12) satisfy the condition

$$|f_{0k}(t,\tilde{z},\tilde{w}) - f_{0k}(t,z,w)| \le 2K \sum_{k=1}^{\infty}(1+\omega_k+\lambda_k) \times (|\tilde{z}_k - z_k| + |\tilde{w}_k - w_k|), \quad k = 1,2,\ldots \tag{7.39}$$

Integrating systems (7.13) and (7.37) over t and subtracting the resulted expressions, we obtain

$$|z_k(t,\varepsilon) - z_{kn}(t,\varepsilon)| \le \frac{\varepsilon}{\omega_k}\int_0^t |f_{0k}(\tau,z(\tau,\varepsilon),w(\tau,\varepsilon)) - f_{0k}^n(\tau,\varepsilon))|d\tau, \quad k = 1,2,\ldots,n,$$

$$|z_k(t,\varepsilon)| \le |\varphi_k| + \frac{\varepsilon}{\omega_k}\int_0^t |f_{0k}(\tau,z(\tau,\varepsilon),w(\tau,\varepsilon))|d\tau, \quad k = n+1,\ n+2,\ldots,$$

$$|w_k(t,\varepsilon) - w_{kn}(t,\varepsilon)| \le \frac{\varepsilon}{\omega_k}\int_0^t |f_{0k}(\tau,z(\tau,\varepsilon),w(\tau,\varepsilon)) - f_{0k}^n(\tau,z_n(\tau,\varepsilon),w_n(\tau,\varepsilon))|\,d\tau, \quad k = 1,2,\ldots,n,$$

$$|w_k(t,\varepsilon)| \le \frac{|\chi_k|}{\omega_k}\int_0^t |f_{0k}(\tau,z(\tau,\varepsilon),w(\tau,\varepsilon))|d\tau, \quad k = n+1,\ n+2,\ldots$$

If we now multiply kth inequality by $(1+\omega_k+\lambda_k)$, $k = 1,2,\ldots$ and then sum up the resulted expressions with respect to k in the interval from one to infinity, then by using the conditions of Theorem 7.1 and condition (7.39) we obtain

$$U_n(t) \le r_n + \varepsilon R_n t + \varepsilon S_n \int_0^t U_n(\tau)d\tau, \tag{7.40}$$

where

$$U_n(t) = \sum_{k=1}^{n}(1+\omega_k+\lambda_k)(|z_k(t,\varepsilon) - z_{kn}(t,\varepsilon)| + |w_k(t,\varepsilon)-$$

$$-w_{kn}(t,\varepsilon)|) + + \sum_{k=n+1}^{\infty}(1+\omega_k+\lambda_k)(|z_k(t,\varepsilon)| + |w_k(t,\varepsilon)|); \tag{7.41}$$

$$r_n = \sum_{k=n+1}^{\infty}(1+\omega_k+\lambda_k)(|\varphi_k| + |\chi_k|\omega_k^{-1}); \tag{7.42}$$

$$R_n = 2\sum_{k=n+1}^{\infty}\frac{1+\omega_k+\lambda_k}{\omega_k}\alpha_k; \tag{7.43}$$

$$S_n = 4K\sum_{k=1}^{n}\frac{1+\omega_k+\lambda_k}{\omega_k}. \tag{7.44}$$

By the Gronwall-Bellman lemma [8] and (7.40), we have

$$U_n(t) \le r_n e^{\varepsilon S_n t} + \frac{R_n}{S_n}(e^{\varepsilon S_n t} - 1), \quad 0 \le t \le T. \tag{7.45}$$

Since r_n and R_n are the remainders of converging series with positive elements, for any $\varepsilon > 0$ there exists n such that $r_n \ge \varepsilon$, $R_n < \varepsilon$. Then we can write

$$U_n(t) \le \varepsilon\left[e^{\varepsilon S_n T}\left(1+\frac{1}{S_n}\right) - \frac{1}{S_n}\right] \equiv \varepsilon P_n(T,\varepsilon), \tag{7.46}$$

where

$$P_n(T,\varepsilon) = \exp\{\varepsilon S_n T\}\left(1+\frac{1}{S_n}\right) - \frac{1}{S_n}. \tag{7.47}$$

Hence

$$|u(x,t,\varepsilon) - u_n(x,t,\varepsilon)| = \Bigg|\sum_{k=1}^{n}[(z_k(t,\varepsilon) - z_{kn}(t,\varepsilon))\cos\omega_k t +$$

$$(w_k(t,\varepsilon) - w_{kn}(t,\varepsilon))\sin\omega_k t]\sin\lambda_k x + \sum_{k=n+1}^{\infty}[z_k(t,\varepsilon)\cos\omega_k t +$$

$$+w_k(t,\varepsilon)\sin\omega_k t]\sin\lambda_k x\Bigg| \le U_n(t).$$

Whence, in view of (7.46), the theorem follows.

Remark 7.1. Since $P_n(t,\varepsilon) \to \infty$ $(n \to \infty)$, the estimate (7.46) does not imply that $u_n(x,t,\varepsilon)$ uniformly converges to the function $u(x,t,\varepsilon)$ as $n \to \infty$. In view of

$$\lim_{\varepsilon\to 0} P_n(T,\varepsilon) = 1, \quad n = \text{const},$$

we can assert that $\varepsilon P_n(T,\varepsilon)$ is proportional to ε for any fixed n and $\varepsilon \ll 1$.

Theorem 7.2. *Suppose that the conditions 1)-3) in Theorem 7.1 are satisfied on the set Ω defined in Theorem 7.1 as well as the following condition*

$$\left|\int_0^t f_{0k}(t, z(t,\varepsilon), w(t,\varepsilon))\sin\omega_k t\,dt\right| \le \alpha_k,$$

$$\left|\int_0^t f_{0k}(t, z(t,\varepsilon), w(t,\varepsilon)) \cos\omega_k t dt\right| \le \beta_k, \quad k = 1, 2, \ldots,$$

the series $\sum_{k=1}^{\infty} \alpha_k$ *and* $\sum_{k=1}^{\infty} \beta_k$ *being convergent.*

Then, for any T $(0 < T < \infty)$, *the statement of Theorem 7.1 holds, with the estimate (7.45) being replaced by*

$$U_n(t) \le (r_n + \varepsilon R_n^*) \exp\{\varepsilon S_n t\}, \quad 0 \le t \le T, \tag{7.48}$$

with

$$R_n^* = \sum_{k=n+1}^{\infty} \frac{1 + \omega_k + \lambda_k}{\omega_k} (\alpha_k + \beta_k). \tag{7.49}$$

Note that condition (1) in Theorem 7.2 is sufficient for a bounded solution of the countable system (7.13) and that of the mixed problem (7.1)–(7.3) to exist for all $t \ge 0$. The estimate (7.48) shows that the choice of n depends on r_n defined by (7.42), which, in its turn, is defined in terms of the Fourier coefficients of the initial functions.

Consider now the countable system (7.49) and assume the following:

a) the Fourier coefficients (7.48), i. e. the right-hand sides of the countable system (7.49), satisfy the condition

$$|f_{0k}(t, \tilde{z}, \tilde{w}) - f_{0k}(t, z, w)| \le \gamma_k \sum_{r=1}^{\infty} (1 + \omega_r + \lambda_r) \left(|\tilde{z}_r - z_r| + |\tilde{w}_r - w_r|\right), \quad k = 1, 2, \ldots,$$

the series $\sum_{k=1}^{\infty} \gamma_k$ being convergent;

b) conditions 2)–4) in Theorem 7.1 or conditions 2) and 3) in Theorem 7.1 and condition 1) in Theorem 7.2 are fulfilled.

Theorem 4.3. *Suppose that conditions a) and b) are satisfied. Then*

$$\lim_{n\to\infty} u_n(x, t, \varepsilon) = u(x, t, \varepsilon), \quad \lim_{n\to\infty} u_{nt}(x, t, \varepsilon) = u_t(x, t, \varepsilon),$$

$$\lim_{n\to\infty} u_{nx}(x, t, \varepsilon) = u_x(x, t, \varepsilon) \quad \forall (x, t) : 0 \le x \le l, \quad 0 \le t \le T,$$

where $u(x, t, \varepsilon)$ *is the classical solution of the mixed problem (7.1)–(7.3) and* $u_n(x, t, \varepsilon)$ *the approximate solution (7.38).*

Proof. In conditions of Theorem 7.3 the estimates (7.45) and (7.48) can be written as follows

$$U_n(t) \le r_n e^{\varepsilon S_n^* t} + \frac{R_n}{S_n^*}(e^{\varepsilon S_n^* t} - 1), \quad 0 \le t \le T,$$

$$U_n(t) \le (r_n + \varepsilon R_n^*) \exp\{\varepsilon S_n^* t\}, \quad 0 \le t \le T,$$

where $S_n^* = 2 \sum_{k=1}^{n} \frac{1 + \omega_k + \lambda_k}{\omega_k} \gamma_k$.

Since the series $\sum_{k=1}^{\infty} (1 + \omega_k + \lambda_k)\omega_k^{-1}\gamma_k$ is convergent and $\lim_{n\to\infty} r_n = 0$, $\lim_{n\to\infty} R_n = 0$, $\lim_{n\to\infty} R_n^* = 0$ we have $\lim_{n\to\infty} U_n(t) = 0$, $0 \le t \le T$.

The series $\sum_{k=1}^{\infty} k^\nu |z_k(t, \varepsilon)|$ and $\sum_{k=1}^{\infty} k^\nu |w_k(t, \varepsilon)|$, $\nu = 0, 1, 2$, are uniformly convergent. Consequently, by using (7.5), we obtain

$$\sum_{k=1}^{\infty} (1 + \omega_k + \lambda_k)(|z_k(t, \varepsilon) - \lim_{n\to\infty} z_{kn}(t, \varepsilon)| + |w_k(t, \varepsilon) -$$

$$- \lim_{n\to\infty} w_{kn}(t,\varepsilon)|) = 0.$$

Hence

$$\lim_{n\to\infty} z_{kn}(t,\varepsilon) = z_k(t,\varepsilon), \quad \lim_{n\to\infty} w_{kn}(t,\varepsilon) = w_k(t,\varepsilon), \quad k = 1,2,\ldots,$$
$$0 \le t \le T.$$

Passing to the limit in (7.38) as $n \to \infty$ yields the statement of the theorem, which completes the proof.

In practice, to verify condition (a) is difficult. It is therefore more natural to use the conditions of Theorems 7.1 and 7.2 for shortening countable systems arisen in solving mixed problems. On the other hand, Theorems 7.1–7.3 show that, with accuracy ε, the main n-frequncy mode of oscillations may exist in systems described by equation (7.1). This mode can be investigated with the help of the truncated system (7.37).

7.5. Determination of the approximate solutions of truncated systems

The truncated system (7.37) is the system of first order ordinary differential equations in standard form. This allows using all the existing methods for solving such systems [9, 48, 86, 91, 109, 157] to determine exact or approximate solutions of system (7.37). If the right-hand sides in system (7.37) are almost periodic in t, the Bogolyubov second theorem [63, 74] or Samoilenko theorem [15] can be used for investigating the structure of solutions. To find approximate solutions, one can use the Bogolyubov-Mitropol'skii averaging method or the Plotnikov-Yarovoi theorems [74]. To find the periodic solition, the method due to Samoilenko and its generalisation [103] can be employed.

7.6. Reduction of the nonlinear equations of hyperbolic type to countable systems

Consider the nonlinear hyperbolic equation

$$u_{tt} - a^2 u_{xx} = cu + \varepsilon f(x,t,u,u_t,u_x,u_{tx},u_{xx}), \tag{7.50}$$

where $\varepsilon > 0$ is a small parameter.

We are interested in the solution of equation (7.50) satisfying the following initial and boundary conditions

$$u(x,0) = \varphi(x), \quad u_t(x,0) = \chi(x), \quad 0 \le x \le l; \tag{7.51}$$

$$L_j\left(u, \frac{\partial u}{\partial x}\right)_{x=j} = 0, \quad j = 0, l, \tag{7.52}$$

where L_j are certain linear homogeneous differential operators with constant coefficients.

The boundary conditions (7.52) correspond to a number of practical cases such as a string or rod with fixed ends [66].

Consider first the nonperturbed equation

$$u_{tt} - a^2 u_{xx} = cu \tag{7.53}$$

with the same initial and boundary conditions (7.51) and (7.52). Representing its solution as a product $u(x,t) = X(x)\,T(t)$ according to the Fourier method, we arrive at the ordinary differential equations

$$T'' + \lambda^2 a^2 T = 0; \tag{7.54}$$

$$X'' + a^2 X = 0 \tag{7.55}$$

(λ is an undefined parameter) and the corresponding boundary conditions with respect to $X(x)$

$$L_j(X, X')|_{x=j} = 0, \quad j = 0, l. \tag{7.56}$$

We assume that the boundary value problem (7.55), (7.56) can be solved in the ordinary way, which leads to the sequence of eigenvalues

$$\lambda_1, \lambda_2, \dots, \lambda_n, \dots \tag{7.57}$$

and the corresponding sequence of eigenfunctions

$$X_1(x), \ X_2(x), \dots, X_n(x), \dots, \tag{7.58}$$

satisfying the orthogonality condition on $[0, l]$

$$\int_0^l X_m(x) X_n(x) dx = 0, \quad m \neq n.$$

Given the eigenvalues (7.57) it is easy to find a sequence of fundamental frequences ω_n occuring in (7.54) from the dispersion equation

$$D(\omega, \lambda_n) \equiv \lambda_n^2 a^2 - \omega^2 - c = 0,$$

The frequences ω_n are assumed to be real. Then $T_n = a_n \cos(\omega_n t + \theta_n)$ determines the fundamental oscillations of equation (7.54) on condition that $\lambda_n^2 a^2 - c > 0$.

So the two-parameter solution of the nonperturbed mixed problem (7.53), (7.51), (7.52) has the form

$$u_n(x, t) = a_n X_n(x) \cos \psi_n, \quad \psi_n = \omega_n t + \theta_n, \quad n = 1, 2, \dots .$$

The last expression determines undamped normal single-frequency oscillations, which take place in the corresponding forms of dynamic equilibrium.

The general solution of the nonperturbed problem (7.53), (7.51), (7.52) takes the form

$$u(x, t, 0) = \sum_{n=1}^{\infty} a_n X_n(x) \cos \psi_n, \tag{7.59}$$

where $\psi_n = \omega_n t + \theta_n$, a_n and θ_n are con stants determined by the initial conditions (7.51).

Suppose that the functions $\varphi(x)$ and $\chi(x)$ occuring in (7.51) can be expanded into the Fourier series by the eigenfunctions (7.58)

$$\varphi(x) = \sum_{n=1}^{\infty} \varphi_n X_n(x), \quad \chi(x) = \sum_{n=1}^{\infty} \chi_n X_n(x), \tag{7.60}$$

where

$$\varphi_n = m_n^{-1} \int_0^l \varphi(x) X_n(x) dx, \quad \chi_n = m_n^{-1} \int_0^l \chi(x) X_n(x) dx,$$

$$m_n = \int_0^l X_n^2(x) dx, \quad n = 1, 2, \dots \tag{7.61}$$

The constants a_n and θ_n are then determined by the expression

$$a_n = \sqrt{\varphi_n^2 + (\chi_n \omega_n^{-1})^2}, \quad \theta_n = -\mathrm{arctg}\, \chi_n (\varphi_n \omega_n)^{-1}, \quad n = 1, 2, \ldots \tag{7.62}$$

Taking into account the form of solution (7.59) of the nonperturbed equation (7.53), we seek the solution of the perturbed equation (7.50) as the series

$$u(x,t,\varepsilon) = \sum_{n=1}^{\infty} \alpha_n(t,\varepsilon) X_n(x) \cos\beta_n(t,\varepsilon), \tag{7.63}$$

where $\beta_n = \omega_n t + h_n(t,\varepsilon)$; $\alpha_n(t,\varepsilon)$ and $h_n(t,\varepsilon)$, $n = 1, 2, \ldots$ are the functions to be found.

Let us introduce the additional relation

$$u_t = -\sum_{n=1}^{\infty} \omega_n \alpha_n(t,\varepsilon) X_n(x) \sin\beta_n(t,\varepsilon). \tag{7.64}$$

Differentiating (7.63) with respect to t and comparing the resulted expression with (7.64), we obtain

$$\sum_{n=1}^{\infty} X_n(x) \left[\frac{d\alpha}{dt} \cos\beta_n + \frac{dh}{dt} \alpha_n \sin\beta_n \right] = 0. \tag{7.65}$$

Computing now the derivatives u_{tt}, u_x, u_{xx}, u_{tx} and combining (7.50), (7.63) and (7.64), we get

$$\sum_{n=1}^{\infty} \omega_n X_n(x) \left[\frac{d\alpha}{dt} \sin\beta_n + \frac{dh}{dt} \alpha_n \cos\beta_n \right] = -\varepsilon f_0(x,t,\alpha,\beta), \tag{7.66}$$

where

$$f_0(x,t,\alpha,\beta) = f\left(x, t, \sum_{n=1}^{\infty} X_n \alpha_n \cos\beta_n, \; -\sum_{n=1}^{\infty} X_n \omega_n \alpha_n \sin\beta_n, \right.$$

$$\left. \sum_{n=1}^{\infty} X_n' \alpha_n \cos\beta_n, \; -\sum_{n=1}^{\infty} X_n' \omega_n \alpha_n \sin\beta_n, \; -\sum_{n=1}^{\infty} \lambda_n^2 X_n \alpha_n \cos\beta_n \right). \tag{7.67}$$

We assume that the function $f_0(x,t,\alpha,\beta)$ can be expanded into the Fourier series by the eigenfunctions $X_n(x)$, $n = 1, 2, \ldots$, i. e.

$$f_0(x,t,\alpha,\beta) = \sum_{n=1}^{\infty} f_{0n}(t,\alpha,\beta) X_n(x), \tag{7.68}$$

where

$$f_{0n}(t,\alpha,\beta) = \frac{1}{m_n} \int_0^l f_0(x,t,\alpha,\beta) X_n(x)\,dx, \quad m_n = \int_0^l X_n^2(x)\,dx. \tag{7.69}$$

Combining (7.65) and (7.66) with (7.68) gives the following system

$$\frac{d\alpha_n}{dt} \cos\beta_n - \frac{dh_n}{dt} \alpha_n \sin\beta_n = 0,$$

$$\frac{d\alpha_n}{dt} \sin\beta_n + \frac{dh_n}{dt} \alpha_n \cos\beta_n = -\frac{\varepsilon}{\omega_n} f_{0n}(t,\alpha,\beta), \quad n = 1, 2, \ldots$$

Whence

$$\frac{d\alpha_n}{dt} = -\frac{\varepsilon}{\omega_n} f_{0n}(t,\alpha,\beta)\sin\beta_n,$$

$$\frac{dh_n}{dt} = -\frac{\varepsilon}{\alpha_n\omega_n} f_{0n}(t,\alpha,\beta)\cos\beta_n, \quad n = 1,2,\ldots \tag{7.70}$$

So to find the solution of the perturbed equation (7.50), one has to solve the countable system of equations (7.70).

In order to ensure that the function (7.63) satisfies the initial conditions (7.51), the boundary conditions for system (7.70) have to be chosen as follows

$$\alpha_n|_{t=0} = a_n, \quad h_n|_{t=0} = \theta_n,$$

where a_n and θ_n are given by (7.62).

Suppose that the following limits exist

$$\lim_{T\to\infty}\frac{1}{T}\int_0^T f_{0n}(t,\alpha,\beta)\sin\beta_n dt = X_{0n}(\alpha,h), \quad n = 1,2,\ldots,$$

$$\lim_{T\to\infty}\frac{1}{T}\int_0^T f_{0n}(t,\alpha,\beta)\cos\beta_n dt = Y_{0n}(\alpha,h), \quad n = 1,2,\ldots,$$

where $\beta = (\omega_1 t + h_1,\ \omega_2 t + h_2,\ldots)$; $\alpha = (\alpha_1,\alpha_2,\ldots)$; $h = (h_1,h_2,\ldots)$. To system (7.70) we can then set in correspondence the averaged system

$$\frac{dx_n}{dt} = -\frac{\varepsilon}{\omega_n} X_{0n}(x,y), \quad x_n(0) = a_n,$$

$$\frac{dy_n}{dt} = -\frac{\varepsilon}{\omega_n x_n} Y_{0n}(x,y), \quad y_n(0) = \theta_n, \quad n = 1,2,\ldots\ . \tag{7.71}$$

Let $(x_n(t), y_n(t))$ be a solution of system (7.71). Then the solution

$$u_I(x,t,\varepsilon) = \sum_{n=1}^{\infty} X_n(x)x_n(t)\cos(\omega_n t + y_n(t))$$

is called the first approximation to the solution of the mixed problem (7.50)–(7.52).

7.7. Investigation of solutions of the equation describing string transverse vibrations in a medium whose resistance is proportional to the velocity in first degree

We first deal with the case where the resistance coefficient is constant. Consider the equation

$$u_{tt} - a^2 u_{xx} = -\varepsilon\mu u_t, \quad \mu = \text{const}, \tag{7.72}$$

with following initial and boundary conditions

$$u(x,0) = \varphi(x), \quad u_t(x,0) = \chi(x), \quad 0 \le x \le l; \tag{7.73}$$

$$u(0,t) = 0, \quad u(l,t) = 0. \tag{7.74}$$

The exact solution of this mixed problem is of the form

$$u(x,t,\varepsilon) = \sum_{n=1}^{\infty} e^{-\frac{1}{2}\varepsilon\mu t} a_n \cos(\omega_n t + \theta_n) \sin \frac{n\pi}{l} x, \tag{7.75}$$

where

$$\omega_n = \frac{1}{2}\sqrt{4\left(\frac{n\pi a}{l}\right)^2 - \mu^2\varepsilon^2}, \quad n = 1, 2, \ldots \tag{7.76}$$

For the series (7.75) and its derivatives to the second order inclusive to be convergent it suffices to require that the derivatives of the initial function $\varphi(x)$ to the second order inclusive be continuous, its third derivative be piecewise continuous and $\varphi(0) = \varphi(l) = 0$, $\varphi''(0) = \varphi''(l) = 0$ hold, and that the initial function $\chi(x)$ be continuously differentiable, its second derivative be piecewise continuous and $\chi(0) = \chi(l) = 0$ hold. Then

$$a_n = \sqrt{\varphi_n^2 + (\chi_n \omega_n^{-1})^{-2}}, \quad \theta_n = -\operatorname{arctg} \chi_n (\varphi_n \omega_n)^{-1}, \quad n = 1, 2, 3, \ldots, \tag{7.77}$$

where φ_n and χ_n are determined by the expressions

$$\varphi_n = \frac{2}{l}\int_0^l \varphi(x) \sin \frac{n\pi x}{l}\, dx,$$

$$\chi_n = \frac{2}{l}\int_0^l \chi(x) \sin \frac{n\pi x}{l}\, dx, \quad n = 1, 2, \ldots \tag{7.78}$$

We shall now investigate the solution of the mixed problem (7.72)–(7.74) by using the asymptotic method presented in §6.6, Chapter 6.

The nonperturbed equation ($\varepsilon = 0$) can be written as

$$u_{tt} - a^2 u_{xx} = 0, \tag{7.79}$$

and the solution of the nonperturbed problem (7.79), (7.73), (7.74) is of the form

$$u(x,t,0) = \sum_{n=1}^{\infty} a_n \cos(\omega_n^{(0)} t + \theta_n) \sin \frac{n\pi x}{l},$$

where $\omega_n^{(0)} = n\pi a/l$, a_n and θ_n, $n = 1, 2, \ldots$, are given by (7.77) and (7.78)

The solution of the perturbed problem (7.72)–(7.74) is sought as the series

$$u(x,t,\varepsilon) = \sum_{n=1}^{\infty} a_n(t,\varepsilon) \cos(\omega_n^{(0)} t + h_n(t,\varepsilon)) \sin \frac{n\pi x}{l}, \tag{7.80}$$

with the additional relation

$$u_t(x,t,\varepsilon) = -\sum_{n=1}^{\infty} \omega_n^{(0)} \alpha_n(t,\varepsilon) \sin(\omega_n^{(0)} t + h_n(t,\varepsilon)) \sin \frac{n\pi x}{l}. \tag{7.81}$$

Since $\varepsilon f \equiv -\varepsilon\mu u_t = \varepsilon\mu \sum_{n=1}^{\infty} \omega_n^{(0)} \alpha_n \sin \beta_n^{(0)} \sin \frac{n\pi x}{l}$ $(\beta_n^{(0)} = \omega_n^{(0)} t + h_n)$, we have $f_{0n} = \mu\omega_n^{(0)} \alpha_n \sin \beta_n^{(0)}$ where f_{0n} are the Fourier coefficients of the function f. Then (7.70) allows us to write the system for determining α_n and h_n,

$$\frac{d\alpha_n}{dt} = -\varepsilon\mu\alpha_n \sin^2 \beta_n^{(0)},$$

$$\frac{dh_n}{dt} = -\varepsilon\mu \sin\beta_n^{(0)} \cos\beta_n^{(0)}, \quad n = 1, 2, \ldots$$

Whence it follows that

$$\underset{t}{M}\{\sin^2\beta_n^{(0)}\} \equiv \lim_{T\to\infty} \frac{1}{T}\int_0^T \sin^2\beta_n^{(0)} dt = \frac{1}{2},$$

$$\underset{t}{M}\{\sin 2\beta_n^{(0)}\} = 0, \; n = 1, 2, \ldots$$

So the truncated system has the form

$$\frac{dx_n}{dt} = -\frac{1}{2}\varepsilon\mu x_n, \; x_n|_{t=0} = a_n, \; \frac{dy_n}{dt} = 0, \; y_n|_{t=0} = \theta_n, \; n = 1, 2, \ldots$$

Whence we derive

$$x_n = a_n \exp\left\{-\frac{1}{2}\varepsilon\mu t\right\}, \quad y_n = \theta_n, \quad n = 1, 2, \ldots$$

Consequently, the first approximation to the solution of the mixed problem (7.72)–(7.74) can be written as follows

$$u_i(x, t, \varepsilon) = \sum_{n=1}^{\infty} e^{-\frac{1}{2}\varepsilon\mu t} a_n \cos(\omega_n^{(0)} t + \theta_n) \sin\frac{\pi n x}{l}.$$

The analysis of the first approximate solution $u_i(x, t, \varepsilon)$ and solution (7.75) of the mixed problem (7.72)–(7.74) shows that the amplitudes of their single-frequency oscillations coincides and decrease in time, only the frequences of single-frequency oscillations being distinct.

Proceeding now with the case where the resistance coefficient is variable, we turn to the equation

$$u_{tt} - a^2 u_{xx} = -\varepsilon\mu(t) u_t \tag{7.82}$$

with the above initial and boundary conditions (7.73) and (7.74) (the function $\mu(t)$ is assumed to be almost periodic).

The solution of the mixed problem (7.82), (7.73) and (7.74) is sought as the series (7.80) with the additional relation (7.81). Since

$$\varepsilon f \equiv -\varepsilon\mu(t) u_t = \varepsilon\mu(t) \sum_{n=1}^{\infty} \omega_n^{(0)} \alpha_n \sin\beta_n^{(0)} \sin\frac{n\pi x}{l},$$

$$\beta_n^{(0)} = \omega_n^{(0)} t + h_n, \quad n = 1, 2, \ldots,$$

we get

$$f_{0n} = \mu(t)\omega_n^{(0)} \alpha_n \sin\beta_n^{(0)}.$$

By using (7.70) we can write down the following equations

$$\frac{d\alpha_n}{dt} = -\varepsilon\alpha_n\mu(t)\sin^2\beta_n^{(0)},$$

$$\frac{dh_n}{dt} = -\varepsilon\mu(t)\sin\beta_n^{(0)}\cos\beta_n^{(0)}, \quad n = 1, 2, \ldots$$

Since the functions $\mu(t)\sin^2\beta_n^{(0)}$ and $\mu(t)\sin 2\beta_n^{(0)}$ are almost periodic in t, their mean values exist

$$\lim_{T\to\infty}\frac{1}{T}\int_0^T \mu(t)\sin^2\beta_n^{(0)}dt = X_n(h_n),$$

$$\lim_{T\to\infty}\frac{1}{T}\int_0^T \mu(t)\sin 2\beta_n^{(0)}dt = Y_n(h_n), \quad n = 1, 2, \ldots$$

The averaged system now takes the form

$$\frac{dx_n}{dt} = -\varepsilon x_n X_n(y_n), \quad x_n|_{t=0} = a_n,$$

$$\frac{dy_n}{dt} = -\frac{1}{2}\varepsilon Y_n(y_n), \quad y_n|_{t=0} = \theta_n, \quad n = 1, 2, \ldots \tag{7.83}$$

Having obtained $y_n, n = 1, 2, \ldots$ from the second equation of system (7.83), we can write

$$x_n(t,\varepsilon) = a_n e^{-\varepsilon\int_0^t X_n(y_n(\varepsilon t))dt}, \quad n = 1, 2, \ldots \tag{7.84}$$

Depending on $\int_0^t X_n(y_n(\varepsilon t))dt$, the approximate solution of the mixed problem (7.82), (7.73), (7.74) acquires various properties. The following cases are possible.

1) If $\int_0^t X_n(y_n(\tau)) \equiv \text{const}$, $n = 1, 2, \ldots$ the first approximation

$$u_t(x,t,\varepsilon) = \sum_{n=1}^{\infty} x_n(t,\varepsilon)\cos(\omega_n^{(0)}t + y_n(\tau))\sin\frac{n\pi x}{l} \tag{7.85}$$

is damped.

2) If $X_n(y_n(t)) = 0$, $n = 1, 2, \ldots$ the first approximation (7.85) is periodic.

3) If at least one of the integrals $\int_0^t X_n(y_n(\varepsilon t))$ is negative for all $t > 0$, the first approximation (7.85) increases as $t \to \infty$.

Example. We shall consider the case that the function $\mu(t) = \cos pt$ is periodic. Equation (7.82) can be rewritten as

$$u_{tt} - a^2 u_{xx} = -\varepsilon \cos pt u_t. \tag{7.86}$$

For the Fourier coefficients f_{0n} we can write

$$f_{0n} = \alpha_n \omega_n \cos pt \sin(\omega_n^{(0)} + h_n),$$

and then

$$\underset{t}{M}\{f_{0n}(t,\alpha,\beta^{(0)})\sin\beta_n^{(0)}\} = \lim_{T\to\infty}\frac{1}{T}\int_0^T \frac{1}{2}[\cos pt - \cos pt\cos 2\beta_n^{(0)}]dt =$$

$$= -\frac{1}{4}\lim_{T\to\infty}\frac{1}{T}\int_0^T [\cos((p + 2\omega_n^{(0)})t + 2h_n) + \cos((p - 2\omega_n^{(0)})t-$$

$$-2h_n)]dt = \begin{cases} 0, & \text{если} \quad p \neq 2\omega_n^{(0)}, \\ -\frac{1}{4}\cos 2h_n, & \text{если} \quad p = 2\omega_n^{(0)}, \quad n = 1, 2, \ldots, \end{cases}$$

$$\underset{t}{M}\{f_{0n}(t, \alpha, \beta^{(0)}) \cos \beta_n^{(0)}\} = \lim_{T\to\infty} \frac{1}{T} \int_0^T \frac{1}{2} \cos pt \sin 2\beta_n^{(0)} dt =$$

$$= \begin{cases} 0, & \text{если} \quad p \neq 2\omega_n^{(0)}, \\ \frac{1}{4}\sin 2h, & \text{если} \quad p = 2\omega_n^{(0)}, \quad n = 1, 2, \ldots, \end{cases}$$

We then proceed in the following manner.

If $p \neq 2\omega_n^{(0)}$, $n = 1, 2, \ldots$ the averaged system (7.83) takes the form

$$\begin{array}{ll} \dfrac{dx_n}{dt} = 0, \quad x_n|_{t=0} = a_n, & x_n(t,\varepsilon) = a_n, \\ & \Leftrightarrow \\ \dfrac{dy_n}{dt} = 0, \quad y_n|_{t=0} = \theta_n, & y_n(t,\varepsilon) = \theta_n, \quad n = 1, 2, \ldots, \end{array}$$

So, in the absence of resonance ($p \neq 2\omega_n^{(0)}$), the first approximation to the solution of equation (7.86) coincides with the solution of the nonperturbed equation (7.79).

For example, if $p = 2\omega_1$, then

$$\frac{dx_1}{dt} = \frac{\varepsilon}{4}\alpha_1 \cos 2y_1, \quad x_1|_{t=0} = a_1,$$

$$\frac{dy_1}{dt} = -\frac{\varepsilon}{4}\sin 2y_1, \quad y_1|_{t=0} = \theta_1,$$

$$\frac{dx_n}{dt} = 0, \quad x_n|_{t=0} = a_n,$$

$$\frac{dy_1}{dt} = 0, \quad y_n|_{t=0} = \theta_n, \quad n = 1, 2, \ldots \tag{7.87}$$

We now proceed with solving system (7.87). We have

$$\frac{dy_1}{dt} = -\frac{\varepsilon}{4}\sin 2y_1 \Leftrightarrow \frac{dy}{\sin 2y_1} = -\frac{\varepsilon}{4}dt \Leftrightarrow \operatorname{tg} y_1 = Ce^{-\frac{\varepsilon t}{2}}.$$

Since

$$y_1|_{t=0} = \theta_n = -\operatorname{arctg}(\chi_1(\omega_1\varphi_1)^{-1})$$

we have

$$\operatorname{tg} y_1 = -\chi_1(\omega_1\omega_1)^{-1}e^{-\frac{\varepsilon t}{2}} \equiv \delta e^{-\frac{\varepsilon t}{2}}. \tag{7.88}$$

In view of (7.88) we can write

$$\cos 2y_1 = \frac{1 - \delta^2 e^{-\varepsilon t}}{1 + \delta^2 e^{-\varepsilon t}}. \tag{7.89}$$

Integration of the first equation of system (7.87) yields

$$x_1 = a_1 \exp\left\{\frac{\varepsilon}{4}\int_0^t \cos 2y_1(\varepsilon t)\, dt\right\}. \tag{7.90}$$

Taking account of (7.89), we can compute the integral

$$\int_0^t \cos 2y_1(\varepsilon t)\,dt = \int_0^t \frac{1-\delta^2 e^{-\varepsilon t}}{1+\delta^2 e^{-\varepsilon t}}\,dt = t + \frac{\varepsilon}{2}\ln\frac{1+\delta^2 e^{-\varepsilon t}}{1+\delta^2}.$$

Substituting it into (7.90) gives

$$\operatorname{tg} y_1 = -\delta e^{-\frac{\varepsilon t}{2}},$$

$$x_1 = a_1 \exp\left\{\frac{\varepsilon t}{4} + \frac{1}{2}\ln\frac{1+\delta^2 e^{-\varepsilon t}}{1+\delta^2}\right\}, \quad \delta = \frac{\chi_1}{\omega_1 \varphi_1}. \tag{7.91}$$

Relations (7.88) and (7.91) show that if $p = 2\omega_1$ (the external frequency p equals the doubled fundamental frequency $2\omega_1$ of the first harmonics), the first approximation

$$u_1(x,t,\varepsilon) = x_1 \cos(\omega_1^{(0)} t + y_1)\sin\frac{\pi x}{l} + \sum_{n=2}^{\infty} x_n \cos(\omega_n^{(0)} t + y_n)\sin\frac{\pi n x}{l}$$

to the solution of the mixed problem (7.86), (7.73), (7.74) increses in time.

7.8. A remark on shortening countable systems obtained when solving nonlinear hyperbolic equations

We shall consider the mixed problem (7.50)–(7.52). Suppose that the boundary value problem (7.55), (7.56) can be solved with respect to $X(x)$. Then we can find the sequence of eigenvalues $\lambda_1, \lambda_2, \ldots, \lambda_n, \ldots$ and corresponding sequence of eigenfunctions $X_1(x), X_2(x) \ldots, X_n(x), \ldots$, are orthogonal on $[0, l]$. Assuming that $\lambda_n^2 a^2 - c > 0$, $n = 1, 2, \ldots$ we have

$$T_n = A_n \cos\omega_n t + B_n \sin\omega_n t,$$

where $\omega_n = \sqrt{\lambda_n^2 a^2 - c}$, A_n and B_n are arbitrary constants, which determine the proper oscillations for equation (7.54).

The general solution of the nonperturbed boundary value problem (7.53), (7.51), (7.52) can be written as the series

$$u(x,t,0) = \sum_{k=1}^{\infty} X_k(x)(A_k \cos\omega_k t + B_k \sin\omega_k t). \tag{7.92}$$

Let us assume that the initial functions $\varphi(x)$ and $\chi(x)$ can be expanded into Fourier series by the eigenfunctions $X_k(x)$, $k = 1, 2, \ldots$, i. e. the expansions in (7.60) hold. Then we obtain

$$A_k = \varphi_k, \quad B_k = \chi_k \omega_k^{-1}, \quad k = 1, 2, \ldots,$$

where φ_k and χ_k are the Fourier coefficients (7.61).

Taking into account the solution (7.92) of the nonperturbed equation (7.53), we seek the solution of the perturbed equation (7.50) as the series

$$u(x,t,\varepsilon) = \sum_{k=1}^{\infty} X_k(x)(z_k(t,\varepsilon)\cos\omega_k t + w_k(t,\varepsilon)\sin\omega_k t) \tag{7.93}$$

with the relation

$$u_t(x,t,\varepsilon) = \sum_{k=1}^{\infty} \omega_k X_k(x)(-z_k(t,\varepsilon)\sin\omega_k t + w_k(t,\varepsilon)\cos\omega_k t) \tag{7.94}$$

added.

Suppose that the function $f_0(x,t,z,w) = f(x,t,u(x,t),\ u_t, u_x, u_{xx}, u_{tx}$ where $u(x,t)$ and $u_t(x,t)$ are given by (7.93) and (7.94), respectively, and u_x, u_{xx}, u_{tx} are the derivatives of $u(x,t)$, can be expanded into the Fourier series by the eigenfunctions $X_k(x)$, $k = 1, 2, \ldots$,

$$f_0(x,t,z,w) = \sum_{k=1}^{\infty} f_{0k}(t,z,w)X_k(x), \tag{7.95}$$

where

$$f_{0k}(t,z,w) = \frac{1}{m_k}\int_0^1 f_0(x,t,z,w)X_k(x)\,dx, \quad m_k = \int_0^l X_k^2(x)dx. \tag{7.96}$$

Proceeding analogously to the calculations in §7.1 of this chapter, we arrive at the following countable system for z_k and w_k, $k = 1, 2, \ldots$:

$$\frac{dz_k}{dt} = -\frac{\varepsilon}{\omega_k} f_{0k}(t,z,w)\sin\omega_k t, \quad z_k(0,\varepsilon) = \varphi_k,$$

$$\frac{dw_k}{dt} = \frac{\varepsilon}{\omega_k} f_{0k}(t,z,w)\cos\omega_k t, \quad w_k(0,\varepsilon) = \chi_k\omega_k^{-1}, \quad k = 1, 2, \ldots \tag{7.97}$$

Besides the countable system in standard form (7.97) we shall consider the so-called truncated system

$$\frac{dz_{kn}}{dt} = -\frac{\varepsilon}{\omega_k} f_{0k}^n(t,z_n,w_n)\sin\omega_k t, \quad z_{kn}(0,\varepsilon) = \varphi_k,$$

$$\frac{dw_{kn}}{dt} = \frac{\varepsilon}{\omega_k} f_{0k}^n(t,z_n,w_n)\cos\omega_n t, w_{kn}(0,\varepsilon) == \chi_k\omega_k^{-1}, \quad k = 1, 2, \ldots, \tag{7.98}$$

where

$$f_{0k}^n(t,z_n,w_n) = f_{0k}(t, z_{1n}, \ldots, z_{nn}, 0,\ 0, \ldots,\ w_{1n}, \ldots, w_{1n},\ 0, 0, \ldots),$$
$$k = 1, 2, \ldots, n.$$

The function

$$u_n(x,t,\varepsilon) = \sum_{k=1}^{n} X_k(x)(z_{kn}\cos\omega_k t + w_{kn}\sin\omega_k t), \tag{7.99}$$

where z_{kn} and w_{kn}, $k = 1, 2, \ldots$ form the solution of the truncated system (7.98), is said to be the approximate solution of the mixed problem (7.50)–(7.52).

We shall prove that the difference $|u(x,t,\varepsilon) - u_n(x,t,\varepsilon)|$ between the exact solution (7.94) and the approximate one (7.99) becomes arbitrarily small on a finite time interval if the number n is properly chosen provided certain conditions hold.

Theorem 7.4. *Suppose that on the set* $\Omega = \{(x,t,u,u_t,u_x,u_{tx},u_{xx}) \in \mathbb{R}^+ : 0 \le x \le l,$ $0 \le t \le T,\ (x,t,u,u_t,u_x,u_{tx},u_{xx}) \in G \subset \mathbb{R}^5\}$ *the following conditions hold*

1) the function $f(x,t,u,u_t,u_x,u_{tx},u_{xx})$ *is continuous with respect to* x *and* t *and satisfies a Lipschitz condition with respect to* u, u_x, u_{tx}, u_{xx} *with a constant* K *that does not depend on* x *and* t;

2) the mixed problem (7.50)–(7.51) possesses the classical solution representable as the series (7.94), where $z_k(0,\varepsilon) = \varphi_k$, $w_k(0,\varepsilon) = \chi_k\omega_k^{-1}$, $k = 1, 2, \ldots$, *the series* $\sum_{k=1}^{\infty} p_k^\nu\varphi_k$, $\nu =$

$0, 1, 2$, and $\sum_{k=1}^{\infty} p_k^{\nu}\chi_k$, $\nu = 0, 1$, are convergent and the series $\sum_{k=1}^{\infty} p_k^{\nu}(|z_k(t,\varepsilon)| + |w_k(t,\varepsilon)|)$, $\nu = 0, 1, 2$ is uniformly convergent (here p_k is either ω_k or λ_k);

3) the function $f(x, t, u(x,t), u_t(x,t), u_x(x,t), u_{tx}(x,t), u_{xx}(x,t))$ can be expanded into the Fourier series (7.95), where $u(x,t)$ is the solution (7.94) of the mixed problem (7.50)-(7.52) and u_t, u_x, u_{tx}, u_{xx} are its partial derivatives;

4) the inequalities

$$\left|\int_0^t f_{0k}(t, z(t,\varepsilon),\ w(t,\varepsilon)) \sin \omega_k t dt\right| \leq b_k,$$

$$\left|\int_0^t f_{0k}(t, z(t,\varepsilon),\ w(t,\varepsilon)) \cos \omega_k t dt\right| \leq c_k, \quad k = 1, 2, \ldots,$$

hold, b_k and c_k, $k = 1, 2, \ldots$, being such that the series $\sum_{k=1}^{\infty} b_k$ and $\sum_{k=1}^{\infty} c_k$ are convergent;

5) the eigenfunctions $X_k(x)$, $k = 1, 2, \ldots$, are bounded, $|X_k(x)| \leq M_k$, $k = 1, 2, \ldots$, and $|X_k'(x)| \leq C\lambda_k$, $C = \text{const}$, $k = 1, 2, \ldots$;

6) the series

$$\sum_{k=1}^{\infty} \omega_k^{-1}(M_k + \omega_k M_k + C\lambda_k\omega_k + \lambda_k^2 M_k)(b_k + c_k)$$

is convergent.

Then for any T, $0 < T < \infty$, there exists a number n such that

$$|u(x,t,\varepsilon) - u_n(x,t,\varepsilon)| < \varepsilon P_n(T,\varepsilon),$$

$$|u_r(x,t,\varepsilon) - u_{nr}(x,t,\varepsilon)| < \varepsilon P_n(T,\varepsilon), \quad r = t,\ x,\ tx,\ xx,$$

where $u(x,t,\varepsilon)$ is the exact solution given by (7.93) and $u_n(x,t,\varepsilon)$ is the approximate solution given by (7.99).

Proof. We observe that the Fourier coefficients (7.96) satisfy the inequality

$$|f_{0k}(t,\tilde{z},\tilde{w}) - f_{0k}(t,z,w)| \leq \frac{KM_k l}{m_k}\sum_{k=1}^{\infty}(M_r + \omega_r M_r + C\lambda_r +$$

$$+ C\lambda_r\omega_r + \lambda_r^2 M_r)(|\tilde{z}_r - z_r| + |\tilde{w}_r - w_r|), \tag{7.100}$$

where $m_k = \int_0^l X_k^2(x)\, dx$.

We first integrate systems (7.97) and (7.98) over t from 0 to t. If we then subtract the resulted expressions, majorise them using (7.100) and the conditions of the theorem, multiply kth inequality by

$$d_k = (M_k + \omega_k M_k + C\lambda_k + C\lambda_k\omega_k + \lambda_k^2 M_k),$$

and sum up, with respect to k from 0 to $+\infty$, the resulted inequalities term by term, we obtain

$$U_n(t) \leq r_n + \varepsilon R_n + \varepsilon S_n \int_0^t U_n(\tau) d\tau, \tag{7.101}$$

where

$$U_n(t) = \sum_{k=1}^{\infty} \alpha_k(|z_k(t,\varepsilon) - z_{kn}(t,\varepsilon)| + |w_k(t,\varepsilon) - w_{kn}(t,\varepsilon)|) +$$

$$+ \sum_{k=n+1}^{\infty} d_k(|z_k(t,\varepsilon)| + |w_k(t,\varepsilon)|);$$

$$r_n = \sum_{k=n+1}^{\infty} d_k(|\varphi_k| + |\chi_k|\omega_k^{-1}); \quad R_n = \sum_{k=n+1}^{\infty} \frac{d_k}{\omega_k}(b_k + c_k);$$

$$S_n = 2Kl \sum_{k=1}^{\infty} \frac{M_k d_k}{m_k \omega_k}.$$

(By the conditions of the theorem, r_n, R_n and S_n are finite for any fixed value of n).

By the Gronwall-Bellman lemma [8], it follows from (7.101) that

$$U_n(t) \leq (r_n + \varepsilon R_n) \exp\{\varepsilon S_n t\} \quad (0 \leq t \leq T). \tag{7.102}$$

Since r_n is the remainder of the convergent series, for any $\varepsilon > 0$ there exists n such that $r_n \leq \varepsilon$. Now (7.102) allows us to write

$$U_n(t) \leq \varepsilon P_n(T,\varepsilon), \tag{7.103}$$

where $P_n(T,\varepsilon) = (1 + R_n)\exp\{\varepsilon S_n T\}$. On the other hand, we have

$$|u(x,t,\varepsilon) - u_n(x,t,\varepsilon)| + |u_t(x,t,\varepsilon) - u_{nt}(x,t,\varepsilon)| + |u_x(x,t,\varepsilon) - u_{nx}(x,t,\varepsilon)| +$$

$$+ |u_{tx}(x,t,\varepsilon) - u_{ntx}(x,t,\varepsilon)| + |u_{xx}(x,t,\varepsilon) - u_{nxx}(x,t,\varepsilon)| \leq$$

$$\leq \sum_{k=1}^{\infty} d_k(|z_k| - z_{kn}| + |w_k - w_{kn}|) + \sum_{k=n+1}^{\infty} d_k(|z_k| + |w_k|) \equiv U_n(t),$$

which being combined with (7.103) proves the theorem.

Note that the condition of Theorem 7.4 are sufficient for the existence of a bounded classical solution of the mixed problem (7.50)–(7.52) for all $t \geq 0$ provided the series $R_n = \sum_{k=n+1}^{\infty} d_k \omega_k^{-1}(b_k + c_k)$ is convergent. The estimate (7.102) shows that in this case the choice of n depends only on r_n. Thus, depending upon r_n, the main n-frequncy mode of oscillations may exist, with accuracy ε, in systems described by equation (7.50). This mode can be investigated with the help of the truncated system (7.98).

7.9. Construction of asymptotic approximations to solutions of linear mixed problems appearing when investigating multi-frequency modes of oscillations

By what was said above, in cases where initial perturbations (as functions occuring in the right-hand sides of initial conditions) are approximated, with the accuracy of the order of magnitude of small parameter, by several eigenfunctions (say, n) of the nonperturbed mixed problem, the perturbed system admits an oscillation mode close to the multi-frequency (n-frequency) one. This is possible when the right-hand side of equation (7.50) does not depend on time t.

Consider the mixed problem

$$u_{tt} - a^2 u_{xx} = cu + \varepsilon f(x, u, u_t, u_x, u_{tx}, u_{xx}); \tag{7.104}$$

$$u(x,0)=\varphi(x),\quad u_t(x,0)=\chi(x),\quad 0\le x\le l; \tag{7.105}$$

$$L_j(u,u_x)|_{x=j}=0,\quad j=0,l, \tag{7.106}$$

where L_j, $j=0,l$ are some linear homogeneous differential operators with constant coefficients.

Suppose that the nonperturbed system ($\varepsilon=0$) possesses a solution representable as the series

$$u(x,t,0)=\sum_{k=1}^{\infty}X_k(x)(A_k\cos\omega_k t+B_k\sin\omega_k t),$$

where $X_k(x)$, $k=1,2,\ldots$ are the eigenfunctions of the boundary value problem

$$X''+\lambda^2X=0,\quad L_j(X,X')|_{x=j}=0,\quad j=0,L,$$

$$\omega_k=\sqrt{\lambda_k^2a^2-c},\quad \lambda_k^2a^2-c>0,\quad k=1,2,\ldots; \tag{7.107}$$

and λ_k, $k=1,2,3,\ldots$ are the eigenvalues of that boundary value problem.

Let us now assume that the initial functions $\varphi(x)$ and $\chi(x)$ occuring in the initial conditions (7.105) can be approximated on $[0,l]$ by the first n eigenfunctions $X_k(x)$, $k=1,2,\ldots,n$, of the boundary value problem (7.107) with an accuracy of the magnitude of ε. Then (7.105) can be represented as

$$u(x,0)=\sum_{k=1}^{n}\varphi_kX_k(x)+\varepsilon y(x),$$

$$u_t(x,0)=\sum_{k=1}^{n}\chi_kX_k(x)+\varepsilon z(x). \tag{7.108}$$

We are now going to find a $2n$-parametric solution of the perturbed mixed problem (7.104), (7.106), (7.108) as the asymptotic expansion

$$u(x,t,\varepsilon)=\sum_{k=1}^{n}X_k(x)a_k(t,\varepsilon)\cos[\psi_k(t,\varepsilon)]+\varepsilon w_1(x,t)+\varepsilon^2w_2(x,t)+\cdots, \tag{7.109}$$

where

$$a_k=\alpha_k+\varepsilon u_{1k}(t,\alpha,h)+\varepsilon^2u_{2k}(t,\alpha,h)+\cdots,$$

$$\psi_k=\omega_kt+h_k+\varepsilon v_{1k}(t,\alpha,h)+\varepsilon^2v_{2k}(t,\alpha,h)+\cdots,\ k=1,2,\ldots,n. \tag{7.110}$$

If we set

$$u_t=-\sum_{k=1}^{n}X_k(x)\omega_ka_k\sin\psi_k+\varepsilon w_{1t}(x,t)+\varepsilon^2w_{2t}(x,t)+\cdots,$$

$$\alpha_k=\alpha_k(\tau),\quad h_k=h_k(\tau),\quad \tau=\varepsilon t,\quad k=1,2,\ldots,n, \tag{7.111}$$

the functions α_k, h_k, v_{ik}, u_{ik} w_i can be determined in the following way. Substituting the expansions from (7.110) into (7.109) and (7.109) we get the asymptotic expansions for u, u_t, u_x. Next, calculating the partial derivatives $(u)_x$, u_{tx}, u_{xx} and taking into account the additional relation (7.111) and equation (7.104), we obtain the chain systems of equations to determine α_k, h_k, u_{ik}, v_{ik}, w_i, $k=1,\ldots,n$, $i=1,2,\ldots$. Then we have

$$u(x,t,\varepsilon)=\sum_{k=1}^{n}X_k(x)[\alpha_k\cos\beta_k+\varepsilon(u_{1k}\cos\beta_k-\alpha_kv_{1k}\sin\beta_k)+$$

$$+\varepsilon^2\left(u_{2k}\cos\beta_k-\alpha_k v_{2k}\sin\beta_k-\frac{1}{2}\alpha_k v_{1k}^2\cos\beta_k-\right.$$

$$\left.-u_{1k}v_{1k}\sin\beta_k\right)+\varepsilon^2\ldots]+\varepsilon w_1(x,t)+\varepsilon^2 w_2(x,t)+\cdots; \tag{7.112}$$

$$u_t(x,t,\varepsilon)=-\sum_{k=1}^{n}X_k(x)\omega_k[\alpha_k\sin\beta_k+\varepsilon(u_{1k}\sin\beta_k+\alpha_k v_{1k}\cos\beta_k)+$$

$$+\varepsilon^2\left(u_{2k}\sin\beta_k+\alpha_k v_{2k}\cos\beta_k-\frac{1}{2}v_{1k}^2\alpha_k\sin\beta_k+u_{1k}v_{1k}\cos\beta_k\right)+$$

$$+\varepsilon^2\ldots]+\varepsilon w_{1t}(x,t)+\varepsilon^2 w_{2t}(x,t)+\cdots; \tag{7.113}$$

$$\beta_k=\omega_k t+h_k;\quad k=1,2,\ldots,n$$

Differentiating (7.12) with respect to t and equating the resulted expression to (7.113), we can write

$$\varepsilon\sum_{k=1}^{n}X_k(x)\left[\frac{d\alpha_k}{d\tau}\cos\beta_k-\frac{dh_k}{d\tau}\alpha_k\sin\beta_k+\frac{\partial u_1}{\partial t}\cos\beta_k+\right.$$

$$\left.+\frac{\partial v_1}{\partial t}\alpha_k\sin\beta_k\right]+\varepsilon^2\sum_{k=1}^{n}X_k(x)\left[\frac{\partial u_{2k}}{\partial t}\cos\beta_k-\right.$$

$$\left.-\frac{\partial v_{2k}}{\partial t}\alpha_k\sin\beta_k+F_{2k}\right]+\varepsilon^2\ldots=0, \tag{7.114}$$

where the functions F_{2k} depend only on α_k, h_k, u_{1k}, v_{1k}, $k=1,2,\ldots,n$ and their first order derivatives with respect to τ, $\tau=\varepsilon t$.

We need to calculate some derivatives,

$$u_{tt}=-\sum_{k=1}^{n}X_k(x)\omega_k^2[\alpha_k\cos\beta_k+\varepsilon(u_{1k}\cos\beta_k-\alpha_k v_{1k}\sin\beta_k)]-$$

$$-\varepsilon\sum_{k=1}^{n}X_k(x)\omega_k\left[\frac{d\alpha_k}{d\tau}\sin\beta_k+\frac{dh_k}{d\tau}\alpha_k\cos\beta_k+\frac{\partial u_{1k}}{\partial t}\sin\beta_k+\right.$$

$$\left.+\frac{\partial v_{1k}}{\partial t}\alpha_k\sin\beta_k\right]-\varepsilon^2\sum_{k=1}^{n}X_k(x)\left[\frac{\partial u_{2k}}{\partial t}\sin\beta_k+\right.$$

$$\left.+\frac{\partial v_{2k}}{\partial t}\alpha_k\cos\beta_k+\Phi_{2k}\right]-\varepsilon^2\cdots+\varepsilon w_{1tt}+\varepsilon^2 w_{2tt}+\cdots, \tag{7.115}$$

where the functions Φ_{2k} depend only on α_k, h_k, u_{1k}, v_{1k}, $k=1,2,\ldots,n$ and their first order derivatives with respect to τ, $\tau=\varepsilon t$;

$$u_x=\sum_{k=1}^{n}X'(x)[\alpha_k\cos\beta_k+\varepsilon(u_{1k}\cos\beta_k-v_{1k}\alpha_k\sin\beta_k)+\varepsilon^2\ldots]+$$

$$+\varepsilon w_{1x}(x,t)+\varepsilon^2 w_{2x}(x,t)+\cdots; \tag{7.116}$$

$$u_{xx}=-\sum_{k=1}^{n}\lambda_k^2 X_k(x)[\alpha_k\cos\beta_k+\varepsilon(u_{1k}\cos\beta_k-v_{1k}\alpha_k\sin\beta_k)+$$

$$+\varepsilon^2\ldots]+\varepsilon w_{1xx}(x,t)+\varepsilon^2 w_{2xx}(x,t)+\cdots; \tag{7.117}$$

$$u_{tx} = -\sum_{k=1}^{n} X'(x)\omega_k[\alpha_k \sin\beta_k + \varepsilon(u_{1k}\sin\beta_k + \alpha_{1k}v_{1k}\cos\beta_k) + $$
$$+\varepsilon^2\ldots] + \varepsilon w_{1tx}(x,t) + \varepsilon^2 w_{2tx}(x,t) + \cdots \tag{7.118}$$

Substituting (7.112), (7.113), (7.115)–(7.118) into equation (7.104) gives

$$-\varepsilon\sum_{k=1}^{n}\omega_k X_k(x)\left[\frac{d\alpha_k}{d\tau}\sin\beta_k + \frac{dh_k}{d\tau}\alpha_k\cos\beta_k + \frac{\partial u_{1k}}{\partial t}\sin\beta_k + \right.$$
$$\left. + \frac{\partial v_{1k}}{\partial t}\alpha_k\cos\beta_k\right] - \varepsilon^2\sum_{k=1}^{n}\omega_k X_k(x)\left[\frac{\partial u_{2k}}{\partial t}\sin\beta_k + \right.$$
$$\left. + \frac{\partial v_{2k}}{\partial t}\alpha_k\cos\beta_k + \Phi_{2k}\right] - \varepsilon^2\cdots + \varepsilon L w_1(x,t) +$$
$$+\varepsilon^2 L w_2(x,t) + \cdots = \varepsilon f(x,t,u,u_t,u_x,u_{tx},u_{xx}), \tag{7.119}$$

where $L\dfrac{\partial^2}{\partial t^2} - a^2 = \dfrac{\partial^2}{\partial x^2} - c$.

Suppose that the function $f(x,t,u,u_t,u_x,u_{tx},u_{xx})$, where $u(x,t)$ and its derivatives are given by (7.112), (7.113), (7.115)–(7.118), can be expanded into the power series by ε

$$f = f_0 + \varepsilon f_1 + \varepsilon^2 f_2 + \cdots \tag{7.120}$$

Substituting (7.120) into (7.119) and comparing the coefficients at equal powers of ε we arrive at the system

$$-\sum_{k=1}^{n}\omega_k X_k(x)\left[\frac{d\alpha_k}{d\tau}\sin\beta_k + \frac{dh_k}{d\tau}\alpha_k\cos\beta_k + \frac{\partial u_{1k}}{\partial t}\sin\beta_k + \right.$$
$$\left. + \frac{\partial v_{1k}}{\partial t}\alpha_k\cos\beta_k\right] + Lw_1 = f_0,$$
$$-\sum_{k=1}^{n}\omega_k X_k(x)\left[\frac{\partial u_{2k}}{\partial t}\sin\beta_k + \frac{\partial v_{2k}}{\partial t}\alpha_k\cos\beta_k + \Phi_{2k}\right] + Lw_2 = f_1, \tag{7.121}$$
$$\cdots\cdots\cdots\cdots\cdots\cdots\cdots\cdots\cdots\cdots$$

Suppose now that the coefficients f_i, $i = 0,1,2,\ldots,n$, of the series (7.120) can be expanded into Fourier series by the eigenfunctions $X_k(x)$, $k = 1,2,\ldots,n$ namely,

$$f_i = \sum_{k=1}^{n} f_{1k}X_k(x) + \sum_{k=n+1}^{d_i} f_{ik}X_k(x), \quad i = 0,1,2,\ldots, \tag{7.122}$$

where

$$f_{ik} = \frac{1}{m_k}\int_0^l f_k X_k(x)\,dx, \quad m_k = f_i = \int_0^l X_k^2(x)\,dx,$$
$$k = 1,2,\ldots,d_i; \quad d_i \le \infty; \quad i == 0,1,2,\ldots, \tag{7.123}$$

By (7.122), it follows from (7.121) that

$$-\omega_k\left[\frac{d\alpha_k}{d\tau}\sin\beta_k + \frac{dh_k}{d\tau}\alpha_k\cos\beta_k + \frac{\partial u_{1k}}{\partial t}\sin\beta_k + \right.$$

$$+\frac{\partial v_{1k}}{\partial t}\alpha_k \cos\beta_k\Big] = f_{0k}, \quad k = 1, 2, \ldots, n,$$

$$Lw_1(x,t) = \sum_{k=n+1}^{d_0} f_{0k}X_k(x),$$

$$-\omega_k\left[\frac{\partial u_{2k}}{\partial t}\sin\beta_k + \frac{\partial v_{2k}}{\partial t}\alpha_k\cos\beta_k + \Phi_2\right] = f_{1k}, \quad k = 1, 2, \ldots, n,$$

$$Lw_2 = \sum_{k=n+1}^{d_1} f_{1k}X_k(x), \tag{7.124}$$

....

Regarding (7.82) as a power series in ε, we can write

$$\sum_{k=1}^{n} X_k(x)\left[\frac{d\alpha_k}{d\tau}\cos\beta_k - \frac{dh_k}{d\tau}\alpha_k\sin\beta_k + \frac{\partial u_{1k}}{\partial t}\cos\beta_k - \right.$$

$$\left. -\frac{\partial v_{1k}}{\partial t}\alpha_k\sin\beta_k\right] = 0,$$

$$\sum_{k=1}^{n} X_k(x)\left[\frac{\partial u_{2k}}{\partial t}\cos\beta_k - \frac{\partial v_{2k}}{\partial t}\alpha_k\sin\beta_k + F_{2k}\right] = 0, \tag{7.125}$$

....

Since the eigenfunctions $X_k(x)$ are linearly independent, (7.125) implies

$$\frac{d\alpha_k}{dt}\cos\beta_k - \frac{dh_k}{d\tau}\alpha_k\sin\beta_k + \frac{\partial u_{1k}}{\partial t}\cos\beta_k - \frac{\partial v_{1k}}{\partial t}\alpha_k\sin\beta_k = 0,$$

$$\frac{\partial u_{2k}}{\partial t}\cos\beta_k - \frac{\partial v_{2k}}{\partial t}\alpha_k\sin\beta_k + F_{2k} = 0, \tag{7.126}$$

....

Combining (7.124) and (7.126) yields the following system for determining the unknown functions α_k, h_k, u_{ik}, v_{ik}, w_i, $k = 1, 2, \ldots, n$; $i = 1, 2, \ldots$:

$$\frac{\partial u_{1k}}{\partial t} = -\frac{1}{\omega_k}f_{0k}(\alpha,\beta)\sin\beta_k - \frac{d\alpha_k}{d\tau},$$

$$\frac{\partial v_{1k}}{\partial t} = -\frac{1}{\alpha_k\omega_k}f_{0k}(\alpha,\beta)\cos\beta_k - \frac{dh_k}{d\tau}, \quad k = 1, 2, \ldots, n,$$

$$Lw_1 = \sum_{k=n+1}^{d_0} f_{0k}(\alpha,\beta)X_k(x),$$

$$\frac{\partial u_{2k}}{\partial t} = -F_{2k}\cos\beta_k - \Phi_{2k}\sin\beta_k - \frac{1}{\omega_k}f_{1k}\sin\beta_k,$$

$$\frac{\partial v_{2k}}{\partial t} = \frac{1}{\alpha_k}F_{2k}\sin\beta_k - \frac{1}{\alpha_k}\Phi_{2k}\cos\beta_k -$$

$$-\frac{1}{\alpha_k\omega_k}f_{1k}\cos\beta_k, \quad k = 1, 2, \ldots, n,$$

$$Lw_2 = \sum_{k=n+1}^{d_1} f_{1k} X_k(x), \tag{7.127}$$

. . . .

Note that f_0 and f_{0k} are explicitly defined, i. e.

$$f_0(x,\alpha,\beta) = f\Bigg(x, \sum_{k=1}^{n} \alpha_k X_k(x)\cos\beta_k,\ -\sum_{k=1}^{n} \omega_k\alpha_k X_k \sin\beta_k,$$

$$\sum_{k=1}^{n} X_k'\alpha_k\cos\beta_k,\ -\sum_{k=1}^{n} X_k'\omega_k\alpha_k\sin\beta_k,\ -\sum_{k=1}^{n}\lambda_k^2 X_k\alpha_k\cos\beta_k\Bigg),$$

$$\beta_k = \omega_k f + h_k, \quad k = 1,2,\ldots,n,$$

$$f_{0k}(\alpha,\beta) = \frac{1}{m_k}\int_0^l f_0(x,\alpha,\beta)X_k(x)dx,$$

$$m_k = \int_0^l X_k^2(x)dx, \quad k = 1,2,\ldots,d_0, \quad d_0 \le \infty. \tag{7.128}$$

Suppose that the mean values

$$\underset{t}{M}\{f_{0k}(\alpha,\beta)\sin\beta_k\} \equiv \lim_{T\to\infty}\frac{1}{T}\int_0^T f_{0k}(\alpha,\beta)\sin\beta_k dt = X_k(\alpha,h),$$

$$\underset{t}{M}\{f_{0k}(\alpha,\beta)\cos\beta_k\} = Y_k(\alpha,\beta), \quad k = 1,2,\ldots,n. \tag{7.129}$$

exist. Let

$$\frac{d\alpha_k}{dt} = -\frac{\varepsilon}{\omega_k}X_k(\alpha,\beta),$$

$$\frac{dh_k}{dt} = -\frac{\varepsilon}{\alpha_k\omega_k}Y_k(\alpha,h), \quad k = 1,2,\ldots,n. \tag{7.130}$$

The functions u_{1k}, v_{1k}, $k = 1,2,\ldots,n$ can then be found from the system

$$\frac{\partial u_{1k}}{\partial t} = -\frac{1}{\omega_k}[f_{0k}(\alpha,\beta)\sin\beta_k - \underset{t}{M}\{f_{0k}(\alpha,\beta)\sin\beta_k\}],$$

$$\frac{\partial v_{1k}}{\partial t} = -\frac{1}{\alpha_k\omega_k}[f_{0k}(\alpha,\beta)\cos\beta_k - \underset{t}{M}\{f_{0k}(\alpha,\beta)\cos\beta_k\}], \tag{7.131}$$

$$k = 1,2,\ldots,n.$$

To determine the initial and boundary conditions for the functions α_k, h_k, u_{ik}, v_{ik}, w_i, $k = 1,2,\ldots,n$; $i = 1,2,\ldots$ we put

$$\alpha_k|_{t=0} = a_k, \quad h_k|_{t=0} = \theta_k, \quad a_k = \sqrt{\varphi_k^2 + (\chi_k\omega_k^{-1})^2},$$

$$\theta_k = -\operatorname{arctg}\chi_k(\omega_k\varphi_k)^{-1},$$

$$u_{ik}|_{t=0} = 0, \quad v_{ik}|_{t=0} = 0, \quad k = 1,2,\ldots,n; \quad i = 1,2,\ldots; \tag{7.132}$$

$$w_i|_{t=0} = 0, \quad w_{it}|_{t=0} = 0, \quad i = 2,3,\ldots,$$

$$w_1|_{t=0} = y(x), \quad w_{1t}|_{t=0} = z(x),$$
$$w_i(0,t) = w_i(l,t) = 0, \quad i = 1, 2, \ldots \tag{7.133}$$

Then one can easily verify that the function $u(x,t,\varepsilon)$ satisfies the initial (7.108) and boundary (7.106) conditions.

The above method for constructing asymptotic solutions of mixed problems applies only if the mean values (7.129) exist; and it can be used if f contains mixed derivatives

$$\frac{\partial^k u}{\partial t \partial x^{k-1}}, \quad k = 1, 2, \ldots, p.$$

7.10. Investigation of single-frequency oscillations for the equation $u_{tt} - a^2 u_{xx} = \varepsilon u^2$

Consider the following mixed problem

$$u_{tt} - a^2 u_{xx} = \varepsilon u^3; \tag{7.134}$$

$$u(0,t) = u(l,t) = 0; \tag{7.135}$$

$$u(x,0) = \varphi(x), \quad u_t(x,0) = \chi(x) \tag{7.136}$$

and suppose that the initial functions $\varphi(x)$ and $\chi(x)$ are given as follows

$$\varphi(x) = \varphi_1 \sin \lambda_1 x + \varepsilon y(x), \quad \chi(x) = \chi_1 \sin \lambda_1 x + \varepsilon z(x), \tag{7.137}$$

where φ_1 and χ_1 are real numbers, $\sin \lambda_1 x$ is the first eigenfunction of the boundary value problem

$$X'' + \lambda^2 X = 0, \quad X(0) = X(l), \quad \lambda_1 = \frac{\pi}{l}.$$

Since the initial functions $\varphi(x)$ and $\chi(x)$ are of the form (7.107), the perturbed systems described by the mixed problem (7.134)–(7.136) admit the single-frequency mode of oscillations. By (7.112), this single-frequency mode is sought in the form of the following expansion

$$u(x,t,\varepsilon) = ((\alpha_1 \cos\beta_1 + \varepsilon u_1 \cos\beta_1 - \varepsilon v_1 \sin\beta_1) + \varepsilon^2 \ldots) \sin \lambda_1 x +$$
$$+ \varepsilon w_1(x,t) + \varepsilon^2 w_2(x,t) + \cdots, \tag{7.138}$$

where $\beta_1 = \omega_1 t + h_1$, $\omega_1 = a\lambda_1 \equiv a\pi/l$; α_1, h_1, u_1, v_1, w_r, $r = 1, 2, \ldots$, are unknown functions.

Since $f = u^3$, the function (7.128) can be written as

$$f_0(x, \alpha_1, \beta_1) = \alpha_1^3 \cos\beta_1^3 \sin^3 \lambda_1 x \equiv \frac{3}{4}\alpha_1^3 \cos\beta_1^3 \sin\lambda_1 x - \frac{1}{4}\alpha_1^3 \cos^3\beta_1 \sin 3\lambda_1 x. \tag{7.139}$$

Whence we find the Fourier coefficients f_{0k}

$$f_{01} = \frac{3}{4}\alpha_1^3 \cos^3\beta_1, \quad f_{02} = 0, \quad f_{03} = -\frac{1}{4}\alpha_1^3 \cos^3\beta_2, \quad f_{0r} = 0, \quad r = 4, 5, \ldots$$

System (7.127) for equation (7.134) can now be written in the form

$$\frac{\partial u_1}{\partial t} = -\frac{3\alpha_1^3}{4\omega_1} \cos^3\beta_1 \sin\beta_1 - \frac{d\alpha_1}{d\tau},$$

$$\frac{\partial v_1}{\partial t} = -\frac{3\alpha_1^3}{4\omega_1}\cos^4\beta_1 - \frac{dh_1}{d\tau}. \tag{7.140}$$

Hence we have

$$\underset{t}{M}\{f_{01}\sin\beta_1\} \equiv \frac{\omega_1}{2\pi}\int_0^{\frac{2\pi}{\omega_1}}\cos^3\beta_1\sin\beta_1 dt = 0,$$

$$\underset{t}{M}\{f_{01}\cos\beta_1\} \equiv \frac{1}{2\pi}\int_0^{2\pi}\cos^4\beta_1 d\beta_1 dt = \frac{3}{8}.$$

In this case system (7.130) takes the form

$$\frac{d\alpha_1}{dt} = 0, \quad \alpha_1|_{t=0} = a_1, \quad \frac{dh_1}{dt} = \frac{9\varepsilon\alpha_1^2}{32\omega_1}, \quad h_1|_{t=0} = \theta_1.w \tag{7.141}$$

Whence, taking account of (7.132), we get

$$\alpha_1 = a_1, \quad a_1 = \sqrt{\varphi_1^2 + \chi_1^2\omega_1^{-1}},$$

$$h_1 = -\frac{9a_1^2\varepsilon t}{32\omega_1} + \theta_1, \quad \theta_1 = -\mathrm{arctg}\chi_1(\omega_1\varphi_1)^{-1}. \tag{7.142}$$

By (7.132) and (7.141), from (7.140) we derive

$$u_1 = \frac{3\alpha_1^3}{128\omega_1^2}[4(\cos 2\beta_1 - \cos 2\theta) + (\cos 4\beta_1 - \cos 4\theta_1)],$$

$$v_1 = -\frac{3\alpha_1^2}{128\omega_1^2}[8(\sin 2\beta_1 - \sin 2\theta_1) + (\sin 4\beta_1 - \sin 4\theta)], \tag{7.143}$$

where $\beta_1 = \omega_1 t + h_1$; α_1 and h_1 are given by (7.142).

Next, $w_1(x,t)$ can be found as a solution of the mixed problem

$$w_{1t} - a^2 w_{1xx} = -\frac{1}{4}\alpha_1^3\cos^3\beta_1\sin 3\lambda_1 x,$$

$$w_1(0,t) = w_1(l,t) = 0, \quad w_1(x,0) = y(x), \quad w_{1t}(x,0) = z(x). \tag{7.144}$$

Note that $\beta_1 = \omega_1 t + h_1$ is completely determined

$$\beta_1 = \left(\omega_1 - \frac{9\varepsilon a_1^2}{32\omega_1}\right)t + \theta_1 \equiv \tilde{\omega}t + \theta_1, \tag{7.145}$$

where $\tilde{\omega} = \omega_1 - \mu\varepsilon$, $\mu = 9a_1^2/32\omega_2$.

Having obtained the solution $w_1(x,t)$ of the mixed problem (7.144) as the series

$$w_1(x,t) = \sum_{k=1}^{\infty} x_k(t)\sin\lambda_k x, \quad \lambda_k = \frac{k\pi}{l}; \quad k = 1, 2, \ldots, \tag{7.146}$$

we can write out the following system

$$\ddot{x}_k + a^2\lambda_k^2 x_k = 0, \qquad k = 1, 2, 4, 5, \ldots,$$

$$\ddot{x}_3 + a^2\lambda_3^2 x_3 = -\frac{1}{4}\alpha_1^3 \cos^3\beta_1.$$

Setting $\omega_k = \lambda_k a$, $k = 1, 2, \ldots$, in this system transforms it into the system

$$\ddot{x}_k + \omega_k^2 x_k = 0, \qquad k = 1, 2, 4, 5, \ldots,$$

$$\ddot{x}_3 + \omega_3^2 x_3 = -\frac{3\alpha_1^3}{16}\cos(\tilde{\omega}_1 t + \theta_1) - \frac{\alpha_1^3}{16}\cos(3\tilde{\omega}_1 t + 3\theta_1). \tag{7.147}$$

The first equation in (7.147) has the solution

$$x_k = A_k \cos\omega_k t + B_k \sin\omega_k t,$$

and the second one

$$x_3(t) = A_3\cos\omega_3 t + B_3 \sin\omega_3 t -$$

$$-\frac{3\alpha_1^3}{16}\frac{\cos\omega_3 t\cos\theta_1 - \cos(\tilde{\omega}_1 t + \theta_1) - \tilde{\omega}_1\omega_3^{-1}\sin\omega_3 t\sin\theta_1}{3\tilde{\omega}_1^2 - \omega_3^2} -$$

$$-\frac{\alpha_1^3}{16}\frac{\cos\omega_3 t\cos 3\theta_1 - \cos(3\tilde{\omega}_1 t + 3\theta_1) - 3\tilde{\omega}_1\omega_3^{-1}\sin\omega_3 t\sin 3\theta_1}{3\tilde{\omega}_1^2 - \omega_3^2}.$$

Since $\omega_1 = \pi a/l$, $\omega_3 = 3\pi a/l$, $\tilde{\omega}_1 = \omega_1 - \mu\varepsilon$, $3\tilde{\omega}_1 = \omega_3 - 3\mu\varepsilon$, $(3\tilde{\omega}_1)^2 - \omega_3^2 = -3\mu\varepsilon(2-\mu\varepsilon)$ we finally get

$$\varepsilon w_1(x,t) = \varepsilon\sum_{k=1}^{\infty}(A_k\cos\omega_k t + B_k\sin\omega_k t)\sin\lambda_k x +$$

$$+\left[\frac{2a_1\omega_1}{27}\frac{\cos\omega_3 t\cos 3\theta_1 - 3\tilde{\omega}_1\omega_3^{-1}\sin\omega_3 t\sin 3\theta_1 - \cos(3\tilde{\omega}_1 t + 3\theta_1)}{2 - 3\mu\varepsilon} -\right.$$

$$\left.-\frac{3a_1^3\varepsilon}{16}\frac{\cos\omega_3 t\cos\theta_1 - \cos(\tilde{\omega}_1 t + \theta_1) - \tilde{\omega}_1\omega_3^{-1}\sin\omega_3 t\sin\theta_1}{\tilde{\omega}_1^2 - \omega_3^2}\right]\sin\lambda_3 x. \tag{7.148}$$

So the first improved approximation to the single-frequency mode of oscillations for the perturbed mixed problem (7.134)–(7.136) has the form

$$u_{1y_n}(x,t,\varepsilon) = (a_1\cos\beta_1 + \varepsilon[u_1\cos\beta_1 - a_1 v_1\sin\beta_1])\sin\lambda_1 x + \varepsilon w_1(x,t),$$

where $\beta_1 = (\omega_1 - 9\varepsilon a_1^2/32\omega_1)t + \theta_1$; u_1, v_1 are defined by (7.143) and $\varepsilon w_1(x,t)$ by (7.148).

7.11. Construction of asymptotic approximations to solutions of nonlinear mixed problems used for investigating single-frequency modes of oscillations with fast and slow variables

Modern problems of the theory of nonlinear oscillations in elastic systems require consideration of nonlinear mixed problems. Therefore the construction of approximate solutions of nonlinear mixed problems, clarification of their structure followed by the analysis and investigation of the corresponding modes of oscillations are of interest for both theory and practice.

We shall consider a nonlinear mixed problem described by the nonlinear equation of the form

$$u_{tt} - a^2 u_{xx} = cu + \varepsilon f(x, \theta, \tau, u, u_t, u_x, u_{tx}, u_{xx}, u_{txx}) + \varepsilon g(\theta, \tau), \tag{7.149}$$

the boundary conditions

$$L_j(u, u_x)|_{x=j} = \varepsilon F_j(\theta, \tau, u|_{x=j}), \qquad j = 0, l, \tag{7.150}$$

close to the linear ones and the initial conditions

$$u(x, 0) = \varphi(x), \qquad u_t(x, 0) = \chi(x). \tag{7.151}$$

Here a is a constant, $\varepsilon > 0$ a small parameter, t time, $\tau = \varepsilon t$; $L_j, j = 0, l$, are certain linear homogeneous differential operators with constant coefficients (see [66]). The righ-hand side of equation (7.149) determines small perturbing (distributed) forces that are 2π-periodic in θ, the instantaneous frequency being equal to $d\theta/dt = \nu(\tau)$.

We first proceed with the nonperturbed boundary value problem

$$u_{tt} - a^2 u_{xx} = cu, \tag{7.152}$$

$$L_j(u, u_x)|_{x=j} = 0, \qquad j = 0, l. \tag{7.153}$$

Searching for its solution as the product $u = X(x)\,T(t)$, we again arrive at the boundary value problem

$$X'' + \lambda^2 X = 0, \quad L_j(X, X')|_{x=j} = 0, \quad j = 0, l. \tag{7.154}$$

Let λ_k, $k = 1, 2, \ldots$, be the eigenvalues of the boundary value problem (7.154) and $X_k(x)$, $k = 1, 2, \ldots$, be the corresponding eigenfunctions. If we assume that $\lambda_k^2 a^2 - c > 0$, $k = 1, 2, \ldots$, the solution of the boundary value problem (7.152), (7.153) can be written as the series

$$u(x, t, 0) = \sum_{k=1}^{\infty} X_k(x) a_k \cos(\omega_k t + \theta_k), \tag{7.155}$$

where $\omega_k = \sqrt{\lambda_k^2 a^2 - c}$, $k = 1, 2, \ldots$ and a_k, θ_k are constants.

Turning to the consideration of the perturbed boundary value problem described by the nonlinear equation (7.149) and nonlinear boundary conditions (7.150) we shall investigate, in the case of resonance, the single-frequency oscillations close to one of the normal oscillations

$$u_k = a_k X_k(x) \cos(\omega_k t + \theta_k) \tag{7.156}$$

of the nonperturbed boundary value problem (7.152), (7.153). Moreover, we allow for the above single-frequency oscillations taking place in the presence of combinative resonance ($fpq\nu(\tau) \approx \omega_k$, where p and q are relatively prime numbers) in the nonlinear system in question; the rise of this kind of resonance is a consequence of perturbing forces in the right-hand sides of equation (7.149) and boundary conditions (7.150). The presence of friction forces in real elastic systems results in proper oscillations with frequences $\omega_n \neq \omega_k$, $n = 1, 2, \ldots k-1, k+1, \ldots$, rapidly vanishing. Therefore if combinative resonance takes place, there are nonstable oscillations close to kth normal oscillation of the nonpertubed system in the corresponding form of dynamic equilibrium.

To be specific, we assume that in the perturbed system the following combinative resonance may occur

$$\frac{p}{q}\nu(\tau) \approx_1 \equiv \sqrt{\lambda_1^2 a^2 - c}, \quad \lambda_1 a^2 - c > 0,$$

in the presence of which a single-frequency mode of nonstable oscillations close to the first normal oscillation of the nonperturbed system (7.152), (7.153) may exist.

The solution of the perturbed mixed problem (7.149)–(7.151) is sought in the following form

$$u(x,t,\varepsilon)=a(t,\varepsilon)X_1(x)\cos[\psi(t,\varepsilon)]+\varepsilon w_1(x,t)+\varepsilon^2 w_2(x,t)+\cdots, \tag{7.157}$$

where

$$a=\alpha+\varepsilon u_1(t,\alpha,h)+\varepsilon^2 u_2(t,\alpha,h)+\cdots,$$
$$\psi=\frac{p}{q}\nu t+h+\varepsilon v_1(t,\alpha,h)+\varepsilon^2 u_2(t,\alpha,h)+\cdots \tag{7.158}$$

We set

$$u_t(x,t,\varepsilon)=-\tilde{\omega}a(t,\varepsilon)X_1(x)\sin[\psi(t,\varepsilon)]+\varepsilon w_{1t}(x,t)+\varepsilon w_{2t}(x,t)+\cdots,$$
$$\alpha=\alpha(\tau),\quad \tilde{\beta}=\tilde{\omega}t+h(\tau),\quad \tilde{\omega}=\frac{p}{q}\nu,\quad \tau=\varepsilon t. \tag{7.159}$$

Expanding $u(x,t,\varepsilon)$, $u_t(x,t,\varepsilon)$ in powers of ε and taking account of the equality $(u)'_t=u_t$ and equation (7.149) we obtain

$$X_1(x)\left[\varepsilon\left(\frac{d\alpha}{d\tau}\cos\tilde{\beta}-\frac{dh}{d\tau}\alpha\sin\tilde{\beta}+\frac{\partial u_1}{\partial t}\cos\tilde{\beta}-\frac{\partial v_1}{\partial t}\alpha\sin\tilde{\beta}\right)+\varepsilon^2\ldots\right]=0,$$

$$\begin{aligned}X_1(x)\{(\omega_1^2-\tilde{\omega}^2)&[\alpha\cos\tilde{\beta}+\varepsilon(u_1\cos\tilde{\beta}--\alpha v_1\sin\tilde{\beta})+\varepsilon^2\ldots]-\\-\varepsilon\tilde{\omega}&\left[\frac{d\alpha}{d\tau}\sin\tilde{\beta}+\frac{dh}{d\tau}\alpha\cos\tilde{\beta}+\frac{\partial u_1}{\partial t}\sin\tilde{\beta}+\frac{\partial v_1}{\partial t}\alpha\cos\tilde{\beta}\right]-\\&-\varepsilon^2\tilde{\omega}\ldots]\}+\varepsilon Lw_1(x,t)+\varepsilon^2 Lw_2(x,t)+\cdots=\\&=\varepsilon f(x,\theta,\tau,u,u_t,u_x,u_{tx},u_{xx},u_{txx})+\varepsilon g(\theta,\tau),\end{aligned} \tag{7.160}$$

where

$$L=\frac{\partial^2}{\partial t^2}-a^2\frac{\partial^2}{\partial x^2}-c.$$

Moreover, the following equality holds

$$\frac{\omega^2-\tilde{\omega}^2}{\tilde{\omega}}=2(\omega-\tilde{\omega})+\frac{(\omega-\omega^2)^2}{\tilde{\omega}}.$$

Letting $\omega_1-\tilde{\omega}=\varepsilon\delta$, we have

$$\omega_1^2-\tilde{\omega}^2=2\varepsilon\delta\tilde{\omega}+\varepsilon^2\delta^2. \tag{7.161}$$

Suppose that the function $f(x,\theta,\tau,u,u_t,u_x,u_{tx},u_{xx},u_{txx})$, where (7.9) is substituted in place of $u(x,t,\varepsilon)$ and the derivatives are replaced with the corresponding derivatives of the function $u(x,t,\varepsilon)$, can be expanded into the power series by ε:

$$f=f_0+\varepsilon f_1+\varepsilon^2 f_2+\cdots \tag{7.162}$$

Comparing the coefficients at equal powers of ε (ε^0 and ε) in (7.160) and using (7.162) we get

$$X_1(x)\left[\frac{d\alpha}{d\tau}\cos\tilde{\beta}-\frac{dh}{d\tau}\alpha\sin\tilde{\beta}+\frac{\partial u_1}{\partial t}\cos\tilde{\beta}-\frac{\partial v_1}{\partial t}\alpha\sin\tilde{\beta}\right]=0,$$

$$X_1(x)\left[2\delta\omega\alpha\cos\tilde{\beta}-\tilde{\omega}\left(\frac{d\alpha}{d\tau}\sin\tilde{\beta}+\frac{dh}{d\tau}\alpha\cos\tilde{\beta}+\frac{\partial u_1}{\partial t}\sin\tilde{\beta}+\right.\right.$$

$$+\alpha\frac{\partial v_1}{\partial t}\cos\tilde{\beta}\Big)\Big] + Lw_1(x,t) = f_0(x,\theta,\tau,\alpha,\tilde{\beta}) + g(\theta,\tau). \tag{7.163}$$

Next, suppose that the coefficient $f_0(x,\theta,\tau,\alpha,\tilde{\beta})$ of the series (7.162) can be expanded into the Fourier series by the eigenfunctions $X_k(x)$, $k = 1,2,\ldots$, of the boundaru value problem (7.154), i. e.

$$f_0(x,\theta,\tau,\alpha,\tilde{\beta}) = f_{01}(\theta,\tau,\alpha,\tilde{\beta})X_1(x) + \sum_{k=2}^{d_0} f_{0k}X_k(x), \tag{7.164}$$

where

$$f_{0k} = \frac{1}{m_k}\int_0^l f_0(x,\theta,\tau,\alpha,\tilde{\beta})X_k(x)dx, \qquad m_k = \int_0^l X_k^2(x)dx,$$

$$k = 1,2,\ldots . d_0; \qquad d_0 \le \infty.$$

Using (7.164) we obtain from (7.163) the following system

$$\frac{\partial u_1}{\partial t} = -\frac{1}{\tilde{\omega}}f_{01}(\theta,\tau,\alpha,\tilde{\beta})\sin\tilde{\beta} + \delta\alpha\sin 2\tilde{\beta} - \frac{d\alpha}{d\tau},$$

$$\frac{\partial v_1}{\partial t} = -\frac{1}{\alpha\tilde{\omega}}f_{01}(\theta,\tau,\alpha,\tilde{\beta})\cos\tilde{\beta} + 2\delta\cos^2\tilde{\beta} - \frac{dh}{d\tau},$$

$$Lw_1 = \sum_{k=2}^{d_0} f_{0k}(\theta,\tau,\alpha,\tilde{\beta})X_k(x) + g(\theta,\tau).$$

Note that $f_0(x,\theta,\tau,\alpha,\tilde{\beta})$ and $f_{0k}(\theta,\tau,\alpha,\tilde{\beta})$, $k = 1,2,3,\ldots$, can easily be written in an explicit form.

Let the mean values

$$\underset{t}{M}\{f_{01}(\theta,\tau,\alpha,\tilde{\beta})\sin\tilde{\beta}\} = Y_1(\tau,\alpha,h),$$

$$\underset{t}{M}\{f_{01}(\theta,\tau,\alpha,\tilde{\beta})\cos\tilde{\beta}\} = Y_2(\tau,\alpha,h).$$

exist.

We set

$$\frac{d\alpha}{d\tau} = -\frac{1}{\tilde{\omega}}Y_1(\tau,\alpha,h), \quad \frac{dh}{dt} = \delta - \frac{1}{\alpha\tilde{\omega}}Y_2(\tau,\alpha,h)$$

or

$$\frac{d\alpha}{dt} = -\frac{\varepsilon q}{p\nu}Y_1(\tau,\alpha,h), \quad \frac{dh}{dt} = \omega_1 - \frac{p\nu}{q} - \frac{\varepsilon q}{\alpha p\nu}Y_2(\tau,\alpha,h). \tag{7.165}$$

The functions u_1 and v_1 are then determined from the system

$$\frac{\partial u_1}{\partial t} = -\frac{1}{\tilde{\omega}}\Big[f_{01}(\theta,\tau,\alpha,\tilde{\beta})\sin\tilde{\beta} - \underset{t}{M}\{f_{01}(\theta,\tau,\alpha,\tilde{\beta})\sin\tilde{\beta}\}\Big] + \delta\alpha\sin 2\tilde{\beta},$$

$$\frac{\partial v_1}{\partial t} = -\frac{1}{\alpha\tilde{\omega}}\Big[f_{01}(\theta,\tau,\alpha,\tilde{\beta})\cos\tilde{\beta} - \underset{t}{M}\{f_{01}(\theta,\tau,\alpha,\tilde{\beta})\cos\tilde{\beta}\}\Big] + \delta\cos 2\tilde{\beta}, \tag{7.166}$$

and $w_1(x,t)$ is a solution of the equation

$$\frac{\partial^2 w_1}{\partial t^2} - a^2\frac{\partial^2 w_1}{\partial x^2} = cw_1 + \sum_{k=2}^{d_0} f_{01}(\theta,\tau,\alpha,\tilde{\beta})X_k(x) + g(\theta,\tau). \tag{7.167}$$

In particular, if we assume that the right-hand sides of the initial conditions (7.151) can be approximated on interval $[0, l]$ by the first eigenfunction $X_1(x)$ with the accuracy of order ε, then the initial conditions (7.151) can be represented in the form

$$u|_{t=0} = \varphi_1 X_1(x) + \varepsilon y(x), \quad u_t|_{t=0} = \chi_1 X_1(x) + \varepsilon z(x), \tag{7.168}$$

where φ_1 and χ_1 are reals.

Using (7.157)-(7.159) and (7.168) we obtain the initial conditions

$$\alpha_1|_{t=0} = a_1, \quad a_1 = \sqrt{\varphi_1^2 + \chi_1^2\omega_1^{-2}},$$

$$h_1|_{t=0} = \theta_1, \quad \theta_1 = -\operatorname{arctg}\chi_1(\omega_1\varphi_1)^{-1}, \quad u_1|_{t=0} = v_1|_{t=0} = 0.$$

The initial and boundary conditions for the function $w_1(x,t)$ are to be chosen as follows

$$w_1(x,0) = y(x), \quad w_{1t}(x,0) = z(x);$$

$$L_j(w_1, w_{1x})|_{x=j} = F_j(\theta, \tau, \alpha X_1(j)\cos\tilde{\beta}), \quad j = 0, l.$$

The first improved approximation for the mixed problem (7.149)-(7.151) thus has the form

$$u_{1yn}(x,t,\varepsilon) = \alpha\cos\tilde{\beta} + \varepsilon(u_1\cos\tilde{\beta} - \alpha v_1 \sin\tilde{\beta}) + \varepsilon w_1(x,t).$$

7.12. A method for constructing asymptotic approximations to solutions of partial differential equations with application to multi-frequency modes of oscillations

If we analyse the method used for the construction of asymptotic approximations to solutions of the nonlinear wave equation (7.104), we observe that the form of solution of the nonperturbed equation ($\varepsilon = 0$) plays an essential part here. In this connection we shall consider the following mixed problem

$$L(M,u) - \rho(M)\frac{\partial^2 u}{\partial t^2} = \varepsilon f(M,\tau,u,u_t,D_x u); \tag{7.169}$$

$$u|_{t=0} = \varphi(M), \quad u_t|_{t=0} = \chi(M), \quad M \in G; \tag{7.170}$$

$$u|_{\Sigma} = \varepsilon F(t,\tau,u)_{M\in\Sigma} t \geq 0, \quad \Sigma = \partial G. \tag{7.171}$$

Here $L(Mu)$ is a linear differential operator, which depends on point M of a region G, function $u(M,t)$ and its derivatives with respect to the coordinates of the point M; G is a bounded region in one-, two- or three-dimensional space; $\Sigma = \partial G$ is its boundary; $\varepsilon > 0$ is a small parameter; f is a nonlinear function of u and its partial derivatives.

Note that the left-hand side of (7.171) may involve a differential operator with constant coefficients, which is applied to the function u. Consider the nonperturbed mixed problem

$$L(M,u) - \rho(M)\frac{\partial^2 u}{\partial t^2} = 0; \tag{7.172}$$

$$u|_{t=0} = \varphi(M), \quad u_t|_{t=0} = \chi(M); \tag{7.173}$$

$$u|_{\Sigma} = 0. \tag{7.174}$$

Suppose that the solution of this problem can be represented as the series

$$u(M,t) = \sum_{k=1}^{\infty}(A_k\cos\sqrt{\lambda_k}t + B_k\sin\sqrt{\lambda_k}t)X_k(M),$$

where $\{\lambda_k\}$ and $\{X_k(M)\}$ are the systems of eigenvalues and eigenfunctions, respectively, of the boundary value problem

$$L(M,X) + \lambda\rho(M)X = 0, \quad X(M)|_{M\in\Sigma} = 0. \tag{7.175}$$

Suppose also that the initial condition (7.170) can be represented as follows

$$u(M,0) = \sum_{k=1}^{n} X_k(M)\varphi_k + \varepsilon y(M),$$

$$u_t(M,0) = \sum_{k=1}^{n} \chi_k X_k(M) + \varepsilon z(M). \tag{7.176}$$

The solution of the perturbed mixed problem (7.169)–(7.171) is sought as the asymptotic expansion

$$u(M,t,\varepsilon) = \sum_{k=1}^{n} X_k(M)a_k(t,\varepsilon)\cos[\psi_k(t,\varepsilon)]+$$

$$+\varepsilon w_1(M,t,\varepsilon) + \varepsilon^2 w_2(M,t,\varepsilon) + \cdots, \tag{7.177}$$

where

$$a_k(t,\varepsilon) = \alpha_k + \varepsilon u_{1k}(t,\alpha,h) + \varepsilon^2 w_{2k}(t,\alpha,h) + \cdots;$$

$$\psi(t,\varepsilon) = \sqrt{\lambda_k}\,t + h_k + \varepsilon v_{1k}(t,\alpha,h) + \varepsilon^2 v_{2k}(t,\alpha,h) + \cdots \tag{7.178}$$

If we set

$$u_t = -\sum_{k=1}^{n} X_k(M)\sqrt{\lambda_k}\,a_k \sin\psi_k + \varepsilon w_{1t} + \varepsilon^2 w_{2t} + \cdots,$$

$$\alpha_k = \alpha_k(\tau), \quad h_k = h_k(\tau), \quad \tau = \varepsilon t, \quad k = 1,2,\ldots,n,$$

then we can write

$$u_{\mathrm{Iy}n}(M,t,\varepsilon) = \sum_{k=1}^{n} X_k(M)[\alpha_k\cos\beta_k + \varepsilon(u_{1k}\cos\beta_k -$$

$$-\alpha_k v_{1k}\sin\beta_k)] + \varepsilon w_1(M,t),$$

for the first improved approximation, where $\beta_k = \sqrt{\lambda_k}\,t + h_k$; α_k, h_k, u_{1k}, v_{1k}, $k = 1,2,\ldots,n$, and $w_1(M,t)$ are determined from the system

$$\frac{\partial u_{1k}}{\partial t} = -\frac{1}{\sqrt{\lambda_k}} f_{0k}(\tau,\alpha,\beta)\sin\beta_k - \frac{d\alpha_k}{d\tau},$$

$$\frac{\partial v_{1k}}{\partial t} = -\frac{1}{\alpha_k\sqrt{\lambda_k}} f_{0k}(\tau,\alpha,\beta)\cos\beta_k - \frac{dh_k}{d\tau}, \quad k = 1,2,\ldots,n;$$

$$L(M,w_1) - \rho(M)\frac{\partial^2 w_1}{\partial t^2} = \sum_{k=n+1}^{d_0} f_{0k}(\tau,\alpha,\beta)X_k(M) \tag{7.179}$$

with the boundary and initial conditions

$$\alpha_k|_{t=0} = a_k, \quad a_k = \sqrt{\varphi_k^2 + \chi_k^2\lambda_k^{-1}},$$

$$h_k|_{t=0} = \theta_k, \quad \theta_k = -\mathrm{arctg}\chi_k(\sqrt{\lambda_k}\,\varphi_k)^{-1},$$

$$u_{1k}|_{t=0} = v_{1k}|_{t=0} = 0, \quad k = 1, 2, \ldots, n,$$

$$w_1|_{t=0} = y(M), \quad w_{1t}|_{t=0} = z(M), \quad M \in G,$$

$$w_1|_\Sigma = F(t, \tau, u|_{\varepsilon=0})_{M\in\Sigma}, \quad t > 0, \quad \tau = \varepsilon t.$$

Here $f_{0k}(\tau, \alpha, \beta)$ are the Fourier coefficients of the function $f_0(M, \tau, \alpha, \beta) = f(M, \tau, u|_{\varepsilon=0}, D_x u|_{\varepsilon=0})$ expanded in the Fourier series by the eigenfunctions $\{X_k(M)\}$ of the boundary value problem (7.175) with weight $\rho(M)$.

The system (7.179) is solvable if the mean values

$$\underset{t}{M}\{f_{0k}(\tau, \alpha, \beta) \sin \beta_k\}, \quad \underset{t}{M}\{f_{0k}(\tau, \alpha, \beta) \cos \beta_k\}, \quad k = 1, 2, \ldots, n.$$

exist.

Note that the general second order equation in the theory of special functions, which appear when solving boundary value problems of the first kind, was studied by Tikhonov and Samarskii in [115]. This book describes the behaviour of solutions of these equations, formulates boundary value problems for partial differential equations, and presents the basic properties of eigenvalues and eigenfunctions as well as the theorem on the expansion of functions into absolutely and uniformly converging series by the eigenfunctions of boundary value problem.

Since the asymptotic method stated above uses the expansion of functions into Fourier series by the eigenfunctions of boundary value problem, we consider an equation for simplest special functions

$$Ly + \lambda \rho y = 0, \quad a < x < h, \quad \rho(x) > 0,$$

$$Ly = \frac{d}{dx}\left(k(x)\frac{dy}{dx}\right) - q(x)y, \quad k(x) > 0, \quad q(x) \geq 0$$

and give the theorem on expansions.

Theorem 7.5. *A function $f(x)$ can be expanded into absolutely and uniformly converging series by the eigenfunctions of a boundary value problem if*

1) the first and second derivatives of $f(x)$ are continuous and piecewise continuous, respectively, for $a < x < b$;

(2) $f(x)$ satisfies the boundary condition; moreover, if $k(a) = 0$, then

$$|f(a)| < \infty \qquad 0 \leq q(x) < \infty,$$

$$f(a) = 0 \qquad q(x) \to \infty \ (x \to \infty).$$

Many problems dealt with above were investigated with the help of the Krylov-Bogolyubov asymptotic methods, for example the problem concerning axially symmetric proper oscillations of circular membrane (see [64, 66] et al.) So the above asymptotic method for constructing approximate solutions can be applied to these problems, providing the first approximations that coincide with those obtained by the Krylov-Bogolyubov asymptotic method.

BIBLIOGRAPHY

References

1. V.E. Abolinya and A.D. Myshkis. (1958) A mixed problem for linear hyperbolic system on the plane. *Uchionye Zap. Latviisk. Univ.*, **20**, p. 87–104 (in Russian).
2. V.E. Abolinya and A.D. Myshkis. (1960) A mixed problem for almost linear hyperbolic system on the plane. *Mat. Sbornik*, **50**, no. 4, p. 423–442 (in Russian).
3. A.A. Andronov, A.A. Vit, and S.E. Khaikin. (1950) *Oscillation Theory*. Moscow, Fizmatgiz, (in Russian).
4. D.V. Anosov. (1960) Averaging in systems of ordinary differential equations with rapidly oscillating solutions. *Izvestiya AN SSSR, Ser. Mat.*, **24**, no. 5, p. 721–742 (in Russian).
5. V.I. Arnol'd. (1963) Small denominators and the stability problem in classical and celestial mechanics. *Uspekhi Mat. Nauk*, **18**, no. 6, p. 92–191 (in Russian).
6. N.A. Artem'yev. (1937) Periodic solutions for a class of partial differential equations. *Izvestiya AN SSSR*, Ser. Mat., no. 1, p. 15–50 (in Russian).
7. Kh. Begnayev and A.N. Filatov. (1970) A property of integro-differential equations. *Dokl. AN Uzbek. SSR*, no. 12, p. 6–8 (in Russian).
8. R. Bellman. (1954) *Stability Theory of Differential Equations*. Moscow, Izdat. Inostr. Lit. (Russian translation).
9. Yu.N. Bibikov. (1981) *A General Course on Ordinary Differential Equations*. Leningrad, Izdat. Leningr. Univ. (in Russian).
10. G.I. Biryuk. (1954) A theorem on existence of almost periodic solutions for certain systems of nonlinear differential equations with small parameter. *Dokl. AN SSSR*, **96**, no. 1, p. 5–7 (in Russian).
11. A.V. Bitsadze. (1981) *Certain Classes of Partial Differential Equations*. Moscow, Nauka. (in Russian).
12. N.N. Bogolyubov. (1945) *On Some Statistical Methods for Mathematical Physics*. Kiev, Izdat. AN Ukr. SSR. (in Russian).
13. N.N. Bogolyubov and D.N. Zubarev. (1955) The method of asymptotic approximation for systems with rotating phase and its application to the motion of charged particles in magnetic field. *Ukr. Mat. Zhurn.* **7**, no. 1, p. 2–17 (in Russian).
14. N.N. Bogolyubov and Yu.A. Mitropol'skii. (1974) Asymptotic Methods in Theory of Nonlinear Oscillations, Moscow, Nauka. (in Russian).
15. N.N. Bogolyubov, Yu.A. Mitropol'skii, and A.M. Samoilenko. (1969) *Method of Accelerated Convergence in Nonlinear Mechanics*. Kiev, Nauk. Dumka. (in Russian).
16. A.D. Bryuno. (1970) A normal form of nonlinear oscillations. In: *Proc. Fifth Intern. Conf. Nonlinear Oscillations*. Kiev, Nauk. Dumka, vol. 1, p. 112–119 (in Russian).
17. M.M. Vainberg and R.I. Kochurovskii. (1959) Towards the variational theory of nonlinear operators and equations. *Dokl. AN SSSR*, **129**, no. 6, p. 1199–1202 (in Russian).
18. M.M. Vainberg. (1972) *The Variational Method and Method of Monotone Operators in Theory of Nonlinear Equations*. Moscow, Nauka. (in Russian).
19. S.A. Vasilishin and V.I. Fodchuk. (1967) Construction of asymptotic solutions for quasilinear partial differential equations with time delay. *Ukr. Mat. Zhurn.*, **19**, no. 4, p. 108–113 (in Russian).
20. A.B. Vasil'yeva and V.F. Butuzov. (1973) *Asymptotic Expansions of Solutions of Singularly Perturbed Equations*. Moscow, Nauka. (in Russian).
21. O. Vejvoda and M. Štedry. (1984) The existence of classical periodic solutions of a wave equation: a relationship between number-theoretical character of period and geometrical properties of solutions. *Differents. Uravn.*, **20**, no. 10. p. 1733–1739 (in Russian).
22. M.I. Vishik. (1972) Solving a system of quasilinear equations of the divergent form under periodic boundary conditions. *Dokl. AN SSSR*, **137**, no. 3, p. 502–505 (in Russian).
23. V.S. Vladimirov. (1967) *Mathematical Physics Equations*. Moscow, Nauka. (in Russian).
24. V.M. Volosov. (1972) *Asymptotic Methods of Investigating Nonlinear Waves in Stratified Medium with Application to Theory of Inner Ocean Waves*. Moscow, Izdat. Mosk. Univ. (in Russian).
25. V.M. Volosov. (1976) Nonlinear waves in nonhomogeneous media: asymptotic methods with applications to oceanography problems. In: *Oscillations in Nonlinear Systems*, Kiev, Inst. Mat. AN Ukr. SSR, p. 3–141 (in Russian).
26. V.M. Volosov and B.I. Morgunov. (1971) *Averaging Method in Theory of Nonlinear Oscillating*

Systems. Moscow, Izdat. Mosk. Univ. (in Russian).
27. I.M. Vul'pe. (1978) Obtainig periodic solutions of nonlinear systems by using the averaging method. *Differents. Uravn.*, **14**, no. 9, p. 1558–1565 (in Russian).
28. E.A. Gayeva, V.G. Kolomiets, and G.P. Khoma. (1979) Averaging in systems of integro-differential equations of hyperbolic type. In: *Fifth All-Union Conf. Qualitative Theory of Differential Equations, Kishinev, 22–24 August 1979, Abstracts*, Kishinev, Shtiintsa, p. 46 (in Russian).
29. A.V. Gaponov, L.A. Ostrovskii, and M.I. Rabinovich. (1970) One-dimensional waves in nonlinear systems with dispersion. *Izvestiya Vuzov. Radiofizika*, **13**, no. 2, p. 163–213 (in Russian).
30. M.I. Gromyak. (1984) Justification of an averaging scheme for hyperbolic system of first order. *Dokl. AN Ukr. SSR*, Ser. A, no. 6, p. 5–7 (in Russian).
31. M.I. Gromyak. (1986) Investigation of generalised periodic solutions of hyperbolic integro-differential equations of second order. *Prepr. 86.14*, Kiev, Inst. Mat. AN Ukr. SSR. (in Russian).
32. M.I. Gromyak. (1986) Justification of an averaging scheme for hyperbolic systems with slow and fast variables: a mixed problem. *Ukr. Mat. Zhurn.*, **38**, no. 5, p. 575–582 (in Russian).
33. M.I. Gromyak, G.P. Khoma, and V.Z. Chornyi. (1986) Periodic solutions of nonlinear wave ordinary integro-differential equations of second order. In: *Questions of the Theory of Asymptotic Methods of Nonlinear Mechanics.* Kiev, Inst. Mat. AN Ukr. SSR, p. 79–84 (in Russian).
34. M.I. Gromyak. (1988) Construction of periodic solutions of wave differential and integro-differential equations of second order. *Ukr. Mat. Zhurn.*, **40**, no. 1, p. 48–53 (in Russian).
35. M.I. Gromyak. (1988) Towards the construction of periodic solutions of wave eqautions of hyperbolic type. In: *Asymptotic Methods in Problems of Mathematical Physics*, Kiev, Inst. Mat. AN Ukr. SSR, p. 69–75 (in Russian).
36. B.P. Demidovich. (1967) *Lectures on Mathematical Theory of Stability.* Moscow, Nauka. (in Russian).
37. Yu.A. Dubinskii. (1968) Quasilinear elliptic and parabolic equations of arbitrary order. *Uspekhi Mat. Nauk*, **23**, no. 1, p. 45–90 (in Russian).
38. N.A. Yevtukha and P.P. Zabreiko. (1985) Samoilenko's method for obtaining periodic solutions of quasilinear differential equations in Banach space. *Ukr. Mat. Zhurn.*, **37**, no. 2, p. 162–168 (in Russian).
39. O.A. Zhautykov. (1964) Application of the averaging method to solution of a partial differential equation ocurring in oscillation theory. In: *Approximate Methods for Solving Differential Equations.* Kiev, Inst. Mat. AN Ukr. SSR, no. 2, p. 52–61 (in Russian).
40. O.A. Zhautykov. (1965) The averaging principle in nonlinear mechanics with application to countable systems of equations. *Ukr. Mat. Zhurn.*, **17**, no. 1, p. 39–46 (in Russian).
41. V.V. Zhikov, S.M. Kozlov, and O.A. Oleinik. (1982) Averaging of parabolic operators with almost periodic coefficients. *Mat. Sbornik*, **117**, no. 1, p. 69–85 (in Russian).
42. P.P. Zabreiko and I.B. Ledovskaya. (1969) Towards the justification of Bogolyubov–Krylov method for ordinary differential equations. *Differents. Uravn.*, **5**, no. 2, p. 240–253 (in Russian).
43. P.P. Zabreiko and Yu.I. Fetisov. (1972) On small parameter methods for hyperbolic equations. *Differents. Uravn.*, **8**, no. 5, p. 823–834 (in Russian).
44. V.A. Il'yin and E.G. Poznyak. (1973) *Foundations of Mathematical Analysis. Vol. 2.* Moscow, Nauka. (in Russian).
45. L.V. Kantorovich and G.P. Akilov. (1977) *Functional Analysis.* Moscow, Nauka. (in Russian).
46. H. Kauderer. (1961) *Nichtlineare Mechanik.* Moscow, Izdat. Inostr. Lit. (Russian translation).
47. M. Kisilevich. (1970) A Bogolyubov type theorem for hyperbolic equation. *Ukr. Mat. Zhurn.*, **22**, no. 3, p. 374–379 (in Russian).
48. E.A. Coddington and N. Levinson. (1958) *Theory of Ordinary Differential Equations.* Moscow, Izdat. Inostr. Lit. (Russian translation).
49. V.G. Kolomiets and G.P. Khoma. (1970) On Bogolyubov-Mitropol'skii averaging principle for a class of second order hyperbolic equations. *Ukr. Mat. Zhurn.*, **22**, no. 3, p. 388–393 (in Russian).
50. M.A. Krasnosel'skii and S.G.Krein. (1955) The averaging principle in nonlinear mechanics. *Uspekhi Mat. Nauk*, **10**, no. 3, p. 147–152 (in Russian).
51. N.M. Krylov and N.N. Bogolyubov. (1937) *Introduction to Nonlinear Mechanics.* Kiev, Izdat. AN Ukr. SSR. (in Russian).
52. R. Courant. (1964) *Partial Differential Equations.* Moscow, Nauka. (Russian translation).
53. J.O. Kurzweil. (1963) Averaging in some special cases of boundary value problems for partial differential equations. ıČas. pro Pestovani Mat., **88**, no. 4, p. 444–456 (in Russian).
54. O.A. Ladyzhenskaya and N.N. Ural'tseva. (1973) *Linear and Quasilinear Equations of Elliptic Type.* Moscow, Nauka. (in Russian).
55. B.M. Levitan. (1973) *Theory of Generalised Translation Operators.* Moscow, Nauka. (in Russian).
56. F.S. Los'. (1950) On averaging principle for differential equations in Hilbert space. *Ukr. Mat. Zhurn.*, **2**, no. 3, p. 87–93 (in Russian).
57. O.B. Lykova. (1966) A generalisation of a theorem of Bogolyubov to the case of Hilbert space. *Ukr. Mat. Zhurn.*, **18**, no. 5, p. 53–59 (in Russian).
58. A.M. Lyapunov. (1950) *General Problem on Motion Stability.* Moscow, Gostekhizdat. (in Russian).
59. I.G. Malkin. (1956) *Certain Problems in Theory of Nonlinear Vibrations.* Moscow, Gostekhizdat. (in

Russian).
60. L.I. Madel'shtam. (1972) *Lectures on Oscillation Theory.* Moscow, Nauka. (in Russian).
61. Z.O. Mel'nik. (1964) A general mixed problem. *Dokl. AN SSSR*, **157**, no. 5, p. 1039–1042 (in Russian).
62. Yu.A. Mitropol'skii. (1964) *Problems of Asymptotic Theory of Nonstationary Oscillations.* Moscow, Nauka. (in Russian).
63. Yu.A. Mitropol'skii. (1971) *Averaging Method in Nonlinear Mechanics.* Kiev, Nauk. Dumka. (in Russian).
64. Yu.A. Mitropol'skii and S.A. Krivosheya. (1980) Asymptotic solution of a class of boundary value problems. *Ukr. Mat. Zhurn.*, **32**, no. 6, p. 846–853 (in Russian).
65. Yu.A. Mitropol'skii and O.B. Lykova. (1973) *Integral Manifolds in Nonlinear Mechanics.* Moscow, Nauka. (in Russian).
66. Yu.A. Mitropol'skii and B.I. Moseyenkov. (1976) *Asymptotic Solutions of Partial Differential Equations.* Kiev, Vyshcha Shkola, (in Russian).
67. Yu.A. Mitropol'skii and A.M. Samoilenko. (1972) Quasiperiodic oscillations in nonlinear systems. *Ukr. Mat. Zhurn.*, **24**, no. 2, p. 179–193 (in Russian).
68. Yu.A. Mitropol'skii and A.M. Samoilenko. (1976) Aymptotic integration of weakly nonlinear systems. *Ukr. Mat. Zhurn.*, **28**, no. 4, p. 493–599 (in Russian).
69. Yu.A. Mitropol'skii, A.M. Samoilenko, and D.I. Martynyuk. (1984) *Systems of Evolutional Equations with Periodic and Conditionally Periodic Coefficients.* Kiev, Nauk. Dumka. (in Russian).
70. Yu.A. Mitropol'skii and A.M. Samoilenko. (1979) Towards the asymptotic expansions in nonlinear mechanics. *Ukr. Mat. Zhurn.*, **31**, no. 1, p. 42–53 (in Russian).
71. Yu.A. Mitropol'skii and G.P. Khoma. (1970) On averaging principle for hyperbolic equations along the characteristics. *Ukr. Mat. Zhurn.*, **22**, no. 5, p. 600–611 (in Russian).
72. Yu.A. Mitropol'skii and G.P. Khoma. (1979) Averaging methods for hyperbolic systems with fast and slow variables: Cauchy problem. *Ukr. Mat. Zhurn.*, **31**, no. 2, p. 149–156 (in Russian).
73. Yu.A. Mitropol'skii and G.P. Khoma. (1979) Averaging methods for hyperbolic systems: mixed problem. *Ukr. Mat. Zhurn.*, **31**, no. 4, p. 396–406 (in Russian).
74. Yu.A. Mitropol'skii and G.P. Khoma. (1983) *Mathematical Justification of Asymptotic Methods of Nonlinear Mechanics.* Kiev, Nauk. Dumka. (in Russian).
75. Yu.A. Mitropol'skii and G.P. Khoma. (1986a) Periodic solutions of second order wave equations. I. *Ukr. Mat. Zhurn.*, **38**, no. 5, p. 593–600 (in Russian).
76. Yu.A. Mitropol'skii and G.P. Khoma. (1986b) Periodic solutions of second order wave equations. II. *Ukr. Mat. Zhurn.*, **38**, no. 6, p. 733–739 (in Russian).
77. Yu.A. Mitropol'skii, G.P. Khoma, and M.I. Gromyak. (1986) Periodic solutions of second order wave differential and integro-differential equations. *Prepr. 86.26*, Kiev, Inst. Mat. AN Ukr. SSR. (in Russian).
78. Yu.A. Mitropol'skii and G.P. Khoma. (1987) Periodic solutions of second order wave equations. III. *Ukr. Mat. Zhurn.*, **39**, no. 3, p. 347–353 (in Russian).
79. Yu.A. Mitropol'skii and G.P. Khoma. (1988) Periodic solutions of second order wave equations. IV. *Ukr. Mat. Zhurn.*, **40**, no. 6, p. 757–763 (in Russian).
80. N.N. Moiseyev. (1969) *Asymptotic Methods of Nonlinear Mechanics.* Moscow, Nauka. (in Russian).
81. V.B. Moseyenkov. (1978) Quasiperiodic solutions of a weakly dissipative nonlinear wave equation. *Ukr. Mat. Zhurn.*, **30**, no. 2, p. 254–257 (in Russian).
82. A. Naife. (1976) *Perturbation Methods.* Moscow, Mir. (Russian translation).
83. A.M. Nakhushev. (1969) Certain boundary value problems for hyperbolic and mixed type equations. *Differents. Uravn.*, **5**, no. 1, p. 44–59 (in Russian).
84. A.M. Nakhushev. (1970) A method for setting well-posed boundary value problems for second order linear hyperbolic equations on the plane. *Differents. Uravn.*, **6**, no. 2, p. 192–195 (in Russian).
85. Yu.I. Naimark and B.V. Gol'berg. (1960) Existence theorems for nonlinear mixed problems. *Dokl. AN SSSR*, **135**, no. 2, p. 313–351 (in Russian).
86. V.V. Nemytskii and V.V. Stepanov. (1949) *Qualitative Theory of Differential Equations.* Moscow, Gostekhizdat. (in Russian).
87. S.M. Nikol'skii. (1973) *A Course on Mathematical Analysis. Vol. 2.* Moscow, Nauka. (in Russian).
88. O.A. Oleinik, G.A. Iosif'yan, and G.P.Panasenko. (1983) Asymptotic expansion of solutions of a system from elasticity theory in perforated regions. *Mat. Statistika*, **120**, no. 1, p. 22–41 (in Russian).
89. K.P. Persidskii. (1959) Countable systems of differential equations and stability of their solutions. *Izvestiya AN Kazakh. SSR.* Ser. Mat. Mekh., no. 7, p. 52–71 (in Russian).
90. I.G. Petrovskii. (1961) *Lectures on Partial Differential Equations.* Moscow, Fizmatgiz. (in Russian).
91. I.G. Petrovskii. (1964) *Lectures on the Theory of Ordinary Differential Equations.* Moscow, Nauka. (in Russian).
92. V.N. Polishchuk and B.I. Ptashnik. (1982) Periodic boundary value problem for linear hyperbolic equations and systems. *Prepr. 82.64*, L'vov, Karpenko Fiziko-Mekhanich. Inst. AN Ukr. SSR. (in Russian).
93. S.I. Pokhozhayev. (1971) Periodic solutions of some nonlinear hyperbolic equations. *Dokl. AN SSSR*,

198, no. 6, p. 1274–1277 (in Russian).
94. L.S. Pontryagin. (1965) *Ordinary Differential Equations.* Moscow, Nauka. (in Russian).
95. A. Poincaré. (1971–1972) *Selected Works. 2 vols.* Moscow, Nauka. (in Russian).
96. M.I. Rabinovich. (1970) Asymptotic method in the theory of nonlinear oscillations in distributed systems. *Dokl. AN SSSR*, **191**, no. 6, p. 1253–1256 (in Russian).
97. M.I. Rabinovich and A.A. Rosenblyum. (1971) Towards the justification of asymptotic methods in the theory of nonlinear distributed system oscillations. *Dokl. AN SSSR*, **199**, no. 3, p. 575–578 (in Russian).
98. I.A. Rudakov. (1985) A problem on free periodic vibrations of string with nonmonotone nonlinearity. *Uspekhi Mat. Nauk*, **241**, no. 1, p. 215–261 (in Russian).
99. G.N. Savin and O.A. Goroshko. (1962) *Dynamics of Thread of Variable Length.* Kiev, Izdat. AN Ukr. SSR. (in Russian).
100. A.M. Samoilenko. (1961) Application of the averaging method to the investigation of oscillations stimulated by instantaneous pulses in second order self-oscillating systems with small parameter. *Ukr. Mat. Zhurn.*, **13**, no. 3, p. 103–109 (in Russian).
101. A.M. Samoilenko. (1962) Concerning a case of continuous dependence of differential equation solutions on parameter. *Ukr. Mat. Zhurn.*, **14**, no. 3, p. 289–299 (in Russian).
102. A.M. Samoilenko. (1971) The averaging method in systems with shocks. *Mat. Fizika*, no. 9, p. 101–117 (in Russian).
103. A.M. Samoilenko and N.I. Ronto. (1976) *Numerical-Analytical Methods of Investigating Periodic Solutions.* Kiev, Vyshcha Shkola. (in Russian).
104. *Oscillation Theory Problem Book.* (1978) Ed. by L.V. Postnikov et al., Moscow, Nauka. (in Russian).
105. A.G. Sveshnikov and A.N. Tikhonov. (1974) *Theory of functions of complex variable.* Moscow, Nauka. (in Russian).
106. Z.F. Sirchenko. (1962) Application of the averaging method to solving partial differential equations. *Ukr. Mat. Zhurn.*, **14**, no.2, p. 222-227 (in Russian).
107. I.V. Skrypnik. (1972) Application of topological methods to equations with monotone operators. *Ukr. Mat. Zhurn.*, **24**, no. 1, p. 69–79 (in Russian).
108. S.L. Sobolev. (1950) *Some Applications of Functional Analysis in Mathematical Physics.* Leningrad, Izdat. Leningr. Univ., 1950 (in Russian).
109. V.M. Starzhinskii and V.A. Yakubovich. (1972) *Linear Differential Equations with Periodic Coefficients.* Moscow, Nauka. (in Russian).
110. V.A. Steklov. (1983) *Basic Problems of Mathematical Physics.* Moscow, Nauka. (in Russian).
111. S.P. Timoshenko. (1959) *Vibrations in Engineering.* Moscow, Fizmatgiz. (in Russian).
112. A.N. Tikhonov. (1948) Concerning dependence of differential equations solutions on small parameter. *Mat. Sbornik*, **22**, no. 2, p. 193–204 (in Russian).
113. A.N. Tikhonov. (1950) Systems of differential equations containing parameters. *Mat. Sbornik*, **27**, no. 2, p. 147–156 (in Russian).
114. A.N. Tikhonov and V.Ya. Arsenin. (1979) *Methods for Solving Ill-Posed Problems.* Moscow, Nauka. (in Russian).
115. A.N. Tikhonov and A.A. Samarskii. (1977) *Equations of Mathematical Physics.* Moscow, Nauka. (in Russian).
116. J. Withem. (1977) *Linear and Nonlinear Waves.* Moscow, Mir. (Russian translation).
117. S.F. Feshchenko, N.I. Shkil', and L.D. Nikolaenko. (1966) *Asymptotic Methods in Theory of Linear Differential Equations.* Kiev, Nauk. Dumka. (in Russian).
118. A.N. Filatov. (1974) *Asymptotic Methods in Theory of Differential and Integro-Differential Equations.* Tashkent, Fan. (in Russian).
119. A.N. Filatov and L.V. Sharova. (1976) *Integral Inequalities and Theory of Nonlinear Oscillations.* Moscow, Nauka. (in Russian).
120. G.M. Fikhtengol'ts. (1969–1970) *A Course on Differential and Integral Calculus.* 3 vols. Moscow, Nauka. (in Russian).
121. V.I. Fushchich. (1978) A new method for investigating group properties of systems of partial differential equations. In: *Group-Theoretic Methods in Mathematical Physics.* Kiev, Inst. Mat. AN Ukr. SSR, p. 5–44 (in Russian).
122. V.I. Fushchich. (1983) *Symmetry of Maxwell Equations.* Kiev, Nauk. Dumka. (in Russian).
123. M.M. Khapayev. (1966) Concerning the averaging method and certain problems related to averaging. *Differents. Uravn.*, **2**, no. 5, p. 600–608 (in Russian).
124. M.M. Khapayev and K.V. Mal'kov. (1986) Concerning a class of methods of investigating the asymptotic behaviour of solutions of differential equations with closed operators. *Differents. Uravn.*, **22**, no. 2, p. 255–267 (in Russian).
125. A.Ya. Khinchin. (1961) *Continued Fractions.* Moscow, Fizmatgiz. (in Russian).
126. G.P. Khoma and V.T. Yatsyuk. (1972) Shortening a countable system of partial differential equations. *Ukr. Mat. Zhurn.*, **24**, no. 3, p. 418–421 (in Russian).
127. G.P. Khoma. (1975) Averaging of first order hyperbolic systems with retarded argument along the characteristics. In: *The Fourth All-Union Conf. on Theory and Application of Differential Equations*

with Deviating Argument, Kiev, 23–26 September 1975. Abstracts, Kiev, Nauk. Dumka, p. 241 (in Russian).

128. G.P. Khoma. (1977) Concerning the averaging method for hyperbolic systems in standard form. In: *Nonlinear Differential Equations in Applied Problems*. Kiev, Inst. Mat. AN Ukr. SSR, p. 194–207 (in Russian).
129. G.P. Khoma. (1978) Averaging in hyperbolic systems in standard form with retarded argument. *Ukr. Mat. Zhurn.*, **30**, no. 1, p. 133–135 (in Russian).
130. G.P. Khoma and E.A. Gayeva. (1978) The averaging method for hyperbolic integro-differential systems with fast and slow variables. **Ukr. Mat. Zhurn.**, **30**, no. 5, p. 696–701 (in Russian).
131. G.P. Khoma. (1979) Partial averaging of hyperbolic systems in standard form. *Differents. Uravn.*, **15**, no. 4, p. 747–750 (in Russian).
132. G.P. Khoma. (1979) Averaging of hyperbolic integro-differential systems with fast and slow variables. In: *Asymptotic Methods of Nonlinear Mechanics*. Kiev, Inst. Mat. AN Ukr. SSR, p. 156–161 (in Russian).
133. G.P. Khoma. (1981a) Certain questions concerning the justification of the Krylov–Bogolyubov asymptotic method. *Prepr. 81.31*, Kiev, Inst. Mat. AN Ukr. SSR. (in Russian).
134. G.P. Khoma. (1981b) Asymptotic methods of investigating wave solutions of second order partial differential equations. *Prepr. 81.32*, Kiev, Inst. Mat. AN Ukr. SSR. (in Russian).
135. G.P. Khoma. (1981c) Asymptotic methods of investigating differential system solutions. *Prepr. 81.35*, Kiev, Inst. Mat. AN Ukr. SSR. (in Russian).
136. G.P. Khoma. (1981d) Concerning the applicability of asymptotic methods to partial differential equations. *Prepr. 81.51*, Kiev, Inst. Mat. AN Ukr. SSR. (in Russian).
137. G.P. Khoma. (1981e) Asymptotic solution of nonlinear hyperbolic equations. In: *The Ninth Intern. Conf. on Nonlinear Oscillations, Kiev, 30 August–6 September 1981. Abstracts*, Kiev, Nauk. Dumka. p. 347–348 (in Russian).
138. G.P. Khoma. (1981f) Obtaining periodic solutions of distributed parameter nonlinear systems of first order. *Ukr. Mat. Zhurn.*, **33**, no. 6, p. 779–786 (in Russian).
139. G.P. Khoma. (1982a) Shortening countable systems. In: *Methods of Nonlinear Mechanics and Their Application*. Kiev, Inst. Mat. AN Ukr. SSR. p. 138–141 (in Russian).
140. G.P. Khoma. (1982b) Asymptotic methods of investigating solutions of differential and integro-differential equations. *Prepr. 82.1*, Kiev, Inst. Mat. AN Ukr. SSR. (in Russian).
141. G.P. Khoma. (1982c) Concerning Banfi-Filatov theorem. *Ukr. Mat. Zhurn.*, **34**, no. 2, p. 253–255 (in Russian).
142. G.P. Khoma. (1982d) Periodic solutions of a wave equation. *Prepr. 82.19*, Kiev, Inst. Mat. AN Ukr. SSR. (in Russian).
143. G.P. Khoma. (1982e) Concerning the investigation of systems of Van der Pol equations in the presence resonance. *Ukr. Mat. Zhurn.*, **34**, no. 5, p. 661–663 (in Russian).
144. G.P. Khoma. (1983) Investigation of nonlinear boundary value problems by asymptotic methods. In: *Republic Conf. on Nonlinear Problems of Mathematical Physics, Donetsk, 12–14 September 1983. Abstracts*, Donestk, p. 134 (in Russian).
145. G.P. Khoma. (1984a) Analytical dependence of hyperbolic equation solutions on parameter. *Ukr. Mat. Zhurn.*, **36**, no. 3, p. 391–394 (in Russian).
146. G.P. Khoma. (1984b) Asymptotic solution of nonlinear hyperbolic equations. In: *Proc. Ninth Intern. Conf. on Nonlinear Oscillations, Kiev, 30 August–6 September 1981.* Kiev, Nauk Dumka., 1984, Vol. 1, p. 391–394 (in Russian).
147. G.P. Khoma. (1984c) Concerning the well-posedness of a mixed problem for a nonlinear wave equation. In: *The Ninth School on Operator Theory in Function Spaces, Ternopol', 13–19 September 1984. Abstracts*, Ternopol', Ternopol' Pedag. Inst. p. 143 (in Russian).
148. G.P. Khoma, M.I. Gromyak, and V.G. Dobrotvor. (1984) Periodic solutions of a nonlinear wave equation: justification of asymptotic methods of nonlinear mechanics for partial differential equations. In: *The Ninth School on Operator Theory in Function Spaces, Ternopol', 13–19 September 1984. Abstracts*, Ternopol', Ternopol' Pedag. Inst. p. 143 (in Russian).
149. G.P. Khoma. (1985a) Concerning the structure of periodic solutions of a second order wave equation. *Prepr. 85.53*, Kiev, Inst. Mat. AN Ukr. SSR. (in Russian).
150. G.P. Khoma. (1985b) The structure of generalised periodic solutions of hyperbolic differential and inegro-differential equations of second order. *Prepr. 85.65*, Kiev, Inst. Mat. AN Ukr. SSR. (in Russian).
151. G.P. Khoma and M.I. Gromyak. (1985) Periodic solutions of hyperbolic integro-differential equations of second order. In: *The Second All-Union Conf. "Lavrent'yevskie Chteniya po Matematike, Mekhanike i Fizike", Kiev, 9–11 September 1985. Abstracts*, Kiev, Nauk. Dumka. (in Russian).
152. G.P. Khoma. (1986) Periodic solutions of wave differential equations of second order. *Prepr. 86.5*, Kiev, Inst. Mat. AN Ukr. SSR. (in Russian).
153. G.P. Khoma and M.I. Gromyak. (1986) Periodic solutions of hyperbolic integro-differential equations of second order. *Ukr. Mat. Zhurn.*, **38**, no. 4, p. 531–534 (in Russian).
154. G.P. Khoma and M.I. Gromyak. (1987) Construction of periodic solutions of second order wave dif-

ferential equations of hyperbolic type. In: *Application of Asymptotic Methods to Theory of Nonlinear Differential Equations.* Kiev, Inst. Mat. AN Ukr. SSR. p. 96–103 (in Russian).

155. G.P. Khoma, M.I. Gromyak, and B.I. Smakula. (1987) Periodic solutions of second order equations with delay. In: *Questions of Integral Manifold Stability for Equations of Mathematical Physics.* Kiev, Inst. Mat. AN Ukr. SSR. p. 97–99 (in Russian).
156. V.A. Chernyatin. (1987) Towards the problem concerning the existence of solutions in a mixed problem for a one-dimensional wave equation. *Vestnik Moskovsk. Univ.* Ser. 1, Matem. i Mekhanika, no. 6, p. 7–16 (in Russian).
157. L.E. El'sgol'ts. (1969) *Differential Equations and Variational Calculus.* Moscow, Nauka. (in Russian).
158. H. Aman and E.Zehnder. (1980) Nontrivial solutions for a class of nonresonance problems and applications to nonlinear differential equations. *Ann. Scuola Norm. Sup. Pisa*, **8**, p. 539–603.
159. A. Bahri and H. Brezis. (1980) Periodic solutions of a nonlinear wave equation. *Proc. Roy. Soc.* Edinburgh. A, **85**, p. 313–320.
160. C. Banfi. (1967) Sul l'approssimazione di procesi nostazionare in mecanica nonlineare. *Bull. Unione Mat. Ital.*, **22**, no. 4, p. 442–451.
161. H. Brezis and L. Nirenberg. (1978a) Characterizations of the ranges of some nonlinear operators and applications to boundary value problems. *Ann. Scuola Norm. Sup. Pisa*, **5**, no. 2, p. 225–326.
162. H. Brezis and L. Nirenberg. (1978b) Forced vibrations for a nonlinear wave equation. *Comm. Pure Appl. Math.*, **31**, no. 1, p. 1–30.
163. H. Brezis, J.M. Coron, and L. Nirenberg. (1980) Free vibrations for a nonlinear wave equation and a theorem of P. Rabinowitz. *Comm. Pure Appl. Math.*, **33**, no. 5, p. 667–684.
164. H. Brezis and J.M. Coron. (1980) Periodic solutions of nonlinear wave equations and Hamiltonian systems. *Amer. J. Math.*, **103**, no. 3, p. 559–570.
165. H. Brezis. (1983) Periodic solutions of nonlinear vibrating string and duality principles. *Bull. Amer. Math. Soc. (N.S.)*, **8**, no. 3, p. 409–426.
166. F.E. Browder. (1964) Nonlinear elliptic boundary value problems. *Amer. J. Math.*, **86**, no. 2, p. 339–357.
167. F.E. Browder. (1965) Nonlinear monotone operators and convex sets in Banach spaces. *Bull. Amer. Math. Soc.*, **71**, no. 5, p. 780–785.
168. J.M. Coron. (1983) Periodic solutions of a nonlinear wave equation without assumption of monotonicity. *Math. Ann.*, **262**, no. 2, p. 273–285.
169. W.S. Hall. (1978) A Rayleigh wave equation – an analysis. *Nonlinear Analysis Theory, Methods and Appl.*, **2**, p. 129–156.
170. H. Lovicarova. (1969) Periodic solutions of weakly nonlinear wave eqaution in one dimension. *Chech. Math. J.*, **19**, no. 2, p. 324–342.
171. J. Mawhin. (1981) Periodic oscillations of some nonlinear wave systems. In: *The Ninth Intern. Conf. on Nonlinear Oscillations, Kiev, 30 August–6 September 1981. Abstracts*, Kiev, Nauk. Dumka. p. 206.
172. G. Minty. (1962) Monotone (nonlinear) operators in Hilbert space. *Duke Math. J.*, **29**, no. 3, p. 341–346.
173. L. Nirenberg. (1981) Variational and topological methods in nonlinear problems. *Bull. Amer. Math. Soc. (N.S.)*, **4**, no. 3, p. 267–302.
174. Nuzeki Hyôzô. (1981) On a representation of solutions of the Cauchy problem for a semilinear hyperbolic system. III. *Mem. Fac. Sci. Koichi Univ.* 3, p. 1–13.
175. P. Rabinowitz. (1967) Periodic solutions of hyperbolic partial differential equations. *Comm. Pure Appl. Math.*, **20**, no. 1, p. 145–205.
176. P. Rabinowitz. (1969) Periodic solutions of nonlinear hyperbolic partial differential equations. *Comm. Pure Appl. Math.*, **22**, no. 1, p. 15–39.
177. P. Rabinowitz. (1971) Time periodic solutions of nonlinear wave equations. *Manuscr. Math.*, **5**, no. 2, p. 165–194.
178. P. Rabinowitz. (1978) Free vibrations for a semilinear wave equation. *Comm. Pure Appl. Math.*, *31*, no. 1, p. 31–68.
179. P. Rabinowitz. (1979) A priori bounds for a semilinear wave eqaution. *Lect. Notes Math.*, vol. 703, p. 340–347.
180. O. Vejvoda. (1959) On the existence and stability of the periodic solution of the second kind of a certain mechanical system. *Chech. Math. J.*, **9**, no. 3, p. 390–415.
181. O. Vejvoda. (1964) Periodic solutions of a nonlinear and weakly nonlinear wave equation in one dimension. *Chech. Math. J.*, **14**, no. 3, p. 341–382.
182. O. Vejvoda and M. Štedry. (1975) Periodic solutions of weakly nonlinear autonomous wave equations. *Chech. Math. J.*, **25** (100), no. 4. p. 536–555.
183. O. Vejvoda. (1981) *Partial Differential Equations: Time-Periodic Solutions.* USA, Sijthoff Noordhoff.

INDEX

Other *Mathematics and Its Applications* titles of interest:

J. Chaillou: *Hyperbolic Differential Polynomials and their Singular Perturbations.* 1979, 184 pp. ISBN 90-277-1032-5

S. Fucik: *Solvability of Nonlinear Equations and Boundary Value Problems.* 1981, 404 pp. ISBN 90-277-1077-5

V.I. Istratescu: *Fixed Point Theory. An Introduction.* 1981, 488 pp. *out of print,* ISBN 90-277-1224-7

F. Langouche, D. Roekaerts and E. Tirapegui: *Functional Integration and Semiclassical Expansions.* 1982, 328 pp. ISBN 90-277-1472-X

N.E. Hurt: *Geometric Quantization in Action. Applications of Harmonic Analysis in Quantum Statistical Mechanics and Quantum Field Theory.* 1982, 352 pp. ISBN 90-277-1426-6

F.H. Vasilescu: *Analytic Functional Calculus and Spectral Decompositions.* 1983, 392 pp. ISBN 90-277-1376-6

W. Kecs: *The Convolution Product and Some Applications.* 1983, 352 pp. ISBN 90-277-1409-6

C.P. Bruter, A. Aragnol and A. Lichnerowicz (eds.): *Bifurcation Theory, Mechanics and Physics. Mathematical Developments and Applications.* 1983, 400 pp. *out of print,* ISBN 90-277-1631-5

J. Aczel (ed.) : *Functional Equations: History, Applications and Theory.* 1984, 256 pp. *out of print,* ISBN 90-277-1706-0

P.E.T. Jorgensen and R.T. Moore: *Operator Commutation Relations.* 1984, 512 pp. ISBN 90-277-1710-9

D.S. Mitrinovic and J.D. Keckic: *The Cauchy Method of Residues. Theory and Applications.* 1984, 376 pp. ISBN 90-277-1623-4

R.A. Askey, T.H. Koornwinder and W. Schempp (eds.): *Special Functions: Group Theoretical Aspects and Applications.* 1984, 352 pp. ISBN 90-277-1822-9

R. Bellman and G. Adomian: *Partial Differential Equations. New Methods for their Treatment and Solution.* 1984, 308 pp. ISBN 90-277-1681-1

S. Rolewicz: *Metric Linear Spaces.* 1985, 472 pp. ISBN 90-277-1480-0

Y. Cherruault: *Mathematical Modelling in Biomedicine.* 1986, 276 pp. ISBN 90-277-2149-1

R.E. Bellman and R.S. Roth: *Methods in Approximation. Techniques for Mathematical Modelling.* 1986, 240 pp. ISBN 90-277-2188-2

R. Bellman and R. Vasudevan: *Wave Propagation. An Invariant Imbedding Approach.* 1986, 384 pp. ISBN 90-277-1766-4

A.G. Ramm: *Scattering by Obstacles.* 1986, 440 pp. ISBN 90-277-2103-3

Other *Mathematics and Its Applications* titles of interest:

C.W. Kilmister (ed.): *Disequilibrium and Self-Organisation.* 1986, 320 pp.
ISBN 90-277-2300-1

A.M. Krall: *Applied Analysis.* 1986, 576 pp.
ISBN 90-277-2328-1 (hb), ISBN 90-277-2342-7 (pb)

J.A. Dubinskij: *Sobolev Spaces of Infinite Order and Differential Equations.* 1986, 164 pp. ISBN 90-277-2147-5

H. Triebel: *Analysis and Mathematical Physics.* 1987, 484 pp.
ISBN 90-277-2077-0

B.A. Kupershmidt: *Elements of Superintegrable Systems. Basic Techniques and Results.* 1987, 206 pp. ISBN 90-277-2434-2

M. Gregus: *Third Order Linear Differential Equations.* 1987, 288 pp.
ISBN 90-277-2193-9

M.S. Birman and M.Z. Solomjak: *Spectral Theory of Self-Adjoint Operators in Hilbert Space.* 1987, 320 pp. ISBN 90-277-2179-3

V.I. Istratescu: *Inner Product Structures. Theory and Applications.* 1987, 912 pp.
ISBN 90-277-2182-3

R. Vich: *Z Transform Theory and Applications.* 1987, 260 pp.
ISBN 90-277-1917-9

N.V. Krylov: *Nonlinear Elliptic and Parabolic Equations of the Second -Order.* 1987, 480 pp. ISBN 90-277-2289-7

W.I. Fushchich and A.G. Nikitin: *Symmetries of Maxwell's Equations.* 1987, 228 pp. ISBN 90-277-2320-6

P.S. Bullen, D.S. Mitrinovic and P.M. Vasic (eds.): *Means and their Inequalities.* 1987, 480 pp. ISBN 90-277-2629-9

V.A. Marchenko: *Nonlinear Equations and Operator Algebras.* 1987, 176 pp.
ISBN 90-277-2654-X

Yu.L. Rodin: *The Riemann Boundary Problem on Riemann Surfaces.* 1988, 216 pp.
ISBN 90-277-2653-1

A. Cuyt (ed.): *Nonlinear Numerical Methods and Rational Approximation.* 1988, 480 pp. ISBN 90-277-2669-8

D. Przeworska-Rolewicz: *Algebraic Analysis.* 1988, 640 pp. ISBN 90-277-2443-1

V.S. Vladimirov, YU.N. Drozzinov and B.I. Zavialov: *Tauberian Theorems for Generalized Functions.* 1988, 312 pp. ISBN 90-277-2383-4

G. Morosanu: *Nonlinear Evolution Equations and Applications.* 1988, 352 pp.
ISBN 90-277-2486-5

Other *Mathematics and Its Applications* titles of interest:

A.F. Filippov: *Differential Equations with Discontinuous Righthand Sides.* 1988, 320 pp. ISBN 90-277-2699-X

A.T. Fomenko: *Integrability and Nonintegrability in Geometry and Mechanics.* 1988, 360 pp. ISBN 90-277-2818-6

G. Adomian: *Nonlinear Stochastic Systems Theory and Applications to Physics.* 1988, 244 pp. ISBN 90-277-2525-X

A. Tesar and Ludovt Fillo: *Transfer Matrix Method.* 1988, 260 pp. ISBN 90-277-2590-X

A. Kaneko: *Introduction to the Theory of Hyperfunctions.* 1989, 472 pp. ISBN 90-277-2837-2

D.S. Mitrinovic, J.E. Pecaric and V. Volenec: *Recent Advances in Geometric Inequalities.* 1989, 734 pp. ISBN 90-277-2565-9

A.W. Leung: *Systems of Nonlinear PDEs: Applications to Biology and Engineering.* 1989, 424 pp. ISBN 0-7923-0138-2

N.E. Hurt: *Phase Retrieval and Zero Crossings: Mathematical Methods in Image Reconstruction.* 1989, 320 pp. ISBN 0-7923-0210-9

V.I. Fabrikant: *Applications of Potential Theory in Mechanics. A Selection of New Results.* 1989, 484 pp. ISBN 0-7923-0173-0

R. Feistel and W. Ebeling: *Evolution of Complex Systems. Selforganization, Entropy and Development.* 1989, 248 pp. ISBN 90-277-2666-3

S.M. Ermakov, V.V. Nekrutkin and A.S. Sipin: *Random Processes for Classical Equations of Mathematical Physics.* 1989, 304 pp. ISBN 0-7923-0036-X

B.A. Plamenevskii: *Algebras of Pseudodifferential Operators.* 1989, 304 pp. ISBN 0-7923-0231-1

N. Bakhvalov and G. Panasenko: *Homogenisation: Averaging Processes in Periodic Media. Mathematical Problems in the Mechanics of Composite Materials.* 1989, 404 pp. ISBN 0-7923-0049-1

A.Ya. Helemskii: *The Homology of Banach and Topological Algebras.* 1989, 356 pp. ISBN 0-7923-0217-6

M. Toda: *Nonlinear Waves and Solitons.* 1989, 386 pp. ISBN 0-7923-0442-X

M.I. Rabinovich and D.I. Trubetskov: *Oscillations and Waves in Linear and Nonlinear Systems.* 1989, 600 pp. ISBN 0-7923-0445-4

A. Crumeyrolle: *Orthogonal and Symplectic Clifford Algebras. Spinor Structures.* 1990, 364 pp. ISBN 0-7923-0541-8

V. Goldshtein and Yu. Reshetnyak: *Quasiconformal Mappings and Sobolev Spaces.* 1990, 392 pp. ISBN 0-7923-0543-4

Other *Mathematics and Its Applications* titles of interest:

I.H. Dimovski: *Convolutional Calculus*. 1990, 208 pp. ISBN 0-7923-0623-6

Y.M. Svirezhev and V.P. Pasekov: *Fundamentals of Mathematical Evolutionary Genetics*. 1990, 384 pp. ISBN 90-277-2772-4

S. Levendorskii: *Asymptotic Distribution of Eigenvalues of Differential Operators*. 1991, 297 pp. ISBN 0-7923-0539-6

V.G. Makhankov: *Soliton Phenomenology*. 1990, 461 pp. ISBN 90-277-2830-5

I. Cioranescu: *Geometry of Banach Spaces, Duality Mappings and Nonlinear Problems*. 1990, 274 pp. ISBN 0-7923-0910-3

B.I. Sendov: *Hausdorff Approximation*. 1990, 384 pp. ISBN 0-7923-0901-4

A.B. Venkov: *Spectral Theory of Automorphic Functions and Its Applications*. 1991, 280 pp. ISBN 0-7923-0487-X

V.I. Arnold: *Singularities of Caustics and Wave Fronts*. 1990, 274 pp. ISBN 0-7923-1038-1

A.A. Pankov: *Bounded and Almost Periodic Solutions of Nonlinear Operator Differential Equations*. 1990, 232 pp. ISBN 0-7923-0585-X

A.S. Davydov: *Solitons in Molecular Systems. Second Edition*. 1991, 428 pp. ISBN 0-7923-1029-2

B.M. Levitan and I.S. Sargsjan: *Sturm-Liouville and Dirac Operators*. 1991, 362 pp. ISBN 0-7923-0992-8

V.I. Gorbachuk and M.L. Gorbachuk: *Boundary Value Problems for Operator Differential Equations*. 1991, 376 pp. ISBN 0-7923-0381-4

Y.S. Samoilenko: *Spectral Theory of Families of Self-Adjoint Operators*. 1991, 309 pp. ISBN 0-7923-0703-8

B.I. Golubov A.V. Efimov and V.A. Scvortsov: *Walsh Series and Transforms*. 1991, 382 pp. ISBN 0-7923-1100-0

V. Laksmikantham, V.M. Matrosov and S. Sivasundaram: *Vector Lyapunov Functions and Stability Analysis of Nonlinear Systems*. 1991, 250 pp. ISBN 0-7923-1152-3

F.A. Berezin and M.A. Shubin: *The Schrödinger Equation*. 1991, 556 pp. ISBN 0-7923-1218-X

D.S. Mitrinovic, J.E. Pecaric and A.M. Fink: *Inequalities Involving Functions and their Integrals and Derivatives*. 1991, 588 pp. ISBN 0-7923-1330-5

Julii A. Dubinskii: *Analytic Pseudo-Differential Operators and their Applications*. 1991, 252 pp. ISBN 0-7923-1296-1

V.I. Fabrikant: *Mixed Boundary Value Problems in Potential Theory and their Applications*. 1991, 452 pp. ISBN 0-7923-1157-4

Other *Mathematics and Its Applications* titles of interest:

A.M. Samoilenko: *Elements of the Mathematical Theory of Multi-Frequency Oscillations*. 1991, 314 pp. ISBN 0-7923-1438-7

Yu.L. Dalecky and S.V. Fomin: *Measures and Differential Equations in Infinite-Dimensional Space*. 1991, 338 pp. ISBN 0-7923-1517-0

W. Mlak: *Hilbert Space and Operator Theory*. 1991, 296 pp. ISBN 0-7923-1042-X

N.Ja. Vilenkin and A.U. Klimyk: *Representation of Lie Groups and Special Functions. Volume 1: Simplest Lie Groups, Special Functions, and Integral Transforms*. 1991, 608 pp. ISBN 0-7923-1466-2

N.Ja. Vilenkin and A.U. Klimyk: *Representation of Lie Groups and Special Functions. Volume 2: Class I Representations, Special Functions, and Integral Transforms*. 1992, 630 pp. ISBN 0-7923-1492-1

N.Ja. Vilenkin and A.U. Klimyk: *Representation of Lie Groups and Special Functions. Volume 3: Classical and Quantum Groups and Special Functions*. 1992, 650 pp. ISBN 0-7923-1493-X
(Set ISBN for Vols. 1, 2 and 3: 0-7923-1494-8)

K. Gopalsamy: *Stability and Oscillations in Delay Differential Equations of Population Dynamics*. 1992, 502 pp. ISBN 0-7923-1594-4

N.M. Korobov: *Exponential Sums and their Applications*. 1992, 210 pp.
ISBN 0-7923-1647-9

Chuang-Gan Hu and Chung-Chun Yang: *Vector-Valued Functions and their Applications*. 1991, 172 pp. ISBN 0-7923-1605-3

Z. Szmydt and B. Ziemian: *The Mellin Transformation and Fuchsian Type Partial Differential Equations*. 1992, 224 pp. ISBN 0-7923-1683-5

L.I. Ronkin: *Functions of Completely Regular Growth*. 1992, 394 pp.
ISBN 0-7923-1677-0

R. Delanghe, F. Sommen and V. Soucek: *Clifford Algebra and Spinor-valued Functions. A Function Theory of the Dirac Operator*. 1992, 486 pp.
ISBN 0-7923-0229-X

A. Tempelman: *Ergodic Theorems for Group Actions*. 1992, 400 pp.
ISBN 0-7923-1717-3

D. Bainov and P. Simenov: *Integral Inequalities and Applications*. 1992, 426 pp.
ISBN 0-7923-1714-9

I. Imai: *Applied Hyperfunction Theory*. 1992, 460 pp. ISBN 0-7923-1507-3

Yu.I. Neimark and P.S. Landa: *Stochastic and Chaotic Oscillations*. 1992, 502 pp.
ISBN 0-7923-1530-8

H.M. Srivastava and R.G. Buschman: *Theory and Applications of Convolution Integral Equations*. 1992, 240 pp. ISBN 0-7923-1891-9

Other *Mathematics and Its Applications* titles of interest:

A. van der Burgh and J. Simonis (eds.): *Topics in Engineering Mathematics*. 1992, 266 pp. ISBN 0-7923-2005-3

F. Neuman: *Global Properties of Linear Ordinary Differential Equations*. 1992, 320 pp. ISBN 0-7923-1269-4

A. Dvurecenskij: *Gleason's Theorem and its Applications*. 1992, 334 pp. ISBN 0-7923-1990-7

D.S. Mitrinovic, J.E. Pecaric and A.M. Fink: *Classical and New Inequalities in Analysis*. 1992, 740 pp. ISBN 0-7923-2064-6

H.M. Hapaev: *Averaging in Stability Theory*. 1992, 280 pp. ISBN 0-7923-1581-2

S. Gindinkin and L.R. Volevich: *The Method of Newton's Polyhedron in the Theory of PDE's*. 1992, 276 pp. ISBN 0-7923-2037-9

Yu.A. Mitropolsky, A.M. Samoilenko and D.I. Martinyuk: *Systems of Evolution Equations with Periodic and Quasiperiodic Coefficients*. 1992, 280 pp. ISBN 0-7923-2054-9

I.T. Kiguradze and T.A. Chanturia: *Asymptotic Properties of Solutions of Nonautonomous Ordinary Differential Equations*. 1992, 332 pp. ISBN 0-7923-2059-X

V.L. Kocic and G. Ladas: *Global Behavior of Nonlinear Difference Equations of Higher Order with Applications*. 1993, 228 pp. ISBN 0-7923-2286-X

S. Levendorskii: *Degenerate Elliptic Equations*. 1993, 445 pp. ISBN 0-7923-2305-X

D. Mitrinovic and J.D. Kečkić: *The Cauchy Method of Residues, Volume 2*. Theory and Applications. 1993, 202 pp. ISBN 0-7923-2311-8

R.P. Agarwal and P.J.Y Wong: *Error Inequalities in Polynomial Interpolation and Their Applications*. 1993, 376 pp. ISBN 0-7923-2337-8

A.G. Butkovskiy and L.M. Pustyl'nikov (eds.): *Characteristics of Distributed-Parameter Systems*. 1993, 386 pp. ISBN 0-7923-2499-4

B. Sternin and V. Shatalov: *Differential Equations on Complex Manifolds*. 1994, 504 pp. ISBN 0-7923-2710-1

S.B. Yakubovich and Y.F. Luchko: *The Hypergeometric Approach to Integral Transforms and Convolutions*. 1994, 324 pp. ISBN 0-7923-2856-6

C. Gu, X. Ding and C.-C. Yang: *Partial Differential Equations in China*. 1994, 181 pp. ISBN 0-7923-2857-4

V.G. Kravchenko and G.S. Litvinchuk: *Introduction to the Theory of Singular Integral Operators with Shift*. 1994, 288 pp. ISBN 0-7923-2864-7

A. Cuyt (ed.): *Nonlinear Numerical Methods and Rational Approximation II*. 1994, 446 pp. ISBN 0-7923-2967-8

Other *Mathematics and Its Applications* titles of interest:

G. Gaeta: *Nonlinear Symmetries and Nonlinear Equations*. 1994, 258 pp.
ISBN 0-7923-3048-X

V.A. Vassiliev: *Ramified Integrals, Singularities and Lacunas*. 1995, 289 pp.
ISBN 0-7923-3193-1

N.Ja. Vilenkin and A.U. Klimyk: *Representation of Lie Groups and Special Functions*. Recent Advances. 1995, 497 pp. ISBN 0-7923-3210-5

Yu. A. Mitropolsky and A.K. Lopatin: *Nonlinear Mechanics, Groups and Symmetry*. 1995, 388 pp. ISBN 0-7923-3339-X

R.P. Agarwal and P.Y.H. Pang: *Opial Inequalities with Applications in Differential and Difference Equations*. 1995, 393 pp. ISBN 0-7923-3365-9

A.G. Kusraev and S.S. Kutateladze: *Subdifferentials: Theory and Applications*. 1995, 408 pp. ISBN 0-7923-3389-6

M. Cheng, D.-G. Deng, S. Gong and C.-C. Yang (eds.): *Harmonic Analysis in China*. 1995, 318 pp. ISBN 0-7923-3566-X

M.S. Livšic, N. Kravitsky, A.S. Markus and V. Vinnikov: *Theory of Commuting Nonselfadjoint Operators*. 1995, 314 pp. ISBN 0-7923-3588-0

A.I. Stepanets: *Classification and Approximation of Periodic Functions*. 1995, 360 pp. ISBN 0-7923-3603-8

C.-G. Ambrozie and F.-H. Vasilescu: *Banach Space Complexes*. 1995, 205 pp.
ISBN 0-7923-3630-5

E. Pap: *Null-Additive Set Functions*. 1995, 312 pp. ISBN 0-7923-3658-5

C.J. Colbourn and E.S. Mahmoodian (eds.): *Combinatorics Advances*. 1995, 338 pp. ISBN 0-7923-3574-0

V.G. Danilov, V.P. Maslov and K.A. Volosov: *Mathematical Modelling of Heat and Mass Transfer Processes*. 1995, 330 pp. ISBN 0-7923-3789-1

A. Laurinčikas: *Limit Theorems for the Riemann Zeta-Function*. 1996, 312 pp.
ISBN 0-7923-3824-3

A. Kuzhel: *Characteristic Functions and Models of Nonself-Adjoint Operators*. 1996, 283 pp. ISBN 0-7923-3879-0

G.A. Leonov, I.M. Burkin and A.I. Shepeljavyi: *Frequency Methods in Oscillation Theory*. 1996, 415 pp. ISBN 0-7923-3896-0

B. Li, S. Wang, S. Yan and C.-C. Yang (eds.): *Functional Analysis in China*. 1996, 390 pp. ISBN 0-7923-3880-4

P.S. Landa: *Nonlinear Oscillations and Waves in Dynamical Systems*. 1996, 554 pp. ISBN 0-7923-3931-2

Other *Mathematics and Its Applications* titles of interest:

A.J. Jerri: *Linear Difference Equations with Discrete Transform Methods.* 1996, 462 pp. ISBN 0-7923-3940-1

I. Novikov and E. Semenov: *Haar Series and Linear Operators.* 1997, 234 pp. ISBN 0-7923-4006-X

L. Zhizhiashvili: *Trigonometric Fourier Series and Their Conjugates.* 1996, 312 pp. ISBN 0-7923-4088-4

R.G. Buschman: *Integral Transformation, Operational Calculus, and Generalized Functions.* 1996, 246 pp. ISBN 0-7923-4183-X

V. Lakshmikantham, S. Sivasundaram and B. Kaymakcalan: *Dynamic Systems on Measure Chains.* 1996, 296 pp. ISBN 0-7923-4116-3

D. Guo, V. Lakshmikantham and X. Liu: *Nonlinear Integral Equations in Abstract Spaces.* 1996, 350 pp. ISBN 0-7923-4144-9

Y. Roitberg: *Elliptic Boundary Value Problems in the Spaces of Distributions.* 1996, 427 pp. ISBN 0-7923-4303-4

Y. Komatu: *Distortion Theorems in Relation to Linear Integral Operators.* 1996, 313 pp. ISBN 0-7923-4304-2

A.G. Chentsov: *Asymptotic Attainability.* 1997, 336 pp. ISBN 0-7923-4302-6

S.T. Zavalishchin and A.N. Sesekin: *Dynamic Impulse Systems.* Theory and Applications. 1997, 268 pp. ISBN 0-7923-4394-8

U. Elias: *Oscillation Theory of Two-Term Differential Equations.* 1997, 226 pp. ISBN 0-7923-4447-2

D. O'Regan: *Existence Theory for Nonlinear Ordinary Differential Equations.* 1997, 204 pp. ISBN 0-7923-4511-8

Yu. Mitropolskii, G. Khoma and M. Gromyak: *Asymptotic Methods for Investigating Quasiwave Equations of Hyperbolic Type.* 1997, 418 pp. ISBN 0-7923-4529-0

R.P. Agarwal and P.J.Y. Wong: *Advanced Topics in Difference Equations.* 1997, 518 pp. ISBN 0-7923-4521-5

N.N. Tarkhanov: *The Analysis of Solutions of Elliptic Equations.* 1997, 406 pp. ISBN 0-7923-4531-2